Springer Atmospheric Sciences

The Springer Atmospheric Sciences series seeks to publish a broad portfolio of scientific books, aiming at researchers, students, and everyone interested in this interdisciplinary field. The series includes peer-reviewed monographs, edited volumes, textbooks, and conference proceedings. It covers the entire area of atmospheric sciences including, but not limited to, Meteorology, Climatology, Atmospheric Chemistry and Physics, Aeronomy, Planetary Science, and related subjects.

Jürgen Steppeler • Jinxi Li

Mathematics of the Weather

Polygonal Spline Local-Galerkin Methods
on Spheres

 Springer

Jürgen Steppeler
Helmholtz-Zentrum Hereon
Climate Service Center Germany (GERICS)
Hamburg, Hamburg, Germany

Jinxi Li
Institute of Atmospheric Physics
Chinese Academy of Sciences
Beijing, China

ISSN 2194-5217   ISSN 2194-5225   (electronic)
Springer Atmospheric Sciences
ISBN 978-3-031-07240-6   ISBN 978-3-031-07238-3   (eBook)
https://doi.org/10.1007/978-3-031-07238-3

This Springer imprint is published by the registered company Springer Nature Switzerland AG
The registered company address is: Gewerbestrasse 11, 6330 Cham, Switzerland

Foreword

Over the last decades, given its increased accuracy and reliability, weather forecast has seen its power to shape our behaviors. Research modeling enjoys a considerable growth. Among the technological backbones that made forecasts possible is the numerical weather prediction (NWP), one of the most celebrated achievements in the field of science and technology for the last half century. NWP is a "model-based" methodology that seeks to integrate ground and space observations with computational techniques. This book is a reader-friendly introduction to the science of developing models of the atmosphere. For readers from relevant fields of study, for example, Earth and Environmental Science, this book may prove to be greatly beneficial in that it gives a general comprehensive introduction to the concepts and mathematical methods used for the creation of atmosphere-related models: ocean, land surface, and even the whole Earth system. Results from this book are also up to date and represent many newest research results.

Developing models of atmosphere is to design numerical schemes/algorithms to solve a set of partial differential equations that describe the physical laws of atmospheric motion and then write the algorithms into compute codes. There are two important things:

- The algorithms should be as accurate as possible to approximate the solutions of the partial differential equations.
- Codes derived from such algorithms must be readily executable and can run efficiently with strong scalability for parallel computers and exaflop computers in the near future.

This book provides essential guides about how to develop models that are both accurate and efficient.

This book introduces the development of atmospheric modeling in an easy-to-follow manner. Starting with simple finite difference procedures for spatial and temporal discretization, the book moves to more advanced schemes. To make life easier for our readers, 1D cases are introduced first with simple examples to explain the essential concepts and ideas. Then it introduces the essentials of schemes of local Galerkin on different spherical grids of the round-ball Earth supplemented by the corresponding algorithms and programming codes.

This book will be helpful for readers to boost their understanding of atmospheric models and their evolution in the future.

IAP, Beijing   Jiang Zhu
June 2020

Preface

Mathematics of the weather (MOW) is a book about the application of mathematics for the construction of numerical models of the atmosphere. The derivation of such mathematical methods is not the topic of this book. For example, the theorem of Pythagoras is not proven here. We are just concerned with the application of this theorem to weather modeling. A number of mathematical principles used here were discovered more recently than Pythagoras's theorem. We just state these results here and guide to their application to modeling. In particular we do not go into lengthy derivations. For such topics, the reader must refer to the mathematical literature or dig into this by themselves. This book is for the engineer, not the lover of mathematical proofs.

Among the huge number of possible numerical schemes, we prefer those which are applicable to realistic modeling. This of course is to a degree the authors' judgment. As on the sphere no regular grids exist, irregular meshes are a point of concern, so are accuracy and efficiency and suitability for multi-processor computers. Suitability on the spherical computational grid is a concern. Among the more modern Local-Galerkin schemes, we concentrate on continuous Galerkin schemes, as some members of this family of methods are rather near to application in realistic models.

After more than 50 years of numerical weather prediction, the authors believe that there is still room for important numerical developments. One indication is the "numerical paradoxon": the developments of atmospheric models in the past 50 years indicate a large increase of forecast quality when going to more accurate schemes in toy models. However, forecast scores for realistic models, such as root mean square errors (RMSE) of temperature, did not increase by going to accurate schemes, such as a higher order of approximation in realistic models. In comparison, the modeling subjects not treated in this book, such as developing refined radiation schemes, did increase forecast scores. Another point is that, for some time, the useful prediction time, measured by RMSE of temperature, stagnates at 6 or 7 days.

This book wants to provide the tools to work on the numerical construction of weather prediction models. Such questions were for the last 20 years discussed in conferences attended mainly by practical model constructors, taking place every

2 years. They had different names: SRNWP (Short-Range Numerical Weather Prediction), MOW, and AMCA (Applied Mathematics and Computation in the Atmosphere) and were co-organized by the authors of this book. A few of these conferences were documented in journals Steppeler (1989, 1996), Steppeler et al. (2001), and Li et al. (2019).

Bad Orb, Germany   Jürgen Steppeler
Beijing, China   Jinxi Li
January 2020

Acknowledgments

The first author dedicates this book to his wife Barbara. The second author would like to express his special thanks to his parents and older sister, whose care and support motivate him to move on and make him want to be a better scientist.

The introductory chapter aims at being understandable to people with only basic mathematical training. The feedback of test readers Hans Jürgen Werth, Wirtheim, and David M. Schultz, University of Manchester, in this respect is appreciated. The sub-chapters on Kalman filters profited from discussions with Roland Potthast, DWD, and Michael Denhard, DWD. The first author's understanding of NWP was much increased by a total of 20 scientific visits of 1 month each at NCAR with hosts Bill Skamarock and Joe Klemp. He profited also from joining the team of two NWP pioneers Winhard Edelmann and Heinz Reiser, DWD. The latter had the foresight to have a specialist of mathematics at DWD's research department, who was not supposed to double as software engineer or similar. DWD's research department was a place of good personal relations between the members. The first author mentions Udo Gärtner, Dieter Fruehwald, Heinz Werner Bitzer, Uli Schättler, and Günther Doms as people contributing to the first author's understanding and development of numerics. The latter, to everyone's grief, died much too early. The technical assistants Annegret Biermann and Yvonne Reiter made the success of numerical conferences at DWD with their personal ambitions. In the opinion of the first author, the mentioned good personal relations were instrumental in DWD's research department remaining and becoming one of the big centers of atmospheric modeling. DWD was among the first national meteorological centers to have operational non-hydrostatic forecasts and made the icosahedral grid practical, a subject on which NCAR at the time had given up. The first author remembers sunny lunchtime walks with John Baumgardner, Los Alamos, along the Main river on occasions of his guest scientist stayed at DWD. We agreed that using his cloud method, non-conserving uniform second-order schemes on the icosahedron are possible, but going beyond this to higher order with conservation was difficult. As seen in Chapter "Platonic and Semi-Platonic Solids", conserving uniform third order schemes on the sphere now seem possible for the construction of NWP models.

Another key subject of the book is cut-cells. For the simplest case of piecewise constant basis functions, they were implemented in the ETA model in Washington D.C. by Fedor Mesinger and Zaviša Janjić. Gallus and Klemp found that this simple method was mathematically faulty and non-convergent. Eventually, this approach was abandoned in Washington D.C. Rather soon DWD could show that a dry model using piecewise linear mountain representations was free of this fault (Steppeler et al. 2002). This means a replacement of the step mountain representation by pyramid mountains. In a visit of 1 month to DWD, Zaviša Janjić helped to include physical parameterizations into this dry model. So, for some time, DWD was the only center to have a fully realistic mesoscale cut-cell model being free of the Gallus and Klemp convergence issues. Udo Gärtner, DWD, was the person encouraging international co-operations at DWD.

Contents

Acronyms

AMCA	Applied Mathematics and Computation in the Atmosphere 2018
CAS	Chinese Academy of Sciences, China
CEE	Compressible Euler Equation
CFD	Computational Fluid Dynamics
CG	Continuous Galerkin
COSMO	COnsortium for Small-scale MOdeling
DG	Discontinuous Galerkin
DWD	Deutscher Wetterdienst, Germany
ECMWF	European Centre for Medium-range Weather Forecast, United Kingdom
EX	EXplicit
FD	Finite Difference
FE	Finite Element
FFT	Fast Fourier Transform
FV	Finite Volume
GC	Galerkin Compiler
HEVI	Horizontally Explicit-Vertically Implicit
HOMME	High-Order Method Modeling Environment, United States
HPE	Primitive Equation
HS	HydroStatic
IAP	Institute of Atmospheric Physics, CAS, China
ICON	Icosahedral Non-Hydrostatic Model, DWD, Germany
IMEX	IMplicit-EXplicit or semi-implicit
LASG	State Key Laboratory of Numerical Modeling for Atmospheric Sciences and Geophysical Fluid Dynamics, IAP, CAS, China
LES	Large Eddy Simulation
L-Galerkin	Local-Galerkin
LM	Lokal Model, DWD, Germany
MOW	Mathematics of the Weather
MPAS	Model for Prediction Across Scales, NCAR, United States
NCAR	National Center for Atmospheric Research, United States

NCAS	National Center for Atmospheric Sciences, United Kingdom
NH	Non-Hydrostatic
NUMA	Non-Hydrostatic Unified Model of the Atmosphere, United States
NWP	Numerical Weather Prediction
ODE	Ordinary Differential Equation
PDE	Partial Differential Equation
QUASAR	QUASi-Arithmetically Rendered
RK	Runge-Kutta
RK4	4th-order Runge-Kutta
RMSE	Root Mean Square Errors
SE	Spectral Element
SISL	Semi-Implicit Semi-Lagrangian
SpEx	Split-Explicit
ST	Spectral Transform
SRNWP	Short-Range Numerical Weather Prediction
SWE	Shallow-Water Equation
WRF	Weather Research and Forecasting model
1D	One-Dimensional
2D	Two-Dimensional
3D	Three-Dimensional

Introduction

Abstract This chapter defines the subject "numerics" of this book, as opposed to other subjects important for numerical weather prediction, such as data assimilation or radiation. The practical engineering approach prevalent in this book is exemplified, as opposed to a mainly theoretical approach. An overview of the different subjects treated is given.

Keywords Discretization · Spherical grids · Polygonal spline methods · Discretized model features · Parameterized model features · Climate-attractor · Lorenz model

Numerical models of the atmosphere (Pedlosky 1987) are an indispensable tool for weather prediction and climate research. The present book concentrates on numerical methods that are currently considered for real-life modeling, meaning that the models use realistic surface conditions and real atmospheric data. It concentrates on the numerical design of such models. This is the mathematical design of the numerical procedures to solve the fluid dynamic equations for moisture, temperature, pressure, and three velocity components. Other important processes, such as radiative transfer, land surface, and ocean interaction with the atmosphere, turbulence parameterization, and the moisture processes transforming the different moisture components into each other, are not treated here. Such processes are normally called "Physical Processes" or "Physical Parameterizations." We stick to this terminology, even though the fluid dynamic and transport processes are no less "physical." Another important component of numerical modeling systems is data assimilation, meaning the assignments of data in the discrete grid using measurements, which are irregularly distributed and of different kinds, such as satellites or rain gauges. This important and difficult subject (Daley 1991) is not fully treated in this book. In section "The Kalman Filter Data Analysis" in chapter "Numerical Tests", just a very short account of the Kalman filter is given, in order to raise the reader's interest.

© Springer Nature Switzerland AG 2022   1
J. Steppeler, J. Li, *Mathematics of the Weather*, Springer Atmospheric Sciences,
https://doi.org/10.1007/978-3-031-07238-3_1

Numerics

The subject of this book is called "numerics." This term indicates that in this field numerical approximations to processes are derived for which the analytic solution is supposed to exist. This terminology comes from early numerical models of the atmosphere, where indeed processes such as radiation or water condensation were rather coarsely represented and the fluid dynamics in comparison was more exactly approximated. We keep using this terminology. Currently, the representation of physical processes has become more accurate. For example, for the representation of radiation, the more advanced schemes can be considered as a discretization of the radiative transfer equation rather than a parameterization. Discretizations are based on approximations of physical laws, for which it may a priori be assumed that they are valid. Parameterizations derive their validity from the fact that they simulate measurements. So the law used with parameterization is potentially established by measurements and numerical experiments. Normally, the set of experiments used to validate discretizations can be smaller than for parameterizations. For example, it is not considered reasonable to investigate if for π the value 3.0 or 3.1415 leads to better results, even though in a special case the less accurate value for π may produce better weather forecasts. This means that for this question, a few experiments are sufficient to see the impact of this change. Such few experiments where statistical significance is not intended are called sensitivity studies.

The field of numerics has given its name to the field, which is called numerical weather prediction (NWP). Much effort has gone into refined numerical procedures, such as approximations of high order. Examples are spectral methods, high-order finite difference methods, and semi-Lagrange methods. For a description of these methods, it is referred to Durran (2010). The present book concentrates on topics which are not extensively treated in Durran (2010) and which play a role in current developments of atmospheric models, and it concentrates on spatial discretization rather than temporal discretization. For time integration, Durran (2010) gives a good overview and describes most methods currently used in atmospheric models. In order that this book is self-contained and can be read without first reading a book on basic numerical methods, such as Durran (2010), chapter "Simple Finite Difference Procedures" will describe the basic principles of some classical numerical procedures, including the fourth-order Runge–Kutta (RK4) time-stepping procedure to be used in this book to test the spatial approximations. Chapter "Simple Finite Difference Procedures" will also treat basic principles for the design of approximations, such as conservation and stability. Not only simple homogeneous schemes will be described, but rather the more general inhomogeneous finite difference schemes. In this book the schemes will be presented for irregular grids. When possible, the schemes are generalized to a form to imply that the approximation order for the case remains the same as known for regular grids. At least, we aim for numerical schemes, which have an approximation order larger than or equal to a target value even at points where the resolution jumps. If in irregular grids the order of approximation for all points is larger than or equal to n^o, it is defined that the scheme has the homogeneous approximation order n^o. For example, the centered difference without conservation is in chapter "Simple Finite Difference Procedures" generalized to a form having an approximation order of two even for points where the resolution jumps, while the ordinary version is reduced to first order for points where the resolution is irregular.

Discretization on Spherical Grids

The rest of the book will treat two areas: Local-Galerkin methods and numerical grids on the sphere. The methods will be described together with test procedures and sample programs. This book could be used for courses in computer applications. The last chapter treats cut-cell methods. This is an emerging subject treating curved boundaries on rather regular grids by assuming that not the whole grid cell is filled by the atmosphere or fluid for boundary cells.

For a discretization, the model area is divided into cells which are intervals in the one-dimensional (1D) case. For more complicated situations, a number of grids will be proposed. On the plane, it is no problem to define regular cell structures. However, cell structures of equal size on the sphere do not exist except for the five Platonic solids described in chapter "Platonic and Semi-Platonic Solids". One way of dealing with this problem is to map the sphere to a plane, as it is done with map projections. The disadvantage is that when just one plane is used, this necessarily leads to singularities, such as at the poles for the latitude–longitude mapping. The global spectral method treats fields on the sphere by spherical harmonics in a similar way as done in quantum mechanics. As these methods are considered difficult for models on computers with a large number of processors, recently models were constructed which divide the atmosphere into a number of patches, such as ten rhomboidal or twenty triangular patches for the icosahedral discretization (see Fig. 1). Inside such patches, rhomboidal or triangular cells can be constructed by subdivision. Figure 1 shows three possibilities of covering the sphere by patches. Chapter "Platonic and Semi-Platonic Solids" will describe some more possibilities. Grids based on the division of the sphere into areas are called polygonal methods. Divisions of the sphere can be obtained by embedding polygons into the sphere and projecting these to the sphere. Chapter "Full and Sparse Hexagonal Grids in the Plane" will give more examples for grids on the sphere and will describe discretization techniques. Polygonal methods on grids such as shown in Fig. 1 were already discussed around 1970 (Sadourny and Morel 1969; Williamson 1968) but it took until 2000 (Baumgardner and Frederickson 1985; Steppeler and Prohl 1996) before they were considered fit for use in realistic atmospheric models. With hindsight, the mathematical problems of early polygonal models on the sphere were associated with map projections and their singularities and a non-uniform numerical approximation order. These problems will be discussed in chapters "2D Basis Functions for Triangular and Rectangular Meshes" and "Full and Sparse Hexagonal Grids in the Plane". Now, for newly developed models, polygonal methods on the

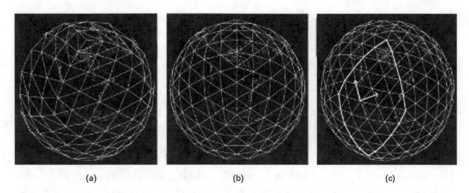

Fig. 1 Examples of spherical grids. The grids are obtained by embedding polygons into the sphere and subdividing them to create a grid (see chapter "Platonic and Semi-Platonic Solids"). The grid cells may be quadrilateral or triangular. Triangular cells may be combined to achieve hexagonal grids, which always involve a few pentagons. The grid is then projected to the surface of the sphere. Two of the figures are projections of Platonic solids, the cube (left) and the icosahedron (right). In realistic models, the icosahedron and the cubed sphere are grids in practical use with realistic models. As indicated by the grid in the middle, there are more possibilities to construct grids, which will be pointed out in chapter "Platonic and Semi-Platonic Solids" (reproduced from Steppeler et al. (2008), © American Meteorological Society, used with permission)

sphere are standard: HOMME (Mishra et al. 2010), ICON (Zängl et al. 2015), MPAS (Park et al. 2014), and NUMA (Giraldo et al. 2013). Not all these methods allow very irregular grids.

Currently, only two polygon based approximations are considered for applications in realistic atmospheric models: the icosahedron and the cubed sphere, which is the cube projected to the sphere. These are Platonic solids. Platonic solids are solids whose surfaces are congruent (identical in shape and size) and regular. Regular surfaces are those whose sides and angles are equal. There exist just five Platonic solids: tetrahedron, cube, octahedron, icosahedron, and dodecahedron. The tetrahedron can be used to derive the Yin-Yang discretization (Kageyama and Sato 2004; Peng et al. 2006), where the sphere is subdivided into just two rhomboidal areas. This will be discussed in chapter "Platonic and Semi-Platonic Solids". Grids are obtained by subdividing the discretization areas. Such grids will necessarily be irregular, except for the very coarse discretizations corresponding to the Platonic solids. More general solids are called semi-Platonic solids (see chapter "Platonic and Semi-Platonic Solids") and the amount of the irregularity of such divisions of the sphere into areas needs to be investigated. In this book, a number of further discretizations, based on semi-Platonic solids, will be analyzed and the tools will be provided to construct grids for a large number of divisions of the sphere into patches (see chapter "Platonic and Semi-Platonic Solids"). Triangular and rhomboidal grids may be constructed. For triangular grids, the patches of grids to be projected to the sphere may be constructed on planes. For rhomboidal grids, a rather convenient way is to construct the grids on bilinear surfaces. Each three points in space determine the plane to construct a patch of a triangular grid. Each four points determine a bilinear surface, on which quasi-regular grids may be constructed easily. For example, the icosahedral grid consists of twenty triangles on which twenty regular patches of triangular grids may be constructed. Combining each two of the triangles to form a rhomboid, each of these ten rhomboids determines a bilinear surface on which quasi-regular rhomboidal grids may be constructed except with the cube, these are not rhomboids in a plane. They can be on two planes with an angle or a bilinear plane. This will be explained in chapter "Platonic and Semi-Platonic Solids". Apart from triangular and rhomboidal cells, hexagonal cells will be discussed. These are often more regular than the corresponding rhomboidal grids and will be described in chapter "Full and Sparse Hexagonal Grids in the Plane" for the plane. The discretizations are done on great circle cells which are obtained by projecting triangular or bilinear rhomboidal grids to the sphere. As an alternative to construct the grids on polygons, the reader can construct the model on a plane or bilinear plane and then use map projections to transfer it to the sphere (Durran 2010). In this book we suggest to use simple Earth centered projection to map triangular of rhomboidal grids to the sphere.

Efficiency of the Computational Grid

The quality of a grid representation depends on its regularity. While it is possible to construct grid cells of equal size on the plane, grids are necessarily irregular on the sphere. Such irregularity of grids is immediately connected to the efficiency of the grid. For example, the regular latitude–longitude grid due to the pole singularity has a very high resolution in east–west direction near the poles. This high resolution in just one direction normally does not benefit the solution. This is a waste of computational resources and nearly regular grids are needed for efficient discretizations. For a different number of polygonal methods on the sphere, the homogeneity of polygonal methods is different. This will be described in chapter "Platonic and Semi-Platonic Solids". As expected, the division of the sphere into small areas leads to more homogeneous grids than with the division into large areas. The Yin-Yang grid, using only 2 deformed rhomboidal patches to cover the sphere, leads to a rather inhomogeneous grid, as compared to the icosahedron. The latter covers the sphere by ten rhomboidal patches (see chapter "Platonic and Semi-Platonic Solids" and Fig. 1).

Another source of inefficiency of grids is grid-anisotropy. This may occur even when the model area is small enough that the grid is locally plane. For example, a square grid allows smallest plane waves with different wavelengths for propagation along the diagonal and parallel to the sides. The relation of these smallest wavelengths in the two directions is for the square grid $1:\sqrt{2}$. As the higher resolution in just one direction is not practically useful, this is a waste of computer resources. Chapters "Full and Sparse Hexagonal Grids in the Plane" and "Platonic and Semi-Platonic Solids" investigate the isotropy of different cell structures. It will come out that hexagonal cells are much more isotropic than squares.

The polygonal methods treated in this book use plane polygons on the plane and great circle polygons for the spheres. As auxiliary constructions to create great circle polygonal cells, grids on bilinear surfaces will be used. Triangular grids can be constructed on plane polygons. There are more complicated surfaces than the sphere on which to solve the atmospheric equations, such as the Earth ellipsoid. The tools to construct grids and discretizations on such more complicated curved surfaces are not treated in this book.

Numerical Methods

To construct finite difference schemes on a system of cells, one method is to assume a functional representation for the fields within the cells. It can be shown that piecewise polynomial representations are the only reasonable assumption for scaled basis functions. Different ways to define such piecewise polynomials are possible. As an alternative to using polynomial coefficients, grid points within a cell can be used as collocation points. In such a way, the time development is equivalent to a system of inhomogeneous finite difference equations associated with the collocation points. Other options to describe schemes are possible and will be described. Such polynomial approximations based on polygonal cells are called polygonal spline methods.

The derivation of inhomogeneous finite difference equations to compute the time development of fields is possible using the classical Galerkin method (see chapter "Simple Finite Difference Procedures"). When used with polygonal cells, this is called the finite element method. An important property of the classical Galerkin methods is grid sparseness with approximation orders greater than 1. This means that not all points of a regular collocation grid are used. In 3D, a great saving of computer time is possible in this way and sparse grids will be discussed in chapter "Finite Difference Schemes on Sparse and Full Grids". Grid sparseness is associated with polynomial representations of order higher than one. It happens naturally on quadrilateral grids, where sparse grids with the classical Galerkin method are called serendipity representation. In chapter "Full and Sparse Hexagonal Grids in the Plane", sparse grids will be investigated also for hexagonal and triangular grid systems.

The more modern versions of the Galerkin method discussed in this book are called L-Galerkin (Herrington et al. 2019; Ullrich et al. 2018; Steppeler 1976; Mishra et al. 2010; Ullrich et al. 2018; Giraldo et al. 2003). These will be described in chapter "Local-Galerkin Schemes in 1D". The grid sparseness found with the classic Galerkin method and associated potential savings of computer time are also present with the more modern L-Galerkin schemes. This will be discussed in chapter "Finite Difference Schemes on Sparse and Full Grids". The change of Galerkin schemes to L-Galerkin schemes is necessary as the original Galerkin procedure is not local, even with explicit time integration. This is a problem for massively parallel computers. The development of the L-Galerkin versions has

Fig. 2 An example of
irregular cells (a)–(e) with
edges shown as black lines
and corners 1–7 where corner
5 is a hanging node

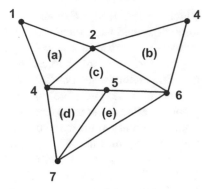

solved this problem. Methods to analyze the usefulness of such grids and L-Galerkin
methods going with these will be described in chapter "Finite Difference Schemes
on Sparse and Full Grids".

In the interior of cells, the fields are represented as polynomials and as such are
continuous and differentiable. In this book the computational methods are based
on the division of the whole computational area into cells. In 2D, the boundary
lines of the cells are called edges. Two or more edges may meet in a point, which
is called a corner point. In Fig. 2, corners and edges are illustrated. It shows five
triangular cells (a)–(e) with seven corner points 1–7. Note that each cell has three
corner points, but the total number seven of corners is not 3×5 as corners may
be shared between different cells. Most corners are end points of edges. However,
corner 5 is on the middle of an edge. Such points are called "hanging nodes."
Grids without hanging nodes are called conforming. Hanging nodes sometimes
require special attention. Consider the situation that at a corner point, two edges
have end points whose angle is different but near to 180 degree. So hanging nodes
may appear as a limiting case of conforming grid nodes. In this book many of the
proposed numerical methods allow the treatment of hanging nodes. Finite difference
methods along edge lines can be treated by treating the hanging node amplitudes as
diagnostic (Steppeler et al. 2008). On corner points and edges as well in the interior
of grid cells, points may be defined carrying amplitudes of fields. Such points
are called collocation points. Finite difference schemes compute time derivatives
at collocation points, and L-Galerkin methods define a functional form of fields
within a cell, using basis functions associated with the cells and grid points. The
time derivatives of collocation grid point values or other amplitudes are obtained by
L-Galerkin procedures, which will be provided in chapter "Local-Galerkin Schemes
in 1D". The regularity requirements of fields at cell interfaces determine very much
the properties of the resulting L-Galerkin schemes and their numerical efficiency.
Some methods admit fields being discontinuous at cell interfaces and these are
called discontinuous Galerkin (DG) methods. For their description, refer to Durran
(2010). This book concentrates on continuous basis functions. When the continuity
of fields at cell interfaces is required, the methods are called continuous Galerkin
(CG) methods. Many versions of such L-Galerkin methods are possible, some of

which are described in chapter "Local-Galerkin Schemes in 1D". For some methods to be described later, such as o3o3, the numerical result of a time step is the same whether performed in grid point space or spectral space. The two versions differ by roundoff error only and can be chosen for numerical efficiency. For other L-Galerkin methods, such as SEs, the numerical result depends on the choice of the collocation grid. Therefore, for this family of methods, the choice of collocation points is called an essential feature. Stronger continuity properties are possible, such as the continuity of the derivatives. Such methods will be treated in chapter "Local-Galerkin Schemes in 1D". Apart from the continuity properties of the fields, the continuity of fluxes is important. This increases the number of possible options. Such methods and some of their properties will be described in chapter "Local-Galerkin Schemes in 1D".

Validations of Numerical Methods Using NWP Models

There are many numerical procedures which could be considered for NWP and we need rational methods to select among them. In this book, we consider methods being suitable for realistic use, that is, with 3D models using realistic initial data. Realistic models predict pressure (p), temperatures (T), and winds ($\mathbf{V}$) as well as a number of moisture components. Further atmospheric contaminants may be modeled. We need verification methods to measure the merits of different schemes. A natural method to compare two numerical approaches is to implement them in a realistic forecast model and compare the results. Before finally using a new numerical approach, such comparisons will have to be done. Figures 3 and 4 give an example from Steppeler et al. (2006). The precipitation output is given for the Lokal Modell (now called COSMO (Steppeler et al. 2006)) developed by Deutscher Wetterdienst (DWD). Figures 3 and 4 compare the standard version of LM with an experimental version, called LMZ. This experimental version uses a different numerical treatment (cut-cells, see sections "A Simple Cut-Cell System Based on the Staggered Low-Order Basis Functions" to "A Conserving Version of the Cut-Cell Scheme" in chapter "Finite Difference Schemes on Sparse and Full Grids") of the lower surface. Figures 3 and 4 give the precipitation and vertical velocity outputs for the two model versions LM and LMZ and the observed precipitation. The output is given for four precipitation sums of 6 hours each, adding up to 24 hrs. It is not the subject of this book to decide which is the better forecast in Fig. 3. Rather, we are interested in developing new numerical schemes and methods to verify them. The comparison of numerical procedures using realistic forecasts is the topic of a large body of literature (Steppeler et al. 2003; Benoit et al. 1997; Dudhia 1993), and the purpose of such research is the improvement of weather or climate predictions.

It is obvious from Figs. 3 and 4 that the different numerical procedures produce quite different forecasts, and this can in principle be used to evaluate the usefulness of numerical procedures. For the example of Figs. 3 and 4, the numerical procedure tested is an alternative numerical treatment of the lower surface. In section "A

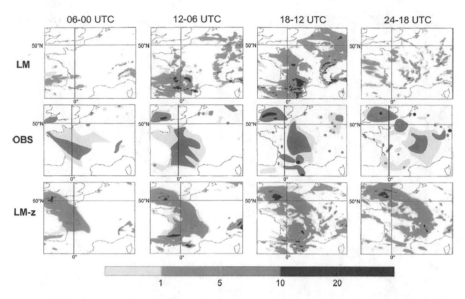

Fig. 3 24-hr precipitation forecasts in mm using the nonhydrostatic 3D model LM. Top: terrain following version, middle: observation, and bottom: cut-cell forecast (reproduced from Steppeler et al. (2006), © American Meteorological Society, used with permission)

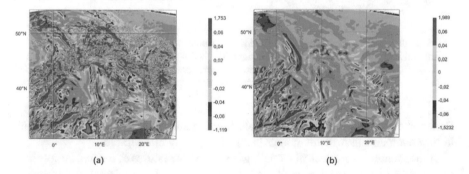

Fig. 4 The vertical velocity belonging to the 6 hr forecast (the second row of results) in m/sec for LM (bottom) and LMZ (top) (reproduced from Steppeler et al. (2006), © American Meteorological Society, used with permission)

Simple Cut-Cell System Based on the Staggered Low-Order Basis Functions" in chapter "Finite Difference Schemes on Sparse and Full Grids", the problem of lower boundary conditions is discussed. Often, one forecast is not enough to draw conclusions. Therefore, Fig. 5 shows the same model comparison for fifty cases of 24 hour forecasts. The root-mean-square error (RMSE) of temperature is shown for the comparison to radio-soundings. While the difference of surface approximations produces a reduction of error in the average of a large number of cases, it is fair to say that the increase of the approximation order (see chapters "Simple Finite

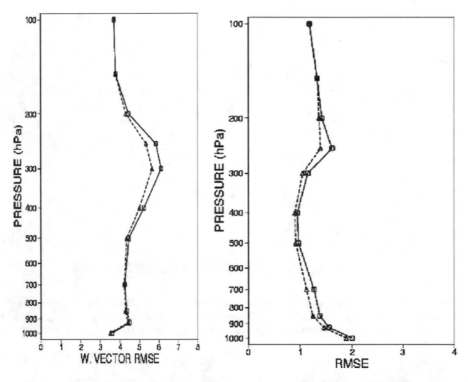

Fig. 5 Radio-sound verifications. (left) RMSE of 24-hour vector wind forecast (m/s) for LM-tf (solid) and LM-z (dashed) as function of height (defined by hydrostatic pressure against radio-sounds). (right) As in left, but for temperature (reproduced from Steppeler et al. (2006), © American Meteorological Society, used with permission)

Difference Procedures" and "Local-Galerkin Schemes in 1D") so far did not result in improvements in realistic models being statistically measurable.

A nice feature of numerical modeling is that many features of the real atmosphere are present in the model as well, and this can be used to understand differences of results of two numerical approaches. For the result concerning precipitation in Fig. 3, it can be shown that the differences of precipitation forecasts are caused by different forecasts to vertical velocities, which are shown in Fig. 4. Both forecasts show the band of ascending motion associated with the front causing the rain. However, already after 6 hr, the vertical velocities are quite different with the two model versions. By means of theoretical meteorology (Pedlosky 1987), it is possible to discuss the realism of such differences. It is also possible to compare to observed vertical velocities. As vertical velocities are difficult to observe directly, indirect observations using other observations are used to compute the vertical velocity field. An example of comparing forecasts of vertical velocities with observations is given in Steppeler (1978) and Steppeler et al. (2013). The difference of the two approximations shown in Fig. 3 is in the approximation of the lower boundary. This condition is that there will be no flow through the lower surface. Approximation principles associated with this boundary condition will be described in chapter "Finite Difference Schemes on Sparse and Full Grids". Research into numerical modeling often involves following up numerical differences of models in the way indicated. In the following chapters this book will concentrate on more simple and basic tests.

Before using a new numerical approach, tests as exemplified above using real data and 3D modeling with realistic initial data are indispensable. Some of such numerical approaches involve classical examples of pure mathematics, such as geometry. Therefore, we are in the strange situation that atmospheric measurements are used to validate mathematical schemes, whose validity is known a priori. This procedure is justified and valid, as we need to know the practical meteorological relevance of a model feature. This means we need to demonstrate if and how much the forecast or climate simulation is improved by a mathematical development. As the validity of many developments discussed in this book is known a priority, it will often be sufficient to investigate the sensitivity of such developments. Often, it is not necessary to do numerical experiments and data validation to the point that statistical significance is achieved. The validity of the numerical developments is known a priori. This can be different for some physical developments, where the choice of a parameter is justified only by numerical results. An example for such parameters is the surface albedo of a discretization cell. The surface albedo in a pixel of say 7×7 km is not constant but in many models is modeled as if it would be constant. The most reasonable value to be assigned can be determined by the best outcome of numerical experiments of the kind shown in Fig. 5. As forecasts at weather centers are done every day, there is a large amount of data available to make such evaluations of parameters significant. Such developments, whose validity is not known a priori but rather need many data for a statistically significant test, are called parameterization. Not all presentations of physical processes in models derive their validity exclusively from a validation of numerical experiments. The progress with modeling physical processes, such as radiation, was related to going from parameterization to discretization. This means that for the inclusion of physical processes into models we need to distinguish between discretized and parameterized physical processes. For the example of radiation, one can imagine that the modeling assumption is made that the albedo of a pixel is constant within each pixel. This is an idealized and not strictly valid assumption, as for typical pixel sizes of pixels in models the albedo may vary within the pixel. This means that while the radiation part of a model may be considered a discretization, the assignment of a pixel albedo may be a parameterization. In this book we are not dealing with parameterizations. The numerical developments treated in this book are discretizations and their validity is known a priori. Therefore the validations of new numerical procedures are often done by a small set of experiments not being large enough to result into statistical significance. The model change is subjected to a sensitivity test and not a statistically significant proof by experiment.

The validation method for numerical forecasting benefits from the large amount of data available in forecast centers where every day forecasts are done. The tuning

Table 1 List of realistic forecast models (Marras et al. 2016)

Model	Country	Institute/University	NH/HS	Type
ETA	USA	NCEP	NH/HS	LAM
GFS	USA	NOAA	HS	GCM
COSMO/LM	GER et al.	DWD	NH	LAM
IFS	GBR	ECMWF	HS	GCM
ICON	GER	MPIfM/DWD	NH/HS	
CAM-FV	USA	NCAR	HS	GCM
CAM-SE	USA	NCAR/SNL	HS	GCM
GRAPES	CHN	CMA	NH	LAM/GCM
MM5	USA	NCAR	NH	LAM
OMEGA	USA	CAP	NH	GCM
HOMME	USA	NCAR	HS	GCM
WRF-ARW	USA	NCAR	NH	LAM
NICAM	JPN	JAMSTEC	NH	LAM/GCM
MPAS	USA	NCAR/LANL	NH	LAM/GCM
NUMA	USA	NPS	NH	LAM/GCM
GONGHO	GBR	MET OFFICE	NH/HS	LAM/GCM
GAMIL	CHN	LASG-IAP	NH/HS	GCM
FAMIL	CHN	LASG-IAP	NH/HS	GCM

of the parameters used with parameterized physical processes is normally done for the modeling systems at forecast centers. As some of such models (e.g., MPAS (Park et al. 2014) for global modeling and WRF for limited area modeling) are public domain, it is possible even for small research groups or single researchers to put numerical developments into models with well-tuned parameterized physics schemes. A list of realistic forecast models is given in Table 1.

The real data validation, as described above for weather forecasting, benefits from large samples, as forecasts are done regularly and use a rather dense measurement system. Therefore, parameter choices for the physical parameterization system can often be determined in a statistically significant way. Models for climate prediction are often derived from weather forecast models (COSMO/ICON, Rocket et al. (2008)). In this way the climate evaluation modes benefit from a well-tuned physics system.

Verifications of Numerical Methods for Climate Modeling

The use of an atmospheric model in climate prediction mode poses problems of validation, which are mentioned only shortly. Models in climate evaluation mode are integrated for long time, being outside the useful prediction range for forecasts. It is known that the prediction of an atmospheric feature such as the temperature at a certain location is valid only for short forecast time. Outside this time range is the chaotic range where forecasts in the classical sense are not valid. Even for very low dimensional schemes, it may happen that there exists a chaotic range (Lorenz 1963). Typically, the chaotic range does not admit arbitrary states. Even though for the atmosphere the state is not predictable, it cannot wander into arbitrary areas of the phase space. It will always be near to a small subset of the phase space. For example, the atmosphere will not reach temperatures outside the climatic range. This subset is called the attractor. Climate evaluations by models try to evaluate the attractor by space and time averaging the model results.

As a simple possibility the climate state can be defined as an average in time. For times being large compared to the averaging interval, we may obtain a state depending on time. There exists the possibility that at a given time the climate state is a linear combination of more than one state. In this way, we obtain the climate attractor as a subset of the phase space of the whole atmosphere. The climate states form a manifold of a dimension small compared to the dimension of the atmospheric phase state. The same large datasets of high-resolution analyzed observations as used for forecasting can be used or its determination. This means that the climate state may contain a lot of spatial detail. This is the case even for the simplest case of a point attractor. If we are interested in the change of climate in time, we have fewer data. The amplitude of the climate state is the variable of interest and there exists only one case, which is the climate of the past. As opposed to this we can make experimental weather forecasts at many dates of the past. If these dates differ by more than ten days, they are outside the useful prediction range and may be considered independent. Model validation for climate purposes has only the one observed climate for validation. So a difficulty of climate validation is that the database is smaller than for weather prediction, as there is only one time trajectory in phase space. The predicted climate signal is typically of low dimension in phase space even though this state may be of high resolution, meaning that it has a lot of regional specification. So the principal value of high-resolution observations or data assimilation is not necessarily of immediate use for climate evaluation. If for example the atmospheric state averaged over a certain time interval is considered to be the climate state, highly variable deviations from it are by definition the non-predictable part of the atmosphere.

The non-climate signal of the system, which is documented by a relative wealth of data, is of interest only in so far as it provides the forcing for the climate part of the atmosphere. In this respect there is an analogy to the turbulent flow in fluid dynamics. This needs a lot of data to be described in detail. However, the turbulent flow does not need to be predicted in detail if the prediction of the laminar flow is the aim. Accuracy of approximations concerning the turbulent part implies an accurate prediction of turbulence details for very short times, which are much shorter than the times of interest for the laminar flow. So we need to know if the forcing provided to the laminar flow comes out correct. If a model of the flow is given, the accuracy can be investigated. In the same way for the atmosphere, it is in principle possible to investigate the accuracy of the forcing of the high-resolution part to the climate signal. Such questions are discussed in Stocker et al. (2013). One way of investigating the errors is to compare model results of different centers to get an indication of the errors involved. The subject of this book is the development of numerical methods. It should be appreciated by the reader that climate applications imply special difficulties of validation as compared to forecasts. The relation of small-scale turbulent state and climate signal is accessible to modeling both in toy and in real-life models. To the knowledge of the authors, investigations into how the forcing for the climate signal can be determined by the transient part of the atmospheric system have not been done. To find such laws for the climate system would be analog to turbulence parameterization in fluid dynamic calculations. In a climate model, the analog to a turbulent motion is the states representing the weather as opposed to the climate state.

In another respect, climate predictions have an advantage over weather prediction. Climate evaluation evaluates the attractor, which is a low dimensional surface or even a curve in a high dimensional phase space. This low dimension of the attractor leads to the effect that sometimes the measurements at one site are sufficient for climate evaluation and its prediction. This is for example the case for pressure measurements at Darvin, Australia. It has been argued that if only a few measurement sites are available for evaluation a low dimension of the attractor is necessary just to be able to speak of a climate system. For the prediction of realistic climate features in very low dimension, see Steppeler (1993).

The simplest situation for climate evaluation is a point attractor, meaning that there is just one state, representing the climatological average. The weather is the deviation from this state. The point attractor may depend on the external forcing, so that the impact of outer conditions on it can be computed by long integrations of atmospheric models. This point attractor could depend on time just by a change of forcing conditions. In order to investigate and verify this, there is only one time trajectory. Note that this kind of attractor is a low dimensional surface in a very high dimensional space, which comprises all relevant amplitudes of the atmosphere dependent on time. This means that each point on the attractor is pointed in a high dimensional space and its determination requires a large number of measurements. To the observer of a climate chart, it appears as a wealth of information and one element of the attractor curve is a map indicating climatic properties of different regions. Such climatic maps may be high resolution. The fact that we are on a low dimensional surface, however, means that the climatic change at one point of the climatic map implies the change at all other points. This has the consequence that a measured time series at two points allows the computation of a whole climate system. This is for example the case for the southern oscillation (Steppeler 1993). So there is a large database to achieve statistically significant conclusions, at least for time intervals where atmospheric measurements are available. In this simple kind of climate evaluation, the famous butterfly, whose action may change the weather development, can be evaluated as follows:

Indeed, if for a forecast we are in the chaotic range of prediction, and not in the predictable range, the small change of initial conditions caused by a butterfly can change the weather forecast after sufficient time totally, meaning within the climatological range. It is unlikely that the butterfly changes the climate. The very idea of a point attractor representing the climate means that the butterfly changes the transient or turbulent part of the atmosphere, not the point attractor. This action is not a change of forcing. A change of climate by small or large causes is possible when the cause is a change of forcing in the point attractor model. There is also the possibility that the climate has its own dynamics. Investigations in this respect are currently in its infancy. This possibility is indicated by the Lorenz toy model (Lorenz 1963) to be described in sections "The Boussinesq Model of Convection Between Heated Plates" and "The Lorenz Paradigmatic Model" in chapter "Simple Finite Difference Procedures". In order for the climate to have its own dynamic, we can assume that there is a low dimensional curve in phase space and this is determined by a few parameters: three for the Lorenz model (Lorenz 1963) and two for the model given in Steppeler (1978). In principle a full atmospheric model in climate mode approximates a large part of the phase space and the weather it computes is a turbulent forcing for the climate manifold. The climate manifold is a low dimensional surface in phase space and positions there can be computed. If the action of the weather on the climate system can be parameterized, meaning that it is written as a function of the parameters determining the climate state. It can be computed in a low dimensional model. This low dimensional climate model can be extracted from the high dimensional climate model (see chapter "Simple Finite Difference Procedures").

The Lorenz model involves three dynamic amplitudes C, L, M which are functions of time t. The derivation of dynamic equations for C, L, M using the Galerkin least square functional is described in chapter "Simple Finite Difference Procedures". This model is a simple example of an unpredictable system. Such unpredictable systems are called chaotic system which means that two initial values of very small difference can result into very different predictions after some time. Even in the limit of the error of the initial values going to 0, the predictability is limited to a finite time interval. The time development of a long integration is shown in Fig. 6. Two branches of the solution can be seen. The solution not at all occupies the whole phase space. The attractor is mainly situated on two "leaves" and transitions between the leaves are possible. The Lorenz model is a paradigm of a chaotic solution. A small error added to the initial value can make the error of prediction large, meaning that the predicted point can be anywhere on the attractor. For more information on this solution, see Lorenz (1963) and Stull (2018). So treating this toy model by the point attractor model, the attractor point would be somewhere in the middle of the curve shown in Fig. 6b. Then the movement of the point on the two branches would be attributed to the turbulent flow part. This is a rather coarse description and another possibility would be to assume that there is an intrinsic climate dynamic. We could assume that the point lies on one of the branches shown in Fig. 6. Outside the two branches shown, a point cannot exist as it would soon join the attractor. We would in the climate dynamic assumption assume that we could add further degrees of freedom which would represent the turbulence in the climate dynamic model. Due to the chaotic nature of the Lorenz model, the point within the attractor cannot be predicted, but the question would be of interest if the transition between the two branches is accessible to prediction. Questions

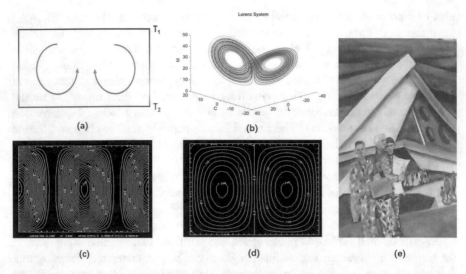

Fig. 6 The convection between plates. (**a**) Symbolic representation and basis function corresponding to C, (**b**) time diagram for C, L, M (Lorenz "Butterfly Solution"), (**c**) and (**d**) flow for points
in different places of the attractor, (**e**) Artist's impression of the Magamecren's world from Artist
Johannes Tittel. The Magamecrens live in a large rectangular cave, allowing voyages of several
days. Settlements in caves are today again considered for the colonization of Mars. This world
was imagined by the Venetian writer and mathematician Giacomo Casanova (Casanova 1964).
The inhabitants of this colorful world communicate by singing pieces of music to each other. The
climate is described by the Lorenz paradigmatic model (see (**a**)–(**d**) which has just three dimensions
in phase space). For the case that the cave is rectangular and has homogeneous surface friction, the
Lorenz model is quantitatively correct and describes real weather and climate in the rectangular
cave. The distribution of rain is given by the predicted upward motion. Long integrations of the
model can be done on a normal PC or smart phone because the model in phase space is only three-
dimensional. The model shows a bifurcation of the climatic solution in the so-called butterfly
diagram (**b**). In the weather interpretation, this means that the locations of rainfall change in the
model. This change is unpredictable. Small errors in the initial values may change the predicted
branch of the climatic solution. This means that the positions of rainfall will be unpredictable for
sufficiently large times. The horizon of predictability of the Lorenz model is finite, meaning that
an arbitrary small error of initial values will not bring the predictability to a finite value

for the validation of numerical schemes would be if a certain numerical procedure
would impact the climatological prediction within the framework of this low and
high dimensional toy model. We could ask how realistic is the low dimensional
Lorenz model, how good is the representation of turbulence implied by the low-
order model, etc. This is not the subject of this book, where we deal just with the
tools of doing this.

In NWP we are using laws of mechanics and thermodynamics, which are
predictable and the uniqueness of the solution implies that with the absence of errors
the solution is determined for an infinite time. This has in former time led to the
concept of "Laplace's Demon", meaning that for a person with infinite knowledge
the world may be known for an infinite time. However, the paradigmatic solution
of this Lorenz concept is changed even for simple systems of classic mechanics. If,

with the Lorenz solution, the initial values are given by an error ϵ, we know that the solution is known only for a finite time $t_{causal}(\epsilon)$. It is now possible to investigate how t_{causal} depends on ϵ. If $t_{causal} < t_0$, then there exists a time horizon, where the future is unknown, even if the initial data are measured with infinite accuracy. This means that the "Laplace's Demon" cannot exist even for simple systems of classic mechanics.

One worrying feature of the paradigm (Lorenz 1963) is that even for this simple model the attractor consists of two branches which are quite different from each other and a point, in the paradigm representing a climate state, can move from one branch to the other, which theoretically could mean a large climate effect for a small cause.

It is shown in chapter "Simple Finite Difference Procedures" that the climate paradigm (Lorenz 1963) was derived from a simple convection model between plates, which physically has a relation to the real climate system. As for the real Earth between north pole and equator, the flow between the plates is determined by a temperature difference and governed by heat transport. The two branches of the Lorenz attractor (see Fig. 6) represent states of different sign of rotation, which can be seen in Fig. 6 by the fact that the amplitude C representing the wind amplitude changes sign between the two "wings of the butterfly." So an inhabitant of an area near one of the plates will see a change of wind direction when a regime change happens between the two parts of the attractor. This inhabitant will experience warm and cold periods. This comes from heat reservoirs shown by different values of the amplitudes L and M that are filled by the flow and may influence the winds, given by amplitude C. The example shows that it may be possible to evaluate the climate not by point values but rather by low dimensional climate dynamics. A high dimensional climate analysis would be difficult to perform. By high dimension here we mean high dimension in phase space, not a high resolution of the climatic states. It is possible and general practice to compute the climate eigenstates with high resolution and obtain climate information up to the local scale. When wanting to compute the climate system's time dependence, the most natural idea is to use a low dimensional system. Current work on climate evaluation is described in Stocker et al. (2013). It is worth mentioning that in the literature of the eighteenth century, a population living between heated plates was described (Casanova 1964), whose inhabitants are called Magamecren. The world of Magamecren can be created by technical means and has been produced at laboratory scales, from which it is known that for some parameter ranges the Lorenz model gives a realistic simulation of the Magamecren world. It is possible that for future settlements on Mars a Magamecren world will be created. As will be pointed out in more detail in chapter "Simple Finite Difference Procedures", the Lorenz model for some parameter ranges gives realistic solutions. It is a paradigmatic model which was derived by discretization of a physically realistic model. Figure 6 shows weather patterns of the Magamecren's world as represented in the Lorenz model.

There is another line of climate research aiming for high-resolution climate information for planning purposes. These high-resolution climate models are done in limited areas, as global high-resolution climate models would cost too much computer time. The aim of such models is to investigate local climate effects when a global scenario is already given or to study local climate effects of local causes. Such a cause may for example be deforestation or the creation of buildings. The verification of numerical schemes for such model applications is somewhat similar to the verification of forecast models and is exemplified in Steppeler et al. (2013). However, even in limited area climate applications integration times above the forecast limit for weather may occur. A review of small-scale climate applications is given in Gutowski et al. (2019).

It is physically plausible that storage systems for heat or moisture, for example, are parameterizable systems relevant for a low dimensional climate system. The parameterizability means that the flow of heat, for example to a storage system, is driven by the heat content itself. Even with just four parameters, it was shown in section "Dynamic Equations of Toy Models" in chapter "Simple Finite Difference Procedures" that experimental data could be reproduced when making the shape of the Galerkin basis realistic (Steppeler 1978). In chapter "Simple Finite Difference Procedures", the Lorenz paradigmatic model is described in more detail. For the numerical researcher, it may be confusing that low dimensional Galerkin models are commonly called low-order models. In this book we stick to the expression "Low-dimension Galerkin models." The word "order" here is reserved to the order of approximation. The numerical tests by fully realistic models are mentioned in the introduction but for the rest of the book we concentrate on more simple tests. It is, however, possible that the future professional studying this book will spend much of his/her professional life in the investigation of tests with realistic models and the physical questions behind this. The current book is just on the methods of numerical model construction.

While the comparison of numerical procedures using realistic models is indispensable, this procedure has a number of drawbacks:

1. To program a numerical procedure for a realistic model is considerable work. Typically, a realistic model has several 10^5 statements, where only a few 10^3 statements concern numerical procedures. The places in the big program to introduce new numerical methods have to be found and modified. Realistic models, such as MPAS, WRF or ICON, are often developed by big forecast centers. A list of realistic models is given in Table 1. Changing the numerics of such models may need the help of a specialist of the target model. The computation time needed on large mainframe computers can be substantial.

2. Often, it is not trivial to pinpoint the exact cause of a difference in forecasts. The huge number of parameters determining the setup of a realistic model have an impact on forecasts which is nonlinear. So the comparison of precipitations shown in Fig. 1 could be different if a different scheme is used to transform the different moisture components used in the model into each other. A case in point is the test of advanced high-order approximations, which as indicated above could not be shown to lead to model improvements proven by RMSE using ensembles in realistic models. For simplified modes, however, the advantages of high order can be demonstrated easily. This difference in performance of high-

order schemes with realistic and simplified models has stunned researchers for decades is called the high-order paradoxon of NWP.

3. For example, the test of stability and the range of stable time steps can in principle be done using realistic models. We just need to find the range of time steps which do not make the model blow up. This would, however, be totally unreasonable. Tests using simplified models, such as used with the von Neumann analysis (see chapter "Simple Finite Difference Procedures"), provide the range of stable time steps with little effort.

In this book, a number of simple models for testing numerical properties of schemes will be presented and used to compare numerical procedures. Such models are called toy models. Many of such toy models were used in a similar way by Durran (2010) with regular resolution. The suggested toy models range from homogeneous advection (see chapters "Simple Finite Difference Procedures" and "Local-Galerkin Schemes in 1D") to a 3D dry model. The choice of procedures presented in this book is determined by their usefulness in current model developments. The dry 3D model presented is nonhydrostatic (see section "Staggered Grid Systems and Their Basis Function Representation" in chapter "Finite Difference Schemes on Sparse and Full Grids") and allows modeling of rather small scales.

A range of new models have been developed recently (such as MPAS, HOMME, or ICON) or are under development. These new developments are made possible by advances in computer technology (within the near future computers with 10^{19} floating point operations per second (exaflop) are expected). Older computers used one processor or just a small number of them. The currently expected boost in computer power will be caused by a large increase in the number of processors. For high-resolution forecasts, these must work together and this causes the need for numerical procedures which allow a large degree of parallelism. Some of the procedures, which performed well with a small number of processors, are no longer suitable. For example, classical Galerkin finite element schemes (see Durran (2010) or chapter "Simple Finite Difference Procedures") perform well with a small number of processors, as there are efficient recursive procedures to solve the matrix equations involved. With many processors, these schemes potentially require much message passing when the number of processors increases. This makes such schemes unsuitable with a large number of processors. A program is called scalable if it is programmed to run n times as fast with n-processors than with one processor.

Scalability for large n is the property to be asked from models being ready for the future with exaflop computers. Generalizations of the classical Galerkin procedure have been developed with the aim of creating scalable models. Such generalized Galerkin procedures are called L-Galerkin (Steppeler and Klemp 2017) and described in Chapter "Local-Galerkin Schemes in 1D". An L-Galerkin method being near to realistic application in the models HOMME and NUMA (Kopera and Giraldo 2011) and Mishra et al. (2010) is the spectral element method (SEM). An emphasis of this book is the description of SEM and other L-Galerkin methods with the view of increasing the efficiency of such methods. Classical Galerkin finite element methods of order two or three naturally used sparse grids (see chapter "Finite Difference Schemes on Sparse and Full Grids"). Sparse grids do not use every point of a regular grid. Because no equations are solved at the unused points a considerable saving of computer time is possible. This book investigates the corresponding concept for L-Galerkin methods. The possible saving depends on the dimension of the problem, the type of discretization cells used, and the order of approximation. According to chapters "Finite Difference Schemes on Sparse and Full Grids" and "Full and Sparse Hexagonal Grids in the Plane", for third-order schemes on hexagonal cells, a saving of computer time by a factor ten appears possible.

While the old model generation used latitude–longitude grids (see Durran (2010)), grids based on Platonic solids are used with many of the modern model developments. Grids currently under development with realistic models are based on the cube or the icosahedron. These grids are known for a long time (Sadourny and Morel 1969; Williamson 1968). The old polygonal developments suffered from grid imprinting which means that the grid is seen in the solution and this is caused by the approximation error being larger at special points of the grid, such as poles. Only recently, it was pointed out that grid imprinting can be avoided (Steppeler et al. 2008; Steppeler and Prohl 1996; Rancic et al. 1996) and grids based on Platonic solids found their place in modeling (Satoh et al. 2014). Apart from Platonic solids also semi-Platonic solids can be used. Cell structures on such grids can be based on triangles, rhomboids, and hexagons (see chapter "2D Basis Functions for Triangular and Rectangular Meshes").

For homogeneous finite differences, the physical processes are often called at every grid point, while for inhomogeneous cells the physical processes may be called only at fewer points (Lander and Hoskins 1997; Steppeler et al. 2019). A method to test such concepts is shown in chapter "Local-Galerkin Schemes in 1D". Due to calling the physics procedures at fewer points, a factor of two in computer resources may be saved while also improving the result.

In summary, this book relies on Durran (2010) for time discretization and simple difference schemes. A few of such concepts are described here, in order to make this book self-contained and make it readable without first having to read other books. For temporal discretization, the tests provided here rely entirely on RK4 (see Durran (2010) and chapter "Simple Finite Difference Procedures"). The space discretizations provided can easily be used with any of the other time discretization methods described in Durran (2010). The simple spatial schemes are generalized and analyzed for the irregular case with a view to obtain a homogeneous order of approximation. Such homogeneous order schemes avoid the problem of grid imprinting and thus are suitable for polygonal and nested grids. L-Galerkin schemes, such as SEM, allow to construct polygonal spline schemes of high homogeneous order in combination with conservation. From the large number of L-Galerkin schemes, some key methods are analyzed. The regularity of splines at corner points is an important element of such polygonal splines. The splines considered in this book are at least continuous in the direction of differentiation. Therefore, the methods presented in this book are CG (DG methods are described in Durran (2010)). Schemes with stronger regularity properties will be described, such as schemes with field or flux representations being differentiable at corner points for fluxes and schemes having differentiable spline representations for the fields.

Like any other book, this book can be read without practical applications. Alternatively, it is possible to use it for courses in practical computations. For this purpose in chapter "Examples of Program", program examples are given. Examples will run on a personal computer or notebook. A compiler and a plot program are necessary.

Simple Finite Difference Procedures

Abstract This chapter gives a few definitions of standard numerical schemes to be used for comparison or as part of the schemes to be described in later chapters. As this book is mainly about spatial discretization, the fourth-order Runge–Kutta scheme is used in tests for time discretization. A rather comprehensive set of possibilities is given in Durran (Numerical methods for fluid dynamics: with applications to geophysics, 2nd edn. Springer, New York, pp. 35–146, 2010), along with an analysis of many standard numerical temporal schemes. This information is given here to make this book self-contained. It is readable with basic knowledge of mathematics and without previous knowledge of numerics and this chapter is overlapping with Durran (Numerical methods for fluid dynamics: with applications to geophysics, 2nd edn. Springer, New York, pp. 35–146, 2010) for the convenience of the reader.

Keywords Dynamic equation · Interior boundaries · Von Neumann analysis · Inhomogeneous finite differences · RK method · Homogeneous approximation order · Order preserving weights

The problems solved in this book can be formulated as systems of conservation laws. Let $F^k(t, \mathbf{r})$ be a set of fields, t being the time variable and $\mathbf{r}$ being a spatial vector of 1D, 2D, or 3D. The index k has the range $k \in (1, k^{dim})$, where k^{dim} indicates the number of different fields used for our model. For example, temperature (T), density (ρ), and three velocity components (u, v, w) for a dry atmospheric model are used. In this case $k^{dim} = 5$. Typical moist models use another five fields to define different moisture components. Then, the equation of motion of these fields is

$$\frac{\partial F^k}{\partial t} = \frac{\partial}{\partial \mathbf{r}} \mathbf{RS}^k(t, \mathbf{r}),\tag{1}$$

where $\mathbf{RS}^k(t, \mathbf{r})$ is a function of the fields F^k. For each k, $\mathbf{RS}^k$ is a 3D vector, called *the flux*, where the components correspond to the three spatial directions x, y, and

© Springer Nature Switzerland AG 2022

J. Steppeler, J. Li, *Mathematics of the Weather*, Springer Atmospheric Sciences,
https://doi.org/10.1007/978-3-031-07238-3_2

z, $\mathbf{r}$ is the vector $\mathbf{r} = (x, y, z)$, and $\frac{\partial}{\partial \mathbf{r}}$ is defined to be the divergence operator. For each 3D vector $\mathbf{RS}(t, \mathbf{r}) = (RS^x, RS^y, RS^z)$, it is defined as $\frac{\partial}{\partial \mathbf{r}}\mathbf{RS}^k(t, \mathbf{r}) = -\nabla \cdot \mathbf{RS} = -\left(\frac{\partial RS^x}{\partial x} + \frac{\partial RS^y}{\partial y} + \frac{\partial RS^z}{\partial z}\right)$. We assume that fluxes are defined in such a way that Eq. (1) is a hyperbolic problem (see Durran (2010)).

Different choices of $F^k(\mathbf{r})$ can represent different physical systems. For example, we may have only one field F^1 and this may represent the density of an atmospheric contaminant, being transported in a given velocity field, meaning the change of the velocity components in time is not computed dynamically. Another possibility is the dry atmosphere, represented for example by three velocity components, temperature, and pressure. Other equivalent choices of fields are possible. For example, one could use the momentum in 3D rather than the velocity components. Durran (2010) gives a number of examples of dynamic systems. Systems used for tests in this book are described in section "Dynamic Equations of Toy Models".

In order to approximate systems of the form Eq. (1) numerically, grid points and cell structures are introduced. In this book, we limit ourselves to the 1D and 2D cases. Let Ω be the area in 2D space where Eq. (1) is valid and denote the boundary of Ω as $\partial \Omega$. Figure 1a shows 2D examples of a rectangular area, where $\partial \Omega$ consists just of the sides of the rectangle. A cell structure is any division of Ω into subareas or cells. These cells must be numbered by an index μ. For any given cell μ, it will be necessary to know the neighbors μ' in order to perform a numerical approximation. When the cells are irregular as shown in Fig. 1b, these neighborhoods are often given by a table of neighboring cells and this cell structure is called *unstructured*. For each cell, the number of neighboring cells can be different. Figure 1a shows the example of a regular division into rectangular cells. Each cell can be indexed by i, j ($i \in 0, 1, 2, \ldots i_{max}$, $j \in 0, 1, 2, \ldots j_{max}$), where we assume the area to be quadratic and the same resolutions in the two directions. In this 2D example, the cell index i, j can be seen as the center point x_i, y_j of a cell. Such grids are called *structured*. For structured grids, no table of neighboring cells is necessary. From Fig. 1a, it is seen that the neighbors of cell i, j are $i-1$, j; i, $j-1$; $i+1$, j and i, $j+1$. To investigate properties of numerical approximations, 1D grids (see Fig. 1d–f) are used, where the cells are intervals of the variable x. As seen from Fig. 1d–f, all 1D grids can be written as structured grids, which explains their popularity for performing preliminary tests of newly designed numerical procedures.

Within each cell grid, points $\mathbf{r}_i$ are assumed. For the fields F^k in Eq. (1), amplitude values F_i^k are assumed. These amplitudes are used to devise numerical procedures and they can also be used to plot the fields. In one dimension, the computational domain is always divided into several intervals bounded by two end points. The center point is defined in the middle of an interval. This center point often carries the amplitudes of the density field. For the dynamic fields (for example, velocity vector), their amplitudes are often positioned at the cell boundaries. Figure 1e gives an example in 1D for grid points being at the cell centers. If densities and velocities are defined at different positions, the cell arrangement is called staggered. Cell boundaries are the black points and the density amplitudes are defined at the white points. Some numerical procedures use more than one

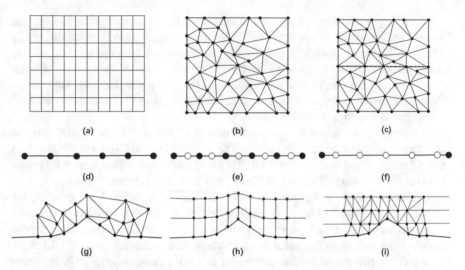

Fig. 1 Examples of the grid structures in NWP models: (**a**) Regular rectangular grid. (**b**) Totally irregular triangular cell structure. This grid requires unstructured programming, using tables of neighboring cells. (**c**) Triangular grid with internal boundary. (**d**–**f**) 1D cell structures with collocation points. Black points indicate cell boundaries or corner points. The black points can but must not be taken as collocation points. Collocation points may be chosen at cell boundaries and center points. This collocation grid is often used with inhomogeneous differences, where the black and white points use different FD formula making the finite difference scheme inhomogeneous. (**e**) Shows two independent collocation points per cell, such as used with a second-degree polynomial representation. (**f**) Shows an interval with six collocation points in one interval, as occasionally used with sixth-order SEs. (**g**)–(**i**) Show cell structures adapting to the lower boundary. (**g**) Is an irregular cell structure accommodating the lower boundary. (**h**) Is more regular than (**i**), but the grid points are not horizontally aligned. The cells in (**f**) are horizontally aligned. The differences of the grid between (**g**), (**h**), and (**i**) potentially lead to different noise structures for the flow along the surface

amplitude per cell. Figure 1e gives the example where amplitudes at cell centers and interval ends are assumed. These grid points assumed in a cell are called *the collocation grid*. Care must be taken to count the number of collocation points, as some of them belong to more than one cell. For the 1D example in Fig. 1e, the center points belong to one cell only, where the points at the edges belong to more than one cell.

For the totally irregular cell structure, a continuous and smooth line may appear, dividing two patches of the grid as indicated in Fig. 1c. Such lines are called *interior boundaries*. Interior boundaries have no physical meaning. They are not physical boundaries. Just the numerical treatment is different at such lines. Such interior boundaries were occasionally found to be a source of numerical error. Even errors larger than the amplitudes of the physical solution and instabilities were found (Thuburn et al. 2001). Another example of interior boundaries occurs with the rhomboidal patches occurring with the icosahedral meshes on the sphere (Fig. 1c in chapter "Introduction"). Currently internal boundaries are sufficiently understood to be no longer a source of problems. However, the spherical grids in Fig. 1 in chapter "Introduction" were considered longtime to be problematic to handle, partly because of internal boundary problems. One important element when avoiding interior boundary problems is the use of methods which do not have a drop of order of approximation at these lines. The concepts of stability and order of approximation will be introduced in the following chapters "Simple Finite Difference Procedures" and "Local-Galerkin Schemes in 1D".

One property of a good solution is that the grid and in particular internal boundaries must not be seen in plots of the results. Section "Shallow Water Tests on the Sphere: Solid Body Rotation, Solid Body Flow, Advection, and Williamson Test No. 6" in chapter "Numerical Tests" will give an example. There can always be special experiments designed to show internal boundaries. For example, the spherical grid shown in Fig. 1c in chapter "Introduction" can be used to compute the height of a water surface using a shallow water surface, where the analytic solution corresponds to the solid body rotation. The numerical results will show some small variations of height, indicating an error of maximum E. On the other hand, the error level $E1$ is the error of physically significant solutions, such as that of meteorological waves. The error level E is not significant and will not show in graphical representations of fields when $E \ll E1$. This experiment has been done and the results are shown in section "Shallow Water Tests on the Sphere: Solid Body Rotation, Solid Body Flow, Advection, and Williamson Test No. 6" in chapter "Numerical Tests". In this case, the amplitude variation caused by the internal boundary error is 7 orders of magnitude smaller than that of the realistic solution. So for the special test solution of the solid body rotation, a very narrow spacing of isolines in plots can make the internal boundaries visible. If, however, the line spacing allows the representation of the physical solution, no grid related effects can be seen. This leads us to the following definition: a solution procedure has a homogeneous order n of approximation when the approximation is of order greater than or equal to n for all points. Special points, such as those at internal boundaries, must be no exception. The order of approximation will be defined in the next sections. Internal boundary effects and more general grid related errors are absent, if the grid is not seen in plots of the solution adjusted to the physical effect to be computed.

A totally different matter is external boundaries, such as the lower boundary in Fig. 1g–i. In Fig. 1a, the external boundary is a straight line. In fact, realistic models, such as those listed in Table 1 in chapter "Introduction", are mostly using horizontal lower boundaries of the model area. Mountains are then introduced in these models with a flat surface by coordinate transformations (Durran 2010). This construction is known as terrain-following coordinate (Phillips 1957). The coordinate transformations can lead to good representations with very well-resolved or smooth mountains. However, in realistic models, mountains normally are not well enough resolved for this procedure to be accurate. This leads to a number of problems which will be discussed in chapter "Finite Difference Schemes on Sparse and Full Grids". Another possibility is to include the lower boundary into the grid definition, as shown in Fig. 1i. If we use straight lines to define cell boundaries, the mountain in Fig. 1g is represented by a piecewise linear spline. On this lower boundary, boundary conditions must be implemented, such as that no mass is going through this boundary. Figure 1h and i shows two options for terrain-following grids. Most existing models use the regular grid point option (h), but this book provides the tool to treat the irregular case (i).

In case that Eq. (1) involves more than one field $F^k(\mathbf{r})$, these different fields will be assumed to be defined using the same cells. However, they do not necessarily have the same collocation points within the cells. If the collocation points for different fields $F^k(\mathbf{r})$ are different, the grid is called *staggered*. Equation (1) is valid in the interior of Ω. At $\partial\Omega$, special constraints are required to express the behavior of the field for obtaining the solution of Eq. (1) which are called *boundary conditions*. Equation (1) requires initial values for $t = t_0$ which is like a boundary condition in time direction, which is applied for one time level only.

In this chapter, we use the simplest test problem of Eq. (1). We assume just one field ($k^{dim} = 1$) and the field is density h. We assume one spatial dimension ($\mathbf{r} = \mathbf{r}(x)$), and for the flux ($\mathbf{RS}(x) = RS^1(x)$) we obtain $RS(x) = -u_0 h(t, x)$. The constant u_0 is the homogeneous advection velocity of the system, and the test problem derived in this way is called *the homogeneous advection problem*. The dynamic equation for homogeneous advection is

$$\frac{\partial h(t, x)}{\partial t} = -\frac{\partial}{\partial x} u_0 h(t, x). \tag{2}$$

For this test problem, we use periodic boundary conditions:

$$h(t, 0) = h(t, L), \tag{3}$$

where the interval $[0, L]$ is the simulation area. The analytic solution of Eqs. (2)–(3) is determined if the initial values of $h(0, x)$ are given. For other t, it is given as

$$h(t, x) = h(0, x - u_0 t - k'L), \tag{4}$$

for each x and k', it is chosen in such a way that $x - u_0 t - k'L \in [0, L]$. The analytic solution given in Eq. (4) can be used to determine the accuracy of numerical solutions of a numerical approximation to Eqs. (2)–(3). Alternatively, it is possible to define $h(t, x)$ for all x using $h(t, x + nL) = h(t, x - n'L) = h(t, x)$ for arbitrary (n, n'). With this definition, Eq. (4) can be simplified to

$$h(t, x) = h(0, x - u_0 t). \tag{5}$$

In order to obtain numerical approximations to Eq. (2), a grid $x_i \in [0, L]$ must be introduced where $i = 0, 1, 2, \ldots, i_{max}$. In this book, we admit an irregular distribution of the grid points. We require that the difference between points must not be too large. Choose values dx_{max} and dx_{min} called maximum and minimum distances and we require from the grid dx_i

$$\begin{cases} x_{i+1} > x_i, \\ x_{i+1} - x_i \leq dx_{max}, \\ x_{i+1} - x_i \geq dx_{min}, \\ x_{i_{max}} - x_0 = L. \end{cases} \tag{6}$$

Except for conditions in Eq. (6), the mesh may be chosen arbitrarily Fig. 1c where the grid is shown in Fig. 1a for the regular cases.

The spatial discretization procedures to be defined in the following have the purpose to obtain approximations to the right-hand side (RHS) of Eq. (2) at the grid points x_i. According to Eq. (2), these will be approximations $h_{a,t,i}$ for the time derivatives $h_{t,i}$ at the points x_i. The subscript a in $h_{a,t,i}$ stands for approximation. $h_{a,t,i}$ will depend on the set h_i of grid point values $h_i = h(t, x_i)$, $h_{a,t,i} = h_{a,t}(\{h_i\})$. The index a is often dropped. After introducing the spatial approximation to be defined in the following sections, we obtain a set of i_{max} ordinary differential equations (ODEs):

$$\frac{dh_i}{dt} = -u_0 h_x(\{h_{i'}\}). \tag{7}$$

The spatial discretization creates for each grid point a differential equation with the variable t only. Therefore, we have an ODE system instead of a PDE system. Eq. (7) is the spatially discrete form of Eq. (2). While h in Eq. (2) is a function of the continuous variable x, h in Eq. (7) depends on the discrete index $i \in \{0, 1, 2, 3, \ldots\}$. The bracket $\{\}$ in Eq. (7) indicates that the time derivative of h at point i does not depend on one h_i but rather a whole set of h_i. For example, in a regular grid with centered FD scheme, we have $-u_0 h_x(\{h_{i'}\}) = -u_0 \frac{h_{i+1} - h_{i-1}}{2dx}$. Durran (2010) gives examples, where the approximation to time derivative depends on the whole set of $h_{i'} : i' \in [0, i_{max}]$. Such schemes are those where the approximate time derivative has a global dependence on $h_{i'}$. An example is the classical (unapproximated) Galerkin scheme (see section "The Classic Galerkin Procedure" in chapter "Local-Galerkin Schemes in 1D"). For the simple schemes and the L-Galerkin schemes in this book, this is not the case and the RHS of Eq. (7) depends only on neighbors of the target point x_i, where the time derivative is approximated:

$$\frac{dh_i}{dt} = -u_0 h_{a,x,i} (h_{i-2}, h_{i-1}, h_i, h_{i+1}, h_{i+2}). \tag{8}$$

To be very precise, in Eqs. (7)–(8), the LHS should be denoted as $\frac{dh_{a,i}}{dt} = h_{a,t,i}$ as these are approximations for the derivatives. For convenience, we always drop the index a. As compared to Eqs. (7), (8) is more local, as it depends for each target point i on only five amplitudes $h_{i'}$, rather than the whole set $\{h_{i'}\}$. Therefore, such approximations are more efficient and more suitable for multi-processor computers than global approximation Eq. (7).

A rather simple and still popular approximation of the derivative in Eq. (2) is the difference of neighbor approximation:

$$\frac{\partial h_i}{\partial t} = -u_0 \frac{h_{i+1} - h_{i-1}}{x_{i+1} - x_{i-1}}. \tag{9}$$

The approximation on the RHS of Eqs. (7), (8), or (9) is called *the spatial discretization* and their derivation will be the purpose of the following sections. Solving the sets of ODEs in Eq. (8) or (7) is called *the time discretization*. A rather complete discussion of time discretizations is given in Durran (2010). In this book we will solve the examples using the fourth-order Runge–Kutta method (RK4), which will be described in the next section. The analysis given in Durran (2010) shows that RK4 is a good spatial approximation providing good accuracy and permitting rather large time steps. While for research models to be discussed here it is a suitable method, real-life models have to be considered for accuracy as related to efficiency. So for real-life models, it may be useful to consider alternatives to RK4, as described in Durran (2010).

A special case is the regular grid:

$$x_i = i dx. \tag{10}$$

Some tests, such as the von Neumann stability analysis, described in section "The Runge–Kutta and Other Time Discretization Schemes", are possible only for regular grids. In this book, tests are performed also for irregular meshes. This is a necessary step toward realistic applications, as the sphere cannot be covered by regular grids.

A special irregular grid is defined as

$$x_i = i dx_i = i dx \cdot (1 + \delta \cdot r_i), \tag{11}$$

where $i = 0, 1, 2, \ldots$, and r_i is a fixed random number between zero and one and when $\delta = 0.0$, the grid is regular. For $\delta = 1.0$ for example, we get an irregular 1D grid to be used with tests.

The accuracy of approximations such as Eq. (8) is often discussed using the term *order of approximation*. The order of approximation defines how fast the error of the approximation converges to 0 with decreasing dx. Let an approximation of the spatial derivative of Eq. (8) be given. Assume a test function where the analytic form of the derivative is known, such as $h(x) = e^{\lambda x}$, and define the grid point values $h_i = h(x_i)$. For grid point x_i, we define the error of approximation for a spatial approximation Eq. (8) at grid point i as

$$err_i = |u_0[h_{x,i} - \lambda h(x_i)]|. \tag{12}$$

In Eq. (12), err_i is dependent on the grid length. The approximation $h_{x,i}$ is said to be of approximation order n, if the error term is of order $n + 1$ meaning that err_i does not go to infinity when x_i is approximated by its neighboring points:

$$\frac{err_i}{max|x_{i+1} - x_i, x_i - x_{i-1}|^{n+1}} < C < +\infty. \tag{13}$$

For practical purposes, it is convenient to investigate the approximation order by plotting the error err_i as a function of grid length using logarithmic scales. Apart from the neighboring point approximation Eq. (9), methods are analyzed for convergence in Fig. 2, which are (a) standard fourth-order FD (o4)

$$h_{t,i} = -u_0 \left(-\frac{1}{3} \frac{h_{i+2} - h_{i-2}}{4dx} + \frac{4}{3} \frac{h_{i+1} - h_{i-1}}{2dx} \right), \tag{14}$$

(b) L-Galerkin o2o3, and (c) second-order spectral element (SE2) using the regular grid Eq. (10) with the spatial derivative of $dx = \frac{1}{8}, \frac{1}{4}, \frac{1}{1}, 1, 2, 4, 8, 16, 32$ and irregular grid Eq. (11). For the case of a regular grid, the difference of neighbor approximation is called *centered FD approximation*. For example, $h_{t,i} = \frac{h_{i+1} - h_{i-1}}{2dx}$. When we use an irregular grid Eqs. (11), (14) must be changed to be

$$h_{t,i} = -u_0 \left(w_i^{-2} h_{i-2} + w_i^{-1} h_{i-1} + w_i h_i + w_i^1 h_{i+1} + w_i^2 h_{i+2} \right), \tag{15}$$

where w_i are for the irregular case computed numerically using the program MOW_GC Huckert and Steppeler (2021). For the special case of a sudden jump of dx, the w_i are given in Table 1.

The order of approximation can then be seen from the steepness of the error curve in Fig. 2. Solid, dotted, and dashed lines are lines of convergence. There are lines of convergence in first and second orders for the three cases. The approximation is of first or second orders if the error curve is parallel to the respective curve. It comes out that for the regular grid, (a) the o4 scheme and (b) o2o3 scheme converge in fourth

Table 1 Values of w_i in Eq. (15) on the irregular grid with resolution jumps

Resolution	w_i^{-2}	w_i^{-1}	w_i	w_i^1	w_i^2
Fine resolution ($dx = 1.0$)	0.083	−0.667	0.0	0.667	−0.083
Resolution jumps	0.1	−0.75	0.167	0.5	−0.017
Resolution jumps	0.167	−1.067	0.75	0.167	−0.017
Resolution jumps	0.152	−0.5	0.083	0.3	−0.036
Coarse resolution ($dx = 2.0$)	0.042	−0.333	0.0	0.333	−0.042
Resolution jumps	0.036	−0.3	0.083	0.5	−0.152
Resolution jumps	0.017	−0.167	−0.75	1.067	−0.167
Resolution jumps	0.017	−0.5	−0.167	0.75	−0.1
Fine resolution ($dx = 1.0$)	0.083	−0.667	0.0	0.667	−0.083

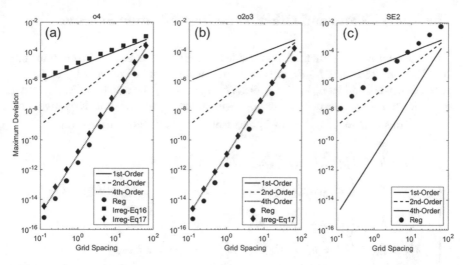

Fig. 2 Convergence speed of o4 (left), o2o3 (middle), and SE2 (right) schemes under irregular and regular grid for different grid spacings: $dx = \frac{1}{8}, \frac{1}{4}, \frac{1}{2}, 1, 2, 4, 8, 16, 32, 64$. The circles represent the error norms for regular grid where the standard o4 scheme defined in Eq. (14) is used for o4, o2o3, and SE2 schemes. The squares are those for irregular grid where the standard o4 schemes defined in Eq. (14) are used and the rhombuses show results for irregular grid where the weighted o4 schemes defined in Eq. (15) are used for o4 and o2o3 schemes (diagram reused from Steppeler et al. (2021))

order and (c) SE2 scheme converges in second order. However, for the irregular grid, the situation changes: the o4 scheme reverts to first order when using the improper approximation Eq. (14) which is suitable for regular grids only. Once the approximation Eq. (15) is applied for o4 in irregular grid, we achieve again fourth-order convergence. The SE2 scheme to be defined in section "Spectral Elements" in chapter "Local-Galerkin Schemes in 1D" keeps second-order convergence with an irregular grid. From construction of o2o3 scheme (defined in chapter "Local-Galerkin Schemes in 1D"), we would expect third-order convergence. However, the order of convergence in o2o3 is demonstrated to be fourth order indicated by Fig. 2, which is faster than expected. This phenomenon is called *super-convergence*. Figure 2c shows the result for SE2, converging in second order. SE2 shows no super-convergence. If the convergence over all points of a cell is demanded, the maximum in Eq. (12) must be taken over the cell, rather than the point i.

In the following, examples of schemes converging in higher order than second order will be given and a modification of the difference of neighbor approximation Eq. (9) will be given, which is second order also for irregular grids.

The Runge–Kutta and Other Time Discretization Schemes

RK4 is a four time-level method, applicable to ODEs such as Eq. (8). To compute the field $h(t + dt, x)$ at time $t + dt$, only one time level t must be known: $h(t, x)$. Let dt be the time step, and let $h_{a,i,t}$ $(h_{i-2}, h_{i-1}, h_i, h_{i+1}, h_{i+2})$ be a spatial discretization according to Eq. (8). We define four approximations $h_{t,i}^k$ for the time derivatives and four versions h_i^k $(k = 0, 1, 2, 3)$ of h_i corresponding to different time levels:

$$
\begin{cases}
h_{t,i}^0 = h_{a,t,i}\left(h_{i-2}^0, h_{i-1}^0, h_i^0, h_{i+1}^0, h_{i+2}^0, t_0\right), h_i^0 = h_i, \\[2mm]
h_{t,i}^1 = h_{a,t,i}\left(h_{i-2}^1, h_{i-1}^1, h_i^1, h_{i+1}^1, h_{i+2}^1, t_0 + \frac{1}{2}dt\right), h_i^1 = h_i^0 + \frac{1}{2}dt \cdot h_{t,i}^0, \\[2mm]
h_{t,i}^2 = h_{a,t,i}\left(h_{i-2}^2, h_{i-1}^2, h_i^2, h_{i+1}^2, h_{i+2}^2, t_0 + \frac{1}{2}dt\right), h_i^2 = h_i^0 + \frac{1}{2}dt \cdot h_{t,i}^1, \\[2mm]
h_{t,i}^3 = h_{a,t,i}\left(h_{i-2}^3, h_{i-1}^3, h_i^3, h_{i+1}^3, h_{i+2}^3, t_0 + dt\right), h_i^3 = h_i^0 + dt \cdot h_{t,i}^2.
\end{cases}
\tag{16}
$$

For the field amplitudes $h_{t+dt,i}$, we obtain

$$
h_{t+dt,i} = h_i + dt \cdot \left(\frac{1}{6}h_{t,i}^0 + \frac{1}{3}h_{t,i}^1 + \frac{1}{3}h_{t,i}^2 + \frac{1}{6}h_{t,i}^3\right).
\tag{17}
$$

An implementation of RK4 in a FORTRAN program is given in section "1D Homogeneous Advection Test". In Eq. (16), we have written the variables for grid point values of h, which will be an important application in this book. Furthermore, Eq. (16) is valid in a more general situation. The h_i could be amplitudes of any kind for which the time development needs to be computed. For example, this could be spectral amplitudes of an L-Galerkin approximation, as shown in chapter "Local-Galerkin Schemes in 1D" or the low dimensional Galerkin approximation to be described in section "The Boussinesq Model of Convection Between Heated Plates".

Durran (2010) gives a rather complete list of alternative time discretizations. Some of these are three time-level schemes. Equations (16) and (17) can be combined with any spatial approximation such as Eq. (9) to compute amplitude sets $h_{t,i}$ at future time levels t. t typically takes values of ndt, n being the number of time steps. Numerical efficiency is best for large time steps dt. When plotting the different time levels of $h_{t,i}$, it may happen that these values become very large. This means the growth of the errors in Eq. (12) is very fast with the time evolving. In fact, they can become larger than any given number after sufficient time steps. This is called *numerical instability* of a numerical scheme. Even a small fluctuation in the initial condition might cause a large deviation or oscillation of the final numerical value from the analytic solution. This phenomenon is not necessarily caused by programming errors. If the solution $h_{t,i}$ remains smaller than a finite number C for all times t, the scheme is called *stable*. For finite difference forms of the simple homogeneous advection equation, Eq. (2), it can be shown in Durran (2010) that the time step of explicit schemes of the form Eq. (8) must be sufficiently small for the scheme to be stable. This can be described by using the dimensionless Courant–Friedrich–Levy number CFL:

$$CFL = dt \frac{u_0}{dx}. \tag{18}$$

Schemes of Eq. (8) have an associated critical CFL number CFL_{crit} (Durran 2010) and the condition of stability is

$$CFL < CFL_{crit}. \tag{19}$$

It should be noted that the theory of stability as will be given in section "The Von Neumann Method of Stability Analysis" requires linear equations of motion and a regular grid. However, the definitions of stability, consistency, and convergence (the two latter terms will be introduced later) are valid in the general case and for irregular resolution. It is possible in the general case to perform forecasts and get information on stability. For the practical model designer, it is common to define a local CFL number using the velocity at a certain point in space. However, it cannot be concluded that the smallest grid length in a surrounding of a point determines the time step. There are cases where this estimate is too pessimistic (Steppeler and Klemp 2017).

RK4 has a critical CFL number of $CFL_{crit} = 2.8$ for regular resolution Eq. (9). For numerical efficiency, it is approximately competitive according to Durran (2010). RK4 is certainly sufficient for the tests performed in this book. For realistic applications, alternative schemes may be considered and different versions of RK schemes are under consideration (Gardner et al. 2018). While the classic RK4 scheme is useful for research model versions, such as treated in this book, operational models need to explore the efficiency of different versions in detail. The realistic WRF model (Park et al. 2014), for example, uses third-order Runge–Kutta (RK3) integration. The differences in efficiency between RK3 and RK4 are marginal. RK4 is more accurate than RK3 and therefore is suitable for scientific investigations into the accuracy of spatial discretizations, which are an interest of this book. For operational applications, if efficiency is a concern, it may be worth to investigate other schemes described in Durran (2010).

Two further FD schemes will be described here, as they are almost universally used in current models, such as those indicated in Table 1 in chapter "Introduction". The implicit time scheme is used for local approximations as defined in Eq. (8). The time scheme is defined as

$$\frac{h_{t+dt,i} - h_{t,i}}{dt} = \frac{1}{2}[RHS(h_{t,i}) + RHS(h_{t+dt,i})]. \tag{20}$$

In Eq. (20), RHS means the RHS of the dynamic equation, Eq. (1). As in Eq. (20), the filed h at the target time $t + dt$ is on RHS of Eq. (20), while the equation is for

$h_{t+dt,i}$, which must be solved for $h_{t+dt,i}$. Such solution is often possible when the dynamic equation is linear. In practice, this method is often applied to some linear terms of the dynamic equation. Such schemes are called *semi-implicit*. We refer to Durran (2010) for the use of semi-implicit schemes in practical modeling which is necessary for efficient discretization of the fast waves (Durran 2010).

The solution procedure often requires approximations, whose performance may need iterations. Such iterations may be heavy on data exchange between processors and so may be a problem for parallel coding and scalability.

In this book implicit approximations are exclusively used in models organized in columns. This means that in the vertical data points line up. For points involving just vertical differentiation, fast direct solvers are available and data exchange between processors is no problem, as the data of each vertical column are stored in the same processor, which is the case with all models listed in Table 1 in chapter "Introduction". When the horizontal discretization is done explicitly, the discretization method is called *horizontally explicit and vertically implicit (HEVI)*.

The implicit time integration is stable for all time steps. It reduces the speed of fast waves, so it cannot be used when the speed of the wave has a physical significance for the weather. So for advection this is not a good choice, as accurate wave velocities for advection are significant for weather prediction. Nonetheless, an implicit treatment of vertical advection has been used and accuracy of vertical advection was sacrificed for efficiency.

Euler forward differencing is another important simple time scheme. For a time scheme $h_{t,i}$, it is defined as

$$\frac{h_{t+dt,i} - h_{t,i}}{dt} = \frac{\partial h_i}{\partial t}. \tag{21}$$

The Euler forward scheme Eq. (21) is only first order in time. It is unstable for discretizations of the advection equation. The Euler forward scheme is used for the time integration of diffusion terms, which will be introduced in section "Diffusion". Many of the most popular time schemes (Durran 2010) used with advection are unstable when applied to diffusion, but Euler forward can be used. This book is about spatial discretization, mainly L-Galerkin methods. For time integration, special attention must be given to the fast waves (Durran 2010; Gardner et al. 2018). Efficiency problems arise from the small time steps which some methods, including RK4 would require. New approaches based on Fourier transformation for time integration are for the moment just experimental (Clancy and Pudykiewicz 2013).

Homogeneous and Inhomogeneous Difference Schemes

The RK4 scheme described in section "The Runge–Kutta and Other Time Discretization Schemes" allows a prediction in time of the field $h(t, x)$ whenever a spatial discretization according to Eq. (7) is given. A simple spatial discretization for Eq. (2) is the centered FD scheme, Eq. (9). It replaces the differentiation to x by the corresponding FD equation. Using section "The Runge–Kutta and Other Time Discretization Schemes" and choosing the grid x_i to be regular and a time step dt, Eq. (21) can be programmed to predict $h_{t+dt,i}$ when $h_{t,i}$ is given. Repeating this process, predictions for longer times can be achieved. Most methods described in 2D and small 3D toy model cases in this book can easily been performed on notebook computers. A compiler and a plot program are needed. The sample programs given in chapter "Examples of Program" are done in MATLAB or FORTRAN. The following chapters will provide examples of more refined spatial discretizations than Eq. (9).

Any distribution of grid points x_i can be used. For tests used in this book, two grids are used, which are defined by the grid distance parameter dx_i.

The regular grid is defined as

$$x_i = i dx_i, i \in [0, i_e],$$ (22)

where $dx_i = dx = const$ and $i_e = 600$ is the end of the interval.

The irregular grid used for testing is defined as

$$x_i = x_{i-1} + dx_i, \text{ for } i \in [1, i_e],$$ (23)

where a 1:2 change of resolution is used in the interval $[1, i_e]$ and

$$\begin{cases} dx_i = dx, & \text{for } i \in [1, 149] \cup [180, i_e], \\ dx_i = 2 \cdot dx, & \text{for } i \in [150, 179]. \end{cases}$$ (24)

Four initial values of different smoothness are used for the tests:

$$h_i = e^{-\frac{(x_i - x_{120})^2}{i^{scale}}}, \text{ for } i^{scale} = 1, 2, 4, 8.$$ (25)

First we use Eq. (9) for regular and irregular grids and the result is shown in Fig. 3. For the choice $dx = 1$ and $dt = 1$, predictions are shown in Figs. 3, 4, 5 and compared to the analytic solution. The grid Eq. (23) with 1:2 change produces a much less accurate solution. For the irregular grid, at the positions where the resolution changes from dx to $2dx$ large inaccuracies arise. This indicates that the centered FD approximation is much more accurate on regular than on irregular grids. The difference in accuracy of the neighbor point approximation apparent in Fig. 3 corresponds to the different order of approximation of this scheme on regular and irregular grids. The scheme Eq. (9) has second-order accuracy on regular grids and this approximation order is reduced to first order for irregular grids. In Fig. 2, the same is shown for o4 schemes (Fig. 2 shows that the order is 4 only on regular grids for o4 and o2o3 schemes. With an appropriate choice of weights, the order

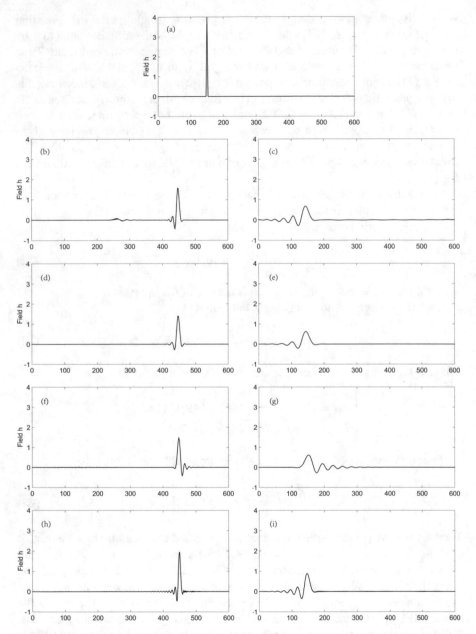

Fig. 3 Runs with the peak initial condition and transport over 300 points with different spacing for $dx = 1$ and $dt = 1$: (**a**) initial values, (**b**) standard o4 spatial difference, (**d**) o3o3 standard difference, (**f**) o3o3 spectral difference, and (**h**) SEM3; transport over 30 000 points: (**c**) standard o4 spatial difference, (**e**) o3o3 standard difference, (**g**) o3o3 spectral difference, and (**i**) SEM3 at all points (reproduced from Steppeler et al. (2019))

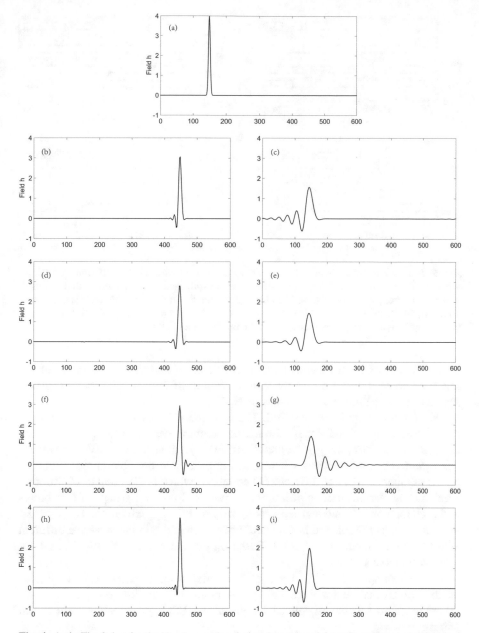

Fig. 4 As in Fig. 3, but for the No. 4 smooth solution (reproduced from Steppeler et al. (2019))

remains 4 with irregular grids, while for a standard choice it goes down to 1). The inaccuracies apparent in Fig. 5 are caused by the irregular resolution. The standard o4 scheme is forth-order only for points where the resolution is regular. Near the jump of resolution, the order of approximation goes down to 1. The standard o4

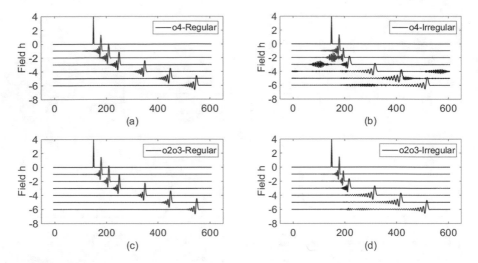

Fig. 5 The solution of a single cell peak for (**a**) o4 scheme with regular cell structure, (**b**) o4 scheme with irregular cell structure, (**c**) o2o3 scheme with regular cell structure, and (**d**) o2o3 scheme with irregular cell structure. The black curves are the initial field, the forecast at $t = 100$, 200, 300, 400 s, and the red and blue curves are the forecast at $t = 30$ and $t = 60$ s at the start and end of the resolution jumps (reproduced from Steppeler et al. (2021))

scheme is not of homogeneous approximation order and this is the cause of the inaccuracies seen in Fig. 5. For the introduction of SE2 and o2o3 schemes, please see chapter "Local-Galerkin Schemes in 1D".

We continue with the regular grid. The centered FD approximation Eq. (9) used the same algebraic form for the difference equations at every point x_i. Just the values of x_i and h_i differ at the grid points. Such FD equations are called *homogeneous FD equations*. Homogeneous schemes are not the only possible choice. Let i^{cell} be the number of types of different FD equations which is the number of different algebraic expressions used at different grid points. So we use groups of i^{cell} points and each member of such a group gets a different FD approximation. The simplest inhomogeneous FD scheme is obtained when even and odd indices have different approximations, meaning $i^{cell} = 2$. Based on this definition, i^{cell} should be 1 for homogeneous schemes.

As an example for $i^{cell} = 2$ at a regular resolution, we choose Eq. (9) as the approximation for even grid point number i. For uneven points i, the inhomogeneous scheme is defined by giving an FD approximation with a real parameter α:

$$\begin{aligned}
\frac{dh_i}{dt} &= -u_0 h_{a,t,i} \\
&= -u_0 \left[(1-\alpha) \frac{h_{i+1} - h_{i-1}}{x_{i+1} - x_{i-1}} + \alpha \left(\frac{dh_{i+1}}{dt} + \frac{dh_{i-1}}{dt} \right) \right].
\end{aligned} \tag{26}$$

Note that $\frac{dh_{i+1}}{dt}$ and $\frac{dh_{i-1}}{dt}$ are already defined in Eq. (21) for uneven i. The inhomogeneous scheme Eq. (26) is second-order due to our assumption that the grid is regular. The second-order approximation is considered desirable. We leave it to the reader to do a test forecast using Eq. (26) with an irregular resolution. For $\alpha = 0$, an acceptable accuracy is obtained for regular resolution only. For this case, we obtain a homogeneous FD scheme.

Equation (26) shows that a large number of consistent inhomogeneous schemes exist. The family of schemes for a range of α has not yet been investigated for accuracy and stability. The importance of such schemes will appear in later chapters when the concepts of high-order differencing and conserving approximations will be investigated. It will turn out that inhomogeneous schemes can be used to combine these two concepts.

Equation (9) and the family of schemes Eq. (26) are just examples for a large number of possible inhomogeneous schemes. We need tools to analyze the stability and other useful properties of such schemes. These are discussed by Durran (2010), and for the purposes of this book a selection of such considerations is described in section "The Von Neumann Method of Stability Analysis". While Durran (2010) discussed widely the case of regular grids, here we put an emphasis on the behavior of schemes on irregular grids.

It is common to have inhomogeneous schemes where some of the amplitudes are treated diagnostically. Let (x_i, x_{i+1}) be the regular grid interval where $dx = x_{i+1} - x_i$. For the interpretation, not only points but also grid intervals are considered. Given a regular set of grid points x_i, there are many ways to define such grid intervals. One way is to define grid points with broken indices:

$$x_{i+\frac{1}{2}} = \frac{1}{2} (x_i + x_{i+1}).$$ (27)

The grid points with half-numbered indices $i + \frac{1}{2}$ are then in the centers of the intervals defined by pairs of natural grid points in the interval (x_i, x_{i+1}). This means that for natural i, the point $x_{i+\frac{1}{2}}$ is in the center of the grid interval (x_i, x_{i+1}) defined in Eq. (27). Often the amplitude value $h_i = h(x_i)$ is then interpreted as the average of the field $h(x)$ in the cell defined in Eq. (27). This grid point is called *the cell center*, as these grid points are defined in the center of the cell.

It is also possible to assume that the grid points x_i and x_{i+1} defining the cells (x_i, x_{i+1}) carry the grid points. This is then called *the corner point*. The name comes from the situation in two and three space dimensions, where these points are the corners of polygonal cells. The natural interpretation of corner grid point representations is that the amplitudes are node points of a linear spline representation. The latter representation is more suitable to generalize toward representations of a high-order approximation.

For inhomogeneous FD schemes, such as defined in Eq. (26), it is convenient to define cells containing more than one point x_i. For example, if the scheme is inhomogeneous with the grid occupancy number i^{cell}, we may define the cells to be (x_i, x_{i+1}) for $i = 0, 1, \ldots, i_e$. Then each cell contains $i^{cell} + 1$ points which could define within the cell a polynomial representation $h(x)$ of a polynomial degree i^{cell}. Note that every corner points belong to two neighboring cells, which means the dimension of the discretization phase sparse is i_e. These considerations will be followed in chapter "Local-Galerkin Schemes in 1D". Note that for L-Galerkin schemes to be defined later, it is convenient to have a regular distribution of points within a cell, even when the cells are irregular.

Consider now the cell centered grid. We define the flux $fl_{i+\frac{1}{2}} = u_0 h_{i+\frac{1}{2}}$ at half-level points $x_{i+\frac{1}{2}}$ as

$$fl_{i+\frac{1}{2}} = u_0 h_{i+\frac{1}{2}} = \frac{1}{2} u_0 \left(h_i + h_{i+1} \right). \tag{28}$$

Then we have as an alternative to Eq. (9) the spatial approximation:

$$\frac{\partial h_i}{\partial t} = -\frac{fl_{i+\frac{1}{2}} - fl_{i-\frac{1}{2}}}{x_{i+\frac{1}{2}} - x_{i-\frac{1}{2}}}. \tag{29}$$

Eq. (29) is an alternative to Eq. (9), which in the special case of a regular grid is equivalent to Eq. (21).

It may be worthwhile to mention that it is possible to have a version of Eq. (9), which is second order also for irregular grids. Introducing this into Eq. (26) will make the associated inhomogeneous scheme also second order. Equation (9) can be written in the following form:

$$\frac{\partial h_i}{\partial t} = -u_0 \frac{\partial h_i}{\partial x} = -u_0(w_i^- h_{i-1} + w_i^+ h_{i+1}), \tag{30}$$

with $w_i^- = -\frac{1}{x_{i+1} - x_{i-1}}$ and $w_i^+ = \frac{1}{x_{i+1} - x_{i-1}}$.

A second-order generalization of Eq. (30) is

$$\frac{\partial h_i}{\partial t} = -\frac{\partial h_i}{\partial x} = -u_0(w_i^- h_{i-1} + w_i^0 h_i + w_i^+ h_{i+1}). \tag{31}$$

The weights w_i are computed numerically, which in this book is done using the program MOW_GC.

For regular grids, the weights are given in Eq. (29) and w_i^0 is 0. For irregular grids corresponding to a jump of resolution by a factor of two and fourth-order schemes, the weights are given in Table 1.

For second order, rather than taking the weights from MOW_GC, an analytic second-order form of Eq. (31) can be used. Write the approximated derivative of h as

$$\frac{\partial h_i}{\partial x} = h_x^{(1)} + h_x^{(2)}, \tag{32}$$

where $h_x^{(1)}$ is given by Eq. (30), including the weights w_i^- and w_i^+.

For the computation of $h_x^{(2)}$, we define

$$h^{(2)} = h_{xx,i} dx',$$ (33)

where $h_{xx,i} = \frac{2\delta h}{\delta x'^2 - dx'^2}$, $\delta x' = x_i - dx'$, $dx' = \frac{1}{2}(x_{i+1} + x_{i-1})$, $\delta h = h_i - \frac{(x_i - x_{i-1})h_{i+1} - (x_{i+1} - x_i)h_{i-1}}{x_{i+1} - xi - 1}$.

Some Further Properties of Finite Difference Schemes

The form Eq. (1) of the dynamic equations is called the flux form of the dynamic equation and the quantities F^k on the LHS define conserved quantities. From the flux form, conservation properties follow immediately. For the special case Eq. (2), considered in this section, the conserved quantity is mass and h in Eq. (2) is the density. Integrate Eq. (2) over the simulation area $[0, L]$ to obtain the total mass $M(t)$:

$$M(t) = \int_0^L h(t, x) dx.$$ (34)

From Eq. (2), we obtain for the time derivative of M:

$$\frac{\partial M(t)}{\partial t} = -\int_0^L \left\{ \frac{\partial}{\partial x} [u_0 h(t, x)] \right\} dx$$

$$= -u_0 [h(t, L) - h(t, 0)] = 0.$$ (35)

Eq. (35) holds because of the periodic condition Eq. (3) and is called the mass conservation equation. The mass defined in Eq. (35) does not depend on time.

It is often required that a conservation property of the analytic equation, in our example the mass Eq. (34), holds also for the FD equation. To check this, it is necessary to obtain FD approximations for mass. In section "Homogeneous and Inhomogeneous Difference Schemes", examples for such FD expressions are given. For the scheme at center point Eq. (27), we define an approximated mass $M_a(t)$ based on the assumption that between all points h is a piecewise linear spline:

$$M_a(t) = \sum_{i=0}^{ie} h_i \left(x_{i+\frac{1}{2}} - x_{i-\frac{1}{2}} \right).$$ (36)

It is easy to show that the discretization Eq. (9) is mass conserving, if mass is defined by Eq. (36). Defining $h_{i+\frac{1}{2}}$ in any way, for example, as $h_{i+\frac{1}{2}} = \frac{h_i + h_{i+1}}{2}$, for

a regular grid ($x_{i+1} - x_i = x_{i+\frac{1}{2}} - x_{i-\frac{1}{2}} = dx$ with dx being independent of i)
Eq. (9) is equivalent to

$$h_{t,i} = -u_0 \frac{h_{i+\frac{1}{2}} - h_{i-\frac{1}{2}}}{dx}. \tag{37}$$

Inserting Eq. (37) into Eq. (36), we obtain

$$\frac{\partial M_a(t)}{\partial t} = \sum_{i=0}^{ie} h_{t,i} dx = -u_0 dx \sum_{i=0}^{ie} \frac{h_{i+\frac{1}{2}} - h_{i-\frac{1}{2}}}{dx} = 0, \tag{38}$$

where the last sign "=" is because of the periodic boundary condition.

Equations (36) to (38) are an example to show how an appropriate definition of mass together with the corresponding FD schemes in space can lead to a mass conserving approximation. Many 3D realistic models make a point of mass conservation in the approximative equations. Other quantities, such as momentum or energy, are often conserved in realistic models.

Obviously, for the more general case of inhomogeneous FD schemes in an irregular grid, the conserving dynamic formula will be more complex than with this simple example of Eq. (36). The basis polygonal spline polynomial methods to be introduced in chapter "Local-Galerkin Schemes in 1D" will lead to such conserving approximations. For the more general system Eq. (1) for each of the equations, a conservation law exists and the basis function approach potentially leads to a conserving approximation. Some of the existing models (Park et al. 2014) currently conserve just mass. The conserved quantities to be used are not uniquely determined, but rather there exists a choice. For a 3D dry meteorological model, three moments, energy, and mass are possible conserved quantities.

Another property of great importance for spatial schemes is their order of approximation. Eq. (13) defines the order of an approximation of the derivative of a function at a point. An FD scheme is called *n-th order* when the approximation of the time derivative at any point is n-th order. While the definition of order is valid for all positive real numbers, in this book we consider only orders to be natural numbers $0, 1, 2, 3, 4, \ldots..$ The order of approximation defines how fast the error of the approximated time derivative is going to 0.

An analytic function is a function given by a convergent power series at a point x which means if for a small surrounding $|dx| < \delta$, it can be represented by

$$h(x + \delta x) = \sum_{i=0}^{+\infty} a_i(x)\delta x^i, \tag{39}$$

where $a_i (i = 0, 1, 2, \ldots)$ are real numbers and the series is convergent to $h(x)$. Examples are trigonometric functions, which are rather popular test functions in NWP models. The definition of the order of approximation of an expression is applicable only for analytic functions. In practice, it is sufficient to use trigonometric functions, such as the *sin* function as test functions. If the order of approximation for a numerical scheme is larger than zero, we call it a consistent scheme. This order of approximation is also called *consistency order*. Increasing the order of approximation of a scheme is often more efficient for achieving accuracy than increasing the resolution. Unfortunately, this improvement of accuracy by increasing the order of approximation could be shown by numerical experiments only with toy models. With real-life models, this could not be observed, which may be seen as an indication of other serious faults of real-life discretizations.

As convergence is difficult to analyze directly, it is useful to apply the convergence theorem (Durran 2010): *a stable and consistent scheme converges*. As consistency of an approximation can often be seen directly, we may expect a scheme to converge if we stay within the stability limit given by the critical CFL number CFL_{crit} for the time step.

A few remarks are given for the practical determination of the order of approximation. When $\frac{\partial h}{\partial x}$ is approximated by n-th approximation order, it is convenient to write the power series Eq. (39) as

$$h(x + \delta x) = \sum_{i=0}^{n} a_i(x)\delta x^i + \sum_{i=n+1}^{+\infty} a_i(x)\delta x^i. \tag{40}$$

If the approximation to be tested involves just one division by dx, the second term in Eq. (40) always implies Eq. (13), and therefore the function system defined by the first term in Eq. (40) is sufficient to investigate. Then, the test for order of approximation can be simplified to

$$\frac{1}{dx^n} \left| \frac{\delta h}{\delta x} - \frac{\partial h}{\partial x} \right| < C < \infty, \tag{41}$$

with $h(x + \delta x) = \sum_{i=0}^{n} a_i(x)\delta x^i$. With $h_{a,x} = \frac{\delta h_a}{\delta x}$ being the derivative approximated by the FD scheme, Eq. (41) implies that only a few test functions $\sum_{i=0}^{n} a_i(x)\delta x^i$ must be used, which are the polynomials of degree n and smaller. This is an $n + 1$ dimensional set of test functions. Also, it is sufficient to use polynomials centered at 0. This means that Eq. (41) must be verified only for the functions at $x = 0$. This is a considerable simplification compared to the tests using an infinite dimensional function system.

After some algebra involving Eqs. (41) and (28), it is verified that for regular grids the centered FD scheme Eq. (9) has approximation order 2. For irregular grids, the approximation order of this scheme is only 1. The classic fourth-order scheme Eq. (14) converges in fourth order for regular grids and the approximation order goes down to first order when the weights are not adjusted. With the weights formulated to achieve fourth order for irregular grids Eq. (15), the convergence is fourth order also with irregular grids. This theoretical result is practically verified in Fig. 2. For regular grids, the form Eq. (29) of the centered FD schemes is arithmetically equivalent to Eq. (9) and therefore is also of approximation order 2. For irregular grids, Eq. (29) turns out to be of first order only. A non-conserving modification of center FDs was given in Eqs. (32)–(33). There exists another modification for Eq. (29) which is second order. This is obtained through replacing the linear interpolation Eq. (28) used in Eq. (29) by an interpolation of at least second order. Assume the following equation at odd values of i:

$$h_{i-\frac{1}{2}} = w_{i+1}^{inter} h_{i+1} + w_i^{inter} h_i + w_{i-1}^{inter} h_{i-1} + w_{i-2}^{inter} h_{i-2}. \tag{42}$$

In order to achieve approximation order 2, the approximation Eq. (42) is more accurate than strictly necessary. The approximation Eq. (42) achieves third order, more accurate than a sufficient second-order requirement. The derivative at point i can then be computed by Eq. (37). The weights w in Eq. (42) can be computed using the Galerkin Compiler MOW_GC Huckert and Steppeler (2021). For regular grids, there is a simpler solution of a second-order interpolation using direct neighbors of $i - 1$ only. Note that for the case of a regular grid, the more complicated formula Eq. (42) with four non-zero w can also be used. However, from the derivations above, it appears that the FD scheme resulting from Eq. (42) is second order. The method of computing fluxes at neighboring points by interpolation and then applying centered FDs leads to a conserving second-order scheme. This allows the construction of second-order conserving schemes on irregular meshes. A number of applications using linear interpolation with irregular meshes means to use Eq. (42) with only two non-zero w. For such schemes, the approximation order drops for irregular parts of the mesh from second to first order. Figure 5 shows an example for the practical effect of such drop of the order of approximation.

The approximation Eq. (26) is derived by approximating h in the vicinity of the target point x_i by a first-degree polynomial. From Fig. 2, it is seen that on irregular meshes this approximation results into first order. Using the high-order interpolation, this approximation increases to second order. For regular grids, it was found that the spatial approximation order is actually 2, being higher than what follows from the interpolation assumption leading to the FD equations. Such increase of the order of approximation is called *super-convergence*. We have seen that for regular grids the centered FD scheme has approximation order 2 due to super-convergence. This order of convergence goes down to first order with irregular grids (see Fig. 2). This is a case where super-convergence to second order depends on a regular grid. In two and three dimensions, this formula is also called finite volume formula and occurs in realistic models. On the sphere, the grids are necessarily irregular. To avoid excessive errors, such models make the grid as regular as possible which is a process called "grid smoothing." Therefore, in this case super-convergence depends on regular resolution. The non-conserving scheme Eq. (32) has order 2 also for irregular grids.

Conserving second-order schemes are obtained by changing the definition of fluxes to second-order approximation in Eq. (29):

$$fl_{i+\frac{1}{2}} = w^i_{-1}h_{i-1} + w^i_0 h_i + w^i_1 h_{i+1} + w^i_2 h_{i+2}. \tag{43}$$

In Eq. (43), the weights can be chosen to achieve third-order accuracy, and for this property they are uniquely determined. For the general irregular case, the w^i_{-2}, ..., w^i_2 can be computed numerically. In this book they are computed by the program MOW_GC. For the special case of a sudden jump in resolution, they are given in Table 1 in chapter "Introduction". For regular resolution with grid length dx, the w are independent of i:

$$w_{-2} = \frac{1}{12dx}, \, w_{-1} = -\frac{2}{3dx}, \, w_1 = \frac{2}{3dx}, \, w_2 = -\frac{1}{12dx}. \tag{44}$$

For second-order approximations, there is more than one possibility to achieve this. The stability of these irregular schemes in general cannot be determined by the von Neumann methods, see Durran (2010) and section "The Von Neumann Method of Stability Analysis" of this book, as this is currently developed for the regular resolution only. Advection on irregular grids can be investigated by prediction experiments. Such experiments give a practical indication of their accuracy. The time step is systematically varied and the range of stable solutions is $dt < dt_{crit}$, where dt_{crit} will determine the critical CFL number. This can also be written as $u_0 \frac{dt}{dx} < CFL_{crit}$. While this method for prediction is immediately applicable to irregular grids, the critical CFL obtained is not very accurate. When the amplification in each time step is rather small and the prediction time is too short, an instability may remain undetected. Also, stability means that the predictions starting from all initial values remain bounded (see Eq. (41)). So, to be sure of stability, predictions would need to be done for all possible initial values. In practice only a small number of initial values will be used. It has turned out that while a test function being a single spectral function is not sufficient, a mixture of waves, such as given by the bell-shaped initial condition use in Fig. 5, will give good results. This follows from the linearity of the test equation and the fact that the bell-shaped function involves many spectral functions. Simple schemes, such as centered FD schemes, often lead to a fast blowup, but with Galerkin schemes it may happen that the eigenvalue has an amount near to 1. So sufficiently long forecasts must be used. It would be better to apply von Neumann method at regular grid first and then obtain more accurate information by long integrations. If the von Neumann methods result into a strong dependence of the absolute value of the eigenvalue on dt near the critical value dt_{crit}, this justifies confidence into the less strict estimation of dt_{crit} by numerical experiments.

When the grid used for testing has large patches of a regular grid, the CFL number is that of the regular grid using the smaller dx (see also Durran (2010)). According to Steppeler and Klemp (2017), the CFL of the homogeneous case is a good approximation for the irregular case, if two of the smallest grid intervals are at least equal and neighboring each other. This means that the CFL of a scheme on an irregular mesh can be estimated as the CFL for regular resolution using the minimum occurring dx (Durran 2010). This is a good approximation as long as the mentioned condition is fulfilled, that 2 neighboring grid intervals have the small size. If just one small dx_i is neighbored by larger dx_i', an example was given by Steppeler and Klemp (2017) that the stability may be much larger. Following Durran (2010), if in a model area the local CFL_i depends on the grid point, the minimum of CFL_i is a lower limit for the CFL of the grid. If the dependence of dx_i of i is not very smooth, the CFL number may be much larger than this minimum.

The Von Neumann Method of Stability Analysis

The von Neumann method for regular grids according to Durran (2010) is also known as "linear analysis." It allows to determine the stability of a spatial discretization for all dx and an arbitrary number of grid points. It also allows to estimate the accuracy of the solution dependent on the scale by which a wave is resolved and find the effective resolution of an FD scheme in this way. It is very simple to perform for homogeneous FD schemes. The method is explained for the example of the centered FD scheme Eq. (2). Assume initial values of the field are possessed by a wave form:

$$h_j = h(x_j) = e^{ik(j \times dx)}. \tag{45}$$

In Eq. (45), i is the imaginary unit, j is the index of the grid point, and k is the wavenumber. The analytic solution of the homogeneous advection equation, Eq. (2), is

$$h(t, x) = e^{-i\omega t} h(0, x) = e^{-i\omega t} e^{ik(j \times dx)}, \tag{46}$$

where the angular frequency is

$$\omega = u_0 k. \tag{47}$$

Note that for the analytic solution ω is purely real and proportional to the wavenumber k. Eq. (47) is called *the analytic dispersion relation*. The Von Neumann analysis checks how good the dispersion relation Eq. (47) is approximated when using the spatial FD scheme, such as Eq. (16). The range of k is limited to the interval $\left(0, \frac{2\pi}{l}\right)$, where l is the wavelength defined as $\frac{1}{k}$. Note that for an approximation in a regular grid, the wavelength has the range $(2dx, +\infty)$.

To obtain the approximated angular frequency ω^a, let us insert the dynamic equation Eq. (2) into Eq. (45) and obtain

$$h_{t,i'} = -i\omega^a(k)h_{i'}, \tag{48}$$

with

$$-i\omega^a(k) = -u_0 \frac{e^{ikdx} - e^{-ikdx}}{2dx} = -iu_0 \frac{sin(kdx)}{dx}. \tag{49}$$

The approximate dispersion relation Eq. (49) deviates from the linear relationship Eq. (47) for larger k. If $k \cdot dx$ approaches $\frac{\pi}{dx}$, the approximate angular frequency goes to 0, which is not the case for the analytic relation Eq. (47).

We define the analytic phase velocity of our test example of homogeneous advection and the numerically approximated phase velocity as follows:

$$\begin{cases} u_{ph} = \dfrac{\omega(k)}{k}, \\ u_{ph}^a = \dfrac{\omega^a}{k} = u_0 \dfrac{sin(kdx)}{kdx}. \end{cases} \tag{50}$$

From Eq. (49), we see that for small k (long wavelengths), the numerically approximated phase velocity approximates the phase velocity. For small wavelengths, there is an interval of k where the usefulness of the approximation becomes questionable and could be explored by numerical experiments of the kind shown in Fig. 5. There is another interval of k where it is clear that the approximation is definitely not useful.

Define the group velocity $u_{gr}(k)$ and its approximated version $u_{gr}^a(k)$:

$$\begin{cases} u_{gr} = \dfrac{\partial \omega}{\partial k} = u_0, \\ u_{gr}^a = \dfrac{\partial \omega^a}{\partial k} = u_0 cos(kdx). \end{cases} \tag{51}$$

The group velocities can be negative, and for the approximation considered above being defined in Eq. (51), there is an interval of large k where this is the case. Such waves of negative group velocities are called *computational modes*. They must definitely be avoided by choosing appropriate initial values. The presence of such computational modes makes the solution irregular and there is no forecast value associated with them. As a general principle of physics, negative group velocities are associated with non-causal behavior (Einstein 1905) and should be avoided. For test equations less simple than Eq. (2), computational modes may appear during the forecast and it is necessary to filter them. Such filtering procedures will be discussed in section "Diffusion".

It should be noted that the theory of group velocities, sometimes called *transfer function analysis* (Sani et al. 1981; Williams and Zienkiewicz 1981; Steppeler 1989;

Schoenstadt 1980), is in the same way applied to communication in telephone lines, and for this application the filtering of computational modes made telephone communication free of accidental noises (Haltiner and Williams 1983). So it becomes clear that some states in grid space are too small to be reasonably predicted. The essential resolution defined in Steppeler (1990) and Ullrich et al. (2018) discards all scales where the prediction is bad. A simple way of defining the essential resolution is to include all wavenumbers with positive group velocities. It is usual to adjust the numerical diffusion (see section "Diffusion") in such a way that the undamped waves have positive group velocities. Kreiss and Oliger (1994) gives an example that the geostrophic adjustment, meaning the reaction of meteorological flows to mountains, is disturbed when waves of negative group velocities occur in a model.

In this book we use the simple definition that the useful part of resolution is given by the useful range of wavenumbers defined by a group velocity being greater than 0. So far no schemes have been found for the advection process where the whole range of discretized wavenumbers belongs to the useful resolution for the advection test problem, though such range may be small for accurate schemes. For the fast waves, it can happen with staggered schemes that all waves have positive group velocity, this is the reason why many models of low approximation order use staggered grids.

The graphic representation of the dispersion relation for Eq. (2) and other schemes are given in Durran (2010). For inhomogeneous FD schemes, the performance of the Von Neumann analysis is more complicated. The von Neumann method for higher-order inhomogeneous schemes will be given in the following chapters and a program for von Neumann analysis is given in chapter "Summary and Outlook".

Dynamic Equations of Toy Models

Toy models are the simple models used for testing numerical schemes rather than using a representation of the real 3D atmosphere. The general form of dynamic equations treated in this book is that of conservation laws Eq. (1). We have seen for the example of homogeneous advection that conservation properties for approximations follow from a conservation law presented with the dynamic equations. Even if an equation is not formally in the form Eq. (1), it may still be arithmetically equivalent and most of the approximations treated in this book will conserve the corresponding fields. The conserved quantity is mass in the example of the test problem of advection treated so far.

The real atmosphere is described by a minimum of five fields which could be three velocity components, pressure, and density for the dry atmosphere. The choice of these fields to describe the dry atmosphere is not unique. For example, model developers could choose momentum instead of the three velocity components. The momentum would be conserved quantities, which is considered an useful choice for dynamic variables.

The treatment of the moist atmosphere will require the introduction of at least one moisture field which could be water vapor. In this case rain could be generated by condensing water vapor and dropping down to the Earth surface as rain droplets. The authors are not aware of any current model being content with such coarse approximations. Refined treatments of moisture in current models require a large number of moisture related fields, such as water vapor, and different fields describing condensed water, such as cloud droplets, rain, snow (Arakawa and Jung 2011). All these fields are normally represented as 3D fields. It is quite common to describe cloud water by refined assumptions on the droplet spectrum, resulting in multiple mode cloud models. Obviously the condensation processes have a potential to create fields which are not very smooth. We have already seen that numerical procedures to predict fields sometimes rely on smoothness properties of the predicted field and this needs to be considered when constructing models.

Atmospheric chemistry prediction models simulate many chemical constituents, in particular for climate and environmental applications. There are potentially hundreds of fields to be treated numerically. The abovementioned physical and chemical processes are not treated in this book. However, all fields mentioned are subject to transport in a velocity field and accurate representations of this process are the subject of this book. The advection process is quite important for atmospheric modeling.

The example Eq. (2) of homogeneous 1D advection is the highly idealized representation of an important physical process, occurring for the prediction of all fields mentioned above. The advection process is conserving when the 3D divergence of the velocity field is 0. We here consider tests where the computational area is a cube without orography inside. The effect of the spherical curvature can be accounted for a coordinate transformation (Durran 2010). In this book, most tests for limited area are done for the plane Earth. Such models can easily be changed to account for spheric effects and the impact of the underlying orography using coordinate transformations (Durran 2010). The same problem will be solved by using cell structures adapted to the geometry in chapters "Finite Difference Schemes on Sparse and Full Grids" and "Platonic and Semi-Platonic Solids".

The equations in this section are given in continuous (unapproximated) form. The introduction of approximations will be the subject for the rest of the book. For tests of advection with a 3D curved velocity field and spatial coordinates x, y, z in a cubic model area, assume a velocity vector $\mathbf{u} = (u, v, w)$ being divergence free:

$$\nabla \cdot \mathbf{u} = \frac{\partial u}{\partial x} + \frac{\partial v}{\partial y} + \frac{\partial w}{\partial z} = 0. \tag{52}$$

Assume a density field $h(x, y, z)$ and define the flux vector $\mathbf{fl} = (fl^x, fl^y, fl^z)$ by

$$\begin{cases} fl^x = -u \cdot h, \\ fl^y = -v \cdot h, \\ fl^z = -w \cdot h. \end{cases} \tag{53}$$

The dynamic equation for the density h is then

$$\frac{\partial h}{\partial t} = -u\frac{\partial h}{\partial x} - v\frac{\partial h}{\partial y} - w\frac{\partial h}{\partial z} = \frac{\partial fl^x}{\partial x} + \frac{\partial fl^y}{\partial y} + \frac{\partial fl^z}{\partial z}, \qquad (54)$$

where the second identity is coming from the divergence free velocity field Eq. (52). Equation (54) means that the dynamic equation, Eq. (54), has the form Eq. (1) of a conservation equation and this means that the density h is a conserved quantity (see section "Some Further Properties of Finite Difference Schemes").

In addition to the dynamic equations Eq. (54), we need to specify boundary conditions. Let the cubic area be defined by the coordinate values $x \in [x_0, x_e]$, $y \in [y_0, y_e]$, $z \in [z_0, z_e]$. Then we assume periodic boundary conditions in x-direction and free-slip conditions for the boundaries $y = y_0$, $y = y_e$, $z = z_0$, $z = z_e$:

$$\begin{cases} h(x_0) = h(x_e), \\ u(x_0) = u(x_e), \\ v(x_0) = v(x_e), \\ w(x_0) = w(x_e), \end{cases} \qquad (55)$$

$$\begin{cases} v(x, y_0, z) = 0, & \text{for } y = y_0, \\ v(x, y_e, z) = 0, & \text{for } y = y_e, \end{cases} \qquad (56)$$

$$\begin{cases} w(x, y, z_0) = 0, & \text{for } z = z_0, \\ w(x, y, z_e) = 0, & \text{for } z = z_e. \end{cases} \qquad (57)$$

The implementation of the periodic boundary condition Eq. (55) in numerical approximations is not considered problematic. For the free-slip boundaries Eqs. (56)–(57), we assume in Eq. (57) that flow is limited by the surface of the rectangle. In sections "A Simple Cut-Cell System Based on the Staggered Low-Order Basis Functions" and "A Conserving Version of the Cut-Cell Scheme", the lower boundary condition and its approximation in atmospheric models will be treated more extensively.

From the 3D equations above, it is possible to derive 2D test problems by putting $v = 0$ or $w = 0$ and requiring h not to depend on the respective space coordinate. Using the same fields u, v, h as used with the 2D version of the advection system, it is possible to write the shallow water equations which are a popular test case involving some dynamics but being less complex than the full 3D nonhydrostatic equations of the atmosphere:

$$\begin{cases} \dfrac{\partial u}{\partial t} = -\mathbf{u} \cdot \nabla u - \dfrac{\partial h}{\partial x} + fv = -u\dfrac{\partial u}{\partial x} - v\dfrac{\partial u}{\partial y} - \dfrac{\partial h}{\partial x} + fv, \\ \dfrac{\partial v}{\partial t} = -\mathbf{u} \cdot \nabla v - \dfrac{\partial h}{\partial y} - fu = -u\dfrac{\partial v}{\partial x} - v\dfrac{\partial v}{\partial y} - \dfrac{\partial h}{\partial y} - fu, \\ \dfrac{\partial h}{\partial t} = -h\nabla \cdot \mathbf{u} - \mathbf{u} \cdot \nabla h = -h\dfrac{\partial u}{\partial x} - h\dfrac{\partial v}{\partial y} - u\dfrac{\partial h}{\partial x} - v\dfrac{\partial h}{\partial y}. \end{cases} \quad (58)$$

where f is the Coriolis parameter.

Note that though Eq. (58) is simplified in comparison to the correct Navier–Stokes equations of the atmosphere, this equation can give important test results of striking similarity to atmospheric flows. The first successful numerical weather prediction by Charney, Fjotoft, and von Neumann (Charney et al. 1950) was done using a shallow water model being a non-divergent version of Eq. (58). Often diffusion terms are added to Eq. (58).

Diffusion

Most fields in an atmospheric model are subject to diffusion. Second-order diffusion is obtained by adding a diffusion term to equations such as Eq. (58). When diffusion alone is presented (meaning to drop the other terms in Eq. (58)), a 1D version for example in y-direction obeys the dynamic equation:

$$\frac{\partial h}{\partial t} = c^{diff}(y)\frac{\partial^2 h}{\partial y^2}. \quad (59)$$

The coefficient $c^{diff}(y)$ is the strength of the diffusion. Terms of a similar form as Eq. (59) are very important for atmospheric modeling and such terms, in a more refined version than Eq. (59), are included to represent the effect of molecular friction and turbulent friction. Such turbulence can occur everywhere in the atmosphere caused by surface roughness in particular. Therefore, the theory and the corresponding software package is often called *boundary layer representation*. 3D second-order diffusion is also used, but operational models often use the 1D versions, as a 3D calculation is numerically very expensive. The computation of turbulence coefficients is part of the physical parameterization package (see chapter "Introduction") and is not the subject of this book. The introduction of the diffusion tendencies into the time-stepping is often done by the numerical developer. In this respect, it must be observed that time integration by explicit RK4 and many other schemes being popular for the advection process are not suitable for diffusion Eq. (59). Euler forward Eq. (21) is a possibility for time-stepping with diffusion operators. For more information and explanations, it is referred to Durran (2010).

In addition to this second-order diffusion operator, which is introduced to represent the physical process of turbulence, models often need diffusion for numerical reasons. For example in Fig. 5b, small-scale computational modes are visible. From Fig. 7, it is seen that the computational modes have a negative group velocity. As this noise is in a small scale, smoothing is suitable to remove it. Second-order diffusion as given in Eq. (59) not only represents the physical process of turbulence but also has a smoothing effect on the solution. For this purpose, a version of Eq. (59) with constant smoothing coefficient is used identical to Eq. (59):

$$\frac{\partial h}{\partial t} = c^{diff} \Delta h, \tag{60}$$

which is a numerical construction and is introduced to remove computational modes where $\Delta h = \frac{\partial^2 h}{\partial x^2}$. This term is not introduced to represent a physical effect but rather to remove errors caused by discretizations of other terms. It is of interest how it acts on small scales, and therefore the numerically discretized form is important. Let us in this section assume that we have fields in grid point space, defined by grid point values h_i for the field $h(x)$. For a regular grid with amplitudes h_i on the grid $x_i = idx$, it is common to use a low-order approximation in order to obtain good smoothing:

$$\Delta h = \frac{1}{dx^2}(h_{i+1} - 2h_i + h_{i-1}). \tag{61}$$

The coefficient c^{diff} in Eq. (59) can be chosen to define the amount of smoothing desired. Eq. (61) is valid for a regular grid and is not constructed to be formally conserving. However, if the mass of an amplitude in the interval $(x_{i-\frac{1}{2}}, x_{i+\frac{1}{2}})$ is defined to be $h_i dx$, then the approximation Eq. (61) is formally mass conserving for constant diffusion coefficient. Eq. (61) is a second-order approximation of the second derivative to x. For irregular meshes, Eq. (61) reverts to first order. It may be worthwhile to give a second-order version of Eq. (61) for an irregular grid. Let $x_{i-1} < x_i < x_{i+1}$ be three grid points for the definition of h_i. Then a second-order version of Eq. (61) is

$$\begin{cases} \delta h = h_i - (w_{i-1} h_{i-1} + w_{i+1} h_{i+1}), \\ \Delta h = -2\alpha \cdot \delta h, \end{cases} \tag{62}$$

with $w_{i-1} = \frac{x_{i+1}-x_i}{x_{i+1}-x_{i-1}}$, $w_{i+1} = \frac{x_i-x_{i-1}}{x_{i+1}-x_{i-1}}$, $\alpha = \frac{2}{x'^2-dx_i^2}$, $x' = x_i - x_m$, $x_m = \frac{1}{2}(x_{i+1} + x_{i-1})$, $dx_i = \frac{1}{2}(x_{i+1} - x_{i-1})$. Note that Eq. (62) reverts to Eq. (60) for the case of a regular grid, meaning $x_i = \frac{1}{2}(x_{i+1} + x_{i-1})$. The scheme Eqs. (60)–(61) removes small-scale errors coming from the discretization of other terms. In an ideal world, we should not use numerical diffusion, though the necessary physical diffusion Eq. (59) represented by a term of a similar form approximates a physical effect. Numerical diffusion is to be considered an error and the aim must be to get along with values for c^{diff} as small as possible. In fact, it is common practice to do some simple tests without diffusion and a method that only runs with heavy diffusion is considered inferior to a model needing less diffusion.

Second-order numerical diffusion is not considered a good choice for a numerical diffusion. If large eddy simulations (LESs) are intended, molecular diffusion must be included, which is represented by constant $c^{diff}(z)$ in 1D, a term similar to Eq. (60). Coefficients depending on x, z arise for the representation of turbulent flow. For molecular diffusion, the coefficient representing molecular diffusion is much smaller than those occurring with turbulence. LES calculations use molecular diffusion and employing high resolution want to calculate the turbulent diffusion by resolving the turbulent motion.

For LES calculations and other models, second-order numerical diffusion is not practical as it is of the same order of magnitude as the physical effect of atmospheric friction. It must be considered as an error as the numerical diffusion term has no physical meaning. With LES, this error has a similar form and magnitude as the physical effect (molecular diffusion) which is to be calculated. Therefore, higher-order numerical diffusion is normally used, such as fourth-order diffusion. This can be obtained by squaring the operator Δ used in Eq. (60). One possibility is to use fourth-order diffusion replacing RHS of Eq. (60) based on Shapiro (1970):

$$\frac{\partial h}{\partial t} = c^{diff} \Delta \Delta h, \tag{63}$$

A simple way to approximate $\Delta \Delta h$ in Eq. (63) is to apply Δ two times to h. A fourth-order diffusion like given in Eq. (63) or similar is called *hyper-diffusion*, in particular in the SE community.

One purpose of a numerical diffusion is to remove the small scales. For a field given at grid points, the smallest-scale field is the $2dx$ wave, given by $h_i = (-1)^i (i = 0, 1, 2, \ldots)$. For a numerical noise filter, we would like to have the $2dx$ component of the field h_i to be reduced. The following is another form of fourth-order diffusion:

$$\Delta^{4,x} = c^{diff} (w^i_{-2} h_i^{-2} + w^i_{-1} h_{i-1} + w^i_0 h_i + w^i_1 h_{i+1} + w^i_2 h_{i+2}), \tag{64}$$

with $w^i_{-2} = w^i_2 = \frac{1}{2}, w^i_{-1} = w^i_1 = 2, w^0 = 3$ for regular resolution. The factor $\frac{1}{dx^4}$, sometimes used, is considered to be included in c^{diff}. For irregular resolution, the weights $w^i_{i'}$ depend on i and can be computed using the Galerkin compiler MOW_GC (Huckert and Steppeler 2021). The upper index x in LHS of Eq. (64) indicates that we are doing a 1D operation in x-direction. In 2D and 3D models, numerical diffusion is normally done for the horizontal coordinate. In the vertical physical diffusion is used. This may become different with very high-resolution models where physical diffusion is done in all three dimensions.

In 2D, with coordinates x and y and a regular grid in 2D, the operator Δ must then have an index x or y (Δ^x or Δ^y) to indicate in which direction it is taken. 2D numerical fourth-order diffusion can then be represented as

$$\frac{\partial h}{\partial t} = c^{diff} (\Delta^x + \Delta^y)(\Delta^x + \Delta^y) h. \tag{65}$$

The following form of 2D numerical fourth-order diffusion is also in use:

$$\frac{\partial h}{\partial t} = c^{diff}(\Delta^x \Delta^x + \Delta^y \Delta^y)h, \tag{66}$$

As numerical diffusion does not represent a physical effect and is included for numerical reasons only, the addition of a second-order numerical diffusion term decreases the order of approximation for the whole dynamic system to one, even when the advection is fourth order and this is a reason why numerical diffusion of second order is normally not considered, but rather fourth- or higher-order diffusion is preferred for the numerical part of diffusion. Therefore, no numerical diffusion of lower than fourth order should be considered. While at the top of NWP models, first-order numerical diffusion is often used in the so-called sponge layer, LESs normally use numerical diffusion of at least fourth order and the advection scheme normally is high order too. Current progress in computer performance allows LESs near solid surfaces. The representation of the curved boundary in such cases must also be correspondingly accurate with LESs. Further information on LES is given in section "Large Eddy Simulation". Also the third-order error associated with second-order advection is too near to the effect to be computed. Therefore, large eddy models require higher than second-order accuracy for the advection scheme.

There always exists the lower boundary, and if the model is not global, we also have lateral boundaries. Very often the diffusion operator, such as the 1D operator in Eq. (63), is applied only for points sufficiently distant to the boundary. This makes sure that all grid points in the stencil are in the model area. For example, for the formula Eq. (63), we can only use points x_i ($i < i_e - 1$), when $i_e dx$ is the boundary of the computational area. There are other options, such as posing boundary conditions. It is also possible to compute one-sided derivatives at the boundary. For the case of a regular (equally distant) grid, the stencil for the points x_{i_e} and x_{i_e-1} is the same as for the point x_{i_e-2}. For the point x_{i_e} and x_{i_e-1}, we obtain the weights for one-sided derivatives by extrapolation. The weights are the same as given for the center point. So the one-sided derivative is the same as applying for x_{i_e} and x_{i_e-1} the same diffusion tendency as for x_{i_e-2}. For lateral boundary conditions, if it is intended to feed results of coarser-resolution models into limited area models, dirty tricks are often considered unavoidable and given in Davies (2014).

Many temporal FD schemes used for the advection process including RK4 are unstable when used with the diffusion process. It is common to use the Euler forward scheme with diffusion operator. Let $h_{t,i}^{diff}$ be the time derivative computed with one of the spatial discretizations of the diffusion operator, Eq. (66). Then the diffusion operator time step is performed as

$$h_i(t + dt) = h_i(t) + h_{i,t}^{diff} dt. \tag{67}$$

In realistic models, diffusion is applied either every time step or every n-th time step. For a computational example shown in Fig. 6, the different smoothing properties are tested by using the peak initial value. The diffusion is applied with 10 steps for

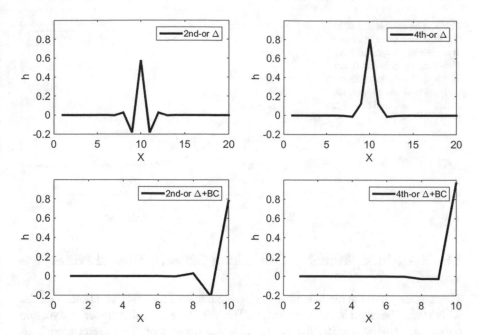

Fig. 6 The diffusion operator applied to the peak initial condition, which has the value 1 at just one point. Top: the application of 10 diffusion steps for $c^{diff} = 0.25$ for second-order diffusion (left) and fourth-order diffusion (right). Bottom: as above, when the maximum is on the boundary of the computational area

fourth order and second order at a regular resolution, and the field h_i for this test is defined to be 0 everywhere except for center point where it is 1 at the top two figures. The bottom two figures of Fig. 6 show the diffusion near the boundary using the extrapolation method as defined above. It is seen that also the boundary amplitude is affected by the smoothing, but not so strong as if this would be a point away from the boundary. In this example the peak is much more surviving the smoothing for fourth order than for second order. Figure 6 shows the diffusion in the presence of a boundary. As expected, there is less reduction of the boundary amplitude than without the wall. Figure 7 is the application of a smoothing step to the test function Eq. (68). This diagram gives an indication what frequencies are damped.

A spectral diagram is obtained by applying the smoothing operator to the field:

$$h_j = e^{2\pi \frac{j}{k} I}, \tag{68}$$

where $k = 2, 3, \ldots$ and $I = \sqrt{-1}$. The diffusion is concentrated near the smallest resolved waves. The spectral diffusion diagram is analog to the dispersion diagram of advection, which will be given in Fig. 7 in chapter "Local-Galerkin Schemes in 1D". The diffusion introduced must be such that the wavenumber area connected to negative group velocities is damped.

Fig. 7 The application of a
diffusion step of fourth order
to a wave of wavenumber k.
The damping factor is shown,
by which the wave solution is
multiplied in one Euler time
step. $c^{coeff} = 1$ and $dt = 1$ are
used

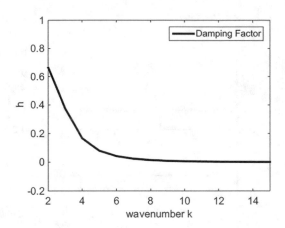

The Boussinesq Model of Convection Between Heated Plates

This section gives the mathematical background for the paradigmatic Lorenz model
which was discussed in the introduction. We introduce a small number of dynamic
parameters (Steppeler 1978) and the three parameters of the Lorenz model. In
particular it is explained here how the Lorenz model gives a rather realistic
representation of the flow between heated plates, in spite of being low dimensional
in phase space.

We set an idealized convection as an example. Convection means the flow
resulting from heating. In this example it is caused by the heat difference between
two plates. We assume a 2D dry incompressible fluid between plates with a
temperature difference. For a sketch of the situation, see Fig. 6 of the introduction.
In Boussinesq approximation, there are two fields: the velocities represented by the
vorticity ξ and temperature T. The prognostic equations of motion are

$$\begin{cases} \dfrac{\partial \xi}{\partial t} = -\left(v\dfrac{\partial \xi}{\partial z} + w\dfrac{\partial \xi}{\partial z} - \dfrac{Ra}{Re^2 Pr}\dfrac{\partial T}{\partial z} - \dfrac{1}{Re}\Delta\xi \right) = F_2, \\[2mm] \dfrac{\partial T}{\partial t} = -\left(v\dfrac{\partial T}{\partial z} + w\dfrac{\partial T}{\partial z} - \dfrac{1}{RePr}\Delta T \right) = F_3, \\[2mm] \dfrac{\partial v}{\partial y} + \dfrac{\partial w}{\partial z} = 0, \end{cases} \tag{69}$$

where Re, Pr, and Ra are the Reynolds number, Prandtl number, and Rayleigh
number, respectively, the velocity components are u and v and the vorticity $\xi = -\frac{\partial w}{\partial y} + \frac{\partial u}{\partial z}$ and

$$v = w = 0, \text{ for } Z = -\frac{1}{2} \text{ and } Z = \frac{1}{2}, \tag{70}$$

where d is the distance of the plates which is assumed to be 1. We treat this a 2D flow, for simplicity. A 3D version (Steppeler 1978) is also possible, which would involve another velocity component u.

$$
\begin{cases}
\tilde{v} = A f^2(y) g_z^3(z), \\
\tilde{w} = -A f_y^1 g^3(z), \\
\tilde{T} = T_0 + \dfrac{1}{2} - Z + T_1 f_y^1(y) g^1(z) + T_2(f^2(y) - c) g^2(z) + T_3 g^2(z), \\
c = \displaystyle\int dy f^2(y) \dfrac{dy}{L}, \\
f^1(y) = \sin\left(\dfrac{2\pi}{L} y\right), \text{ for } y \in [0, L], \\
f^2(y) = \cos\left(\dfrac{2\pi}{L} y\right), \text{ for } y \in [0, L], \\
g^1(z) = \dfrac{1}{4} - z^2, \\
g^2(z) = \left(\dfrac{1}{4} - X^2\right) z, \\
g^3(z) = \left(\dfrac{1}{4} - Z^2\right)^2.
\end{cases}
\tag{71}
$$

Equation (71) is used as a linear function system. It is called linear because the dependence of the test function on the amplitudes A, T_1, T_2, T_3 is linear. This linear dependence on the dynamic amplitudes means that the classic Galerkin method can be used to derive dynamic equations for these amplitudes. This solution procedure with process adapted basis functions is different from the solution using an FD scheme in a grid, which adapted in the rest of this book. FD schemes can resolve any field within a range of scales. In contrast to this Eq. (71) uses basis functions which are chosen to represent the given physical situation of conventional flow between heated plates. A Galerkin method using specially adapted basis functions is called an adaptive Galerkin method. A grid point solution is given in Ogura and Yagihashi (1970), which uses a number of about 1000 amplitudes to model this physical situation. The adapted Galerkin method uses only four dynamic amplitudes to obtain a rather realistic simulation, being comparable to the grid point simulation which uses more computational resources. For the Lorenz paradigmatic model, this is reduced to 3 amplitudes. Note that the physical situation of convection between plates is an idealized picture of the atmospheric situation. Therefore, this adaptive Galerkin approach can potentially be used to parameterize convection and large-scale atmospheric flow. The latter leads to the Lorenz model which will be discussed in the next chapter.

From the functional ansatz, Eq. (71), and the dynamic equations, Eq. (69), the dynamic equations are obtained by the Galerkin procedure. This means that the equations, Eq. (69), are multiplied with all basis functions used in Eq. (71) and integrated over x and z. This results into the following dynamic equations:

$$
\begin{cases}
\dfrac{\partial A}{\partial t} = \dfrac{Ra}{Re^2 Pr}\alpha_1^4 T_1 + \dfrac{1}{Re}\alpha_2^4 A, \\[2mm]
\dfrac{\partial T_1}{\partial t} = \alpha_1^5 A T_2 + \alpha_2^5 A + \dfrac{\alpha_3^5}{RePr}T_1 + \alpha_5^4 A T_3, \\[2mm]
\dfrac{\partial T_2}{\partial t} = \alpha_1^6 A T_1 + \dfrac{\alpha_2^6}{RePr}T_2, \\[2mm]
\dfrac{\partial T_3}{\partial t} = \alpha_1^7 A T_1 + \dfrac{\alpha_2^7}{RePr}T_3.
\end{cases}
\tag{72}
$$

For the parameters $\alpha^{i,k}$ in Eq. (72), we obtain by evaluating the Galerkin integrals of the basis functions in Eq. (71) as described in section "Dynamic Equations of Toy Models":

$$
\begin{cases}
\alpha_1^4 = -1.010315, \\
\alpha_2^4 = -23.008418, \\
\alpha_1^5 = -0.002976, \\
\alpha_2^5 = -0.214285, \\
\alpha_3^5 = -19.800878, \\
\alpha_4^5 = -0.035714, \\
\alpha_1^6 = -3.266959, \\
\alpha_2^6 = -81.203510, \\
\alpha_1^7 = -4.900438, \\
\alpha_2^7 = -41.900007,
\end{cases}
\tag{73}
$$

where the above equation has always the stationary solution $A = T_1 = T_2 = T_3 = 0$.

For

$$
Ra > \dfrac{\alpha_3^5 \alpha_2^4}{\alpha_2^5 \alpha_1^4},
\tag{74}
$$

there is a connective solution given by the parameters:

$$
\begin{cases}
T_2 = -\left(\alpha_2^5 \gamma + \dfrac{\alpha_3^5}{Re\,Pr}\right) / (\gamma(\alpha_1^5 + \alpha_4^5 \gamma_1)), \\[2ex]
\gamma = -\dfrac{\alpha_1^4 Ra}{\alpha_2^4 Re\,Pr}, \\[2ex]
\gamma_1 = -\dfrac{\alpha_2^6 \alpha_1^7}{\alpha_1^6 \alpha_2^7}, \\[2ex]
T_3 = T_2 \gamma_1 \\[2ex]
T_1 = \sqrt{\left(-\dfrac{\alpha_2^6 T_2}{\alpha_1^6 \gamma\, Re\,Pr}\right)}, \\[2ex]
A = T_1 \gamma.
\end{cases}
\tag{75}
$$

The temperature fields corresponding to the solution with $Re = 513, Pr = 0.713$, and $Ra = 8010.0$ are given in Fig. 8. A comparison of this stationary solution with FD computations and experimental data is given in Steppeler (1978).

Low dimension solutions have a potential significance for convection in shallow layers of the atmosphere. This is normally too small scale to be resolved in models of grid length larger than 50 km. Under the name "gray area of convection" (Gao et al. 2017; Honnert et al. 2020), this is a recognized problem for climate modeling. In Steppeler (1978), low dimensional modeling was proposed to include such shallow convection in climate models.

The non-stationary solutions of Eq. (72) give rise to the paradigmatic Lorenz low dimensional model, which will be described in the next section.

Fig. 8 The temperature field (isotherms) for the stationary case computed by the method of infinitesimal functionals (reproduced from Steppeler (1978))

The Lorenz Paradigmatic Model

The Lorenz paradigmatic model (see introduction) is a model of three dynamic amplitudes coming from a discretization of the model of convection between plates as described in section "The Boussinesq Model of Convection Between Heated Plates". Thus the Lorenz model is not an invented equation but rather is obtained by discretization and the results can verify positively with experimental results, even when only three dynamic amplitudes are used. This verification was done for a set of parameters where experimental data are available. As is the case with all numerical models, realistic simulations are not obtained for all parameter ranges.

It has the amplitudes $C(t)$, $L(t)$, $M(t)$ and dynamic equations of motion are

$$
\begin{cases}
\dfrac{\partial C}{\partial t} = \sigma (L - C), \\[2mm]
\dfrac{\partial L}{\partial t} = r \cdot C - L - C \cdot M, \\[2mm]
\dfrac{\partial M}{\partial t} = q \cdot C \cdot L - b \cdot M,
\end{cases}
\tag{76}
$$

where parameters leading to the chaotic solutions (shown in the introduction) are $\sigma = 10$, $b = 8$, $r = 28$, and $q = 2$.

To derive these solutions from the low dimensional Galerkin solution Eq. (69) in section "The Boussinesq Model of Convection Between Heated Plates", we assume $\dot{T}_3$ in Eq. (72) to be 0, leaving only the first three equations of Eq. (72). Then we can translate between Eqs. (69) and (76) assuming the following correspondence between the amplitudes of Eqs. (72) and (76):

$$
\begin{cases}
C = A \cdot c, \\
L = T_1 \cdot l, \\
M = T_2 \cdot m.
\end{cases}
\tag{77}
$$

Using the ansatz Eq. (77) and comparing Eq. (76) with Eq. (72), we obtain as the translation between the variables A, T_1, T_2 of Eq. (46) and C, L, M of Eq. (76):

$$
\begin{cases}
l = \dfrac{\alpha_3^5}{Re \cdot Pr}, \\[3mm]
\sigma = \dfrac{Ra}{Re} \left(\dfrac{\alpha_1^4}{\alpha_3^5} \right), \\[3mm]
c = -\dfrac{\alpha_2^4 \alpha_3^5}{\alpha_1^4 Ra}, \\[3mm]
m = \dfrac{\alpha_2^5}{c}, \\[3mm]
q = \dfrac{\alpha_1^6}{c \cdot l}, \\[3mm]
b = -\dfrac{\alpha_2^6}{Re \cdot Pr}.
\end{cases}
\tag{78}
$$

The wavelength of the rolls is used just as given with the stationary solution above. Eqs. (77)–(78) allow to compute A, T_1, and T_2 from C, L, M and in this way compute the physical states according to Eq. (72) of the trajectories in C, L, M space. Using a high-resolution model of the same flow, it is in principle possible for the toy model (Eq. (69), as described in Lorenz (1963)), to answer questions like if the implied turbulence parameterization in the low dimension model is realistic. For the toy model, such questions are less difficult to treat than for a realistic model, where such questions are in principle also possible. The difficulty of realistic models is that many parameterizations are necessary. One advantage of this toy paradigm model is that a rather complete solution may be obtained, as a 2D rather fine-resolution Boussinesq solution will fit even in personal computers. The model may also be made more realistic by including simple models of heat transfer and heat storage in the plates. This would then be an analogy to the real atmosphere, which is completely known, including turbulence. This model would then be suitable to investigate the existence of climate states and a possible realism of the Lorenz paradigmatic model.

From Eq. (78), we can compute the non-dimensional parameters Ra, Re, Pr, when assuming that the model Eq. (72) is emulated as Eq. (76) with $\sigma = 10, b = 8, r = 28, q = 2$:

$$
\begin{cases}
Re \cdot Pr = \dfrac{\alpha_2^6}{b}, \\[2ex]
Ra = \sigma \dfrac{\alpha_3^5}{\alpha_1^4} Re, \\[2ex]
q = \dfrac{1}{Ra \cdot Re \cdot Pr} \dfrac{\alpha_3^5 \alpha_2^4 \alpha_3^5}{\alpha_1^4}.
\end{cases}
\tag{79}
$$

from which the non-dimensional parameters are computed:

$$
\begin{cases}
Ra = b \dfrac{\alpha_3^5 \alpha_2^4 \alpha_3^5}{\alpha_1^4 q \cdot \alpha_2^6}, \\[2ex]
Re = Ra \dfrac{\alpha_1^4}{\sigma \alpha_3^5}, \\[2ex]
Pr = \dfrac{\alpha_2^6}{b \cdot Re}.
\end{cases}
\tag{80}
$$

In Eq. (78), this will allow to compute the parameters L, C, M, and according to Eq. (77) the dynamic parameters A, T_1 and T_2 can be computed when C, L, M are given. Equation (71) allows to compute the 2D field functions determining the velocities and temperatures. This means that for the time development of the butterfly solution, it is connected to the time development of the 2D fields of temperature and velocities. The physical meaning of the time development can be explored. While this does not necessarily mean that the time development is realistic, the model may be modified to become so. From general considerations reported in the introduction for considering climate as the laminar part of a turbulent flow, we may expect that a relevant low dimensional sub-manifold in phase space exists. Currently, it is not in a fully reliable way known if such short time predictable low dimensional sub-systems exist for the real atmosphere. It is also not very well known if eventual bifurcations of such systems exist and if beyond the relative low range predictability of the low dimensional subset this system is of relevance for an eventual analysis of climate bifurcations. Climate evaluation for the future is based on the assumption that in the phase space a low dimensional predictable subset exists. Otherwise just a few long range observations would not be sufficient to analyze climate change. Current climate evaluation model runs often analyze a point ("point attractor") in phase space and its dependence on parameters, such as CO^2. An example for the analysis of climate beyond the point attractor assumption is the southern oscillation, where attempts were made to analyze a low dimensional attractor (Steppeler 1993).

The analysis of low dimensional subsets of the phase space for climate analysis purposes is in its infancy. The Lorenz paradigmatic model offers the possibility to analyze such systems rather completely and with small technical resources. Such Lorenz models are not based on an invented equation. They are rather based on a discretized physical system. For people, such as the Magamecrens, who live in a rectangular cave, they describe the real weather.

Local-Galerkin Schemes in 1D

Abstract This chapter describes Local-Galerkin (L-Galerkin) methods, which are generalizations of the classic Galerkin procedure. L-Galerkin methods are local and therefore more practical on multiprocessing computers. A well-known example is the spectral element (SE) method, but there exist a large number of alternatives, named o*nom* or the third-degree method. o*nom* approximates the fields in order n and the fluxes in order m. All methods described here use piecewise continuous polynomial spaces and therefore are continuous Galerkin (CG) methods. An important consideration with o*nom* and the classic Galerkin methods is the sparseness of the grid, which is alternatively called *the serendipity grid* which was already used with the classic Galerkin method. This means that not all points of a regular grid are used and this offers a considerable potential saving of computer time.

Keywords Galerkin method · Local Galerkin method · Spectral elements · Finite elements · L-Galerkin methods · Irregular resolution · Distributions

In this chapter, we consider numerical approaches using piecewise polynomial representations in 1D. We assume a discretization over a 1D domain $\Omega = (-\infty, +\infty)$, and the domain is divided into the sum of intervals $\Omega_i = (x_i, x_{i+1})$ with the interval length:

$$dx_{\Omega_i} = x_{i+1} - x_i, \text{ for } i = 0, 1, 2, ... \tag{1}$$

If the domain is divided regularly, then $dx_{\Omega_i} = dx$ for $i = 0, 1, 2, ...$. For a grid point x_i, the dynamic field is represented as a polynomial for each grid interval Ω_i that is a pixel for our discretization. As in Chapter "Simple Finite Difference Procedures", we call the x_i at the interval ends *corner points*. Let n^{ord} be the degree of the polynomials used. Then amplitudes are needed to define a polynomial function in each interval. If field values are used as amplitudes, then the point is called *collocation point*. Otherwise, amplitudes are called *spectral amplitudes*.

J. Steppeler, J. Li, *Mathematics of the Weather*, Springer Atmospheric Sciences,
https://doi.org/10.1007/978-3-031-07238-3_3

Let us consider the example of third-degree polynomials in this chapter. We introduce a number of collocation points in each interval whose grid point values determine the polynomial. A third-degree polynomial (n^{ord}=3) is defined by four grid point values. If we assume that field values are located at two corner points, two more points must carry amplitudes of the polynomial in Ω_i. We indicate these by broken indices $i + \frac{1}{3}$ and $i + \frac{2}{3}$. Within each interval Ω_i, we assume a regular distribution of collocation points for most applications:

$$\begin{cases} x_{i+\frac{1}{3}} = x_i + \dfrac{1}{3}dx_{\Omega_i}, \\ x_{i+\frac{2}{3}} = x_i + \dfrac{2}{3}dx_{\Omega_i}. \end{cases} \qquad (2)$$

In a similar way, the representation by piecewise second-degree polynomials (n^{ord}=2) required the collocation point:

$$x_{i+\frac{1}{2}} = x_i + \frac{1}{2}dx_{\Omega_i}. \qquad (3)$$

With o*nom*, dx_{Ω_i} may depend on i. However, the collocation grid is regular within each cell Ω_i, even when the corner points x_i are irregular. The global distribution of the collocation grid points $x_0, x_{\frac{1}{3}}, x_{\frac{2}{3}}, x_1, x_{1+\frac{1}{3}}, \ldots$ is regular for an irregular corner grid x_i. For the SE approach, an irregular distribution of collocation points within each cell Ω_i will be necessary (see Section "Spectral Elements" in Chapter "Simple Finite Difference Procedures").

The grid cells may be arbitrary. It could go to a very high resolution in some parts of the model domain. For the purpose of convergence studies, we assume an approximately homogeneous grid that means that the grid distances dx_{Ω_i} vary only within a certain range. We assume an approximate grid scale dx^0 with:

$$\frac{1}{2}dx^0 < dx_{\Omega_i} < \frac{3}{2}dx^0. \qquad (4)$$

The limitation Eq. (4) is used only to discuss the order of an approximation. The L-Galerkin methods to be discussed admit arbitrary grids, but their efficiency and accuracy for certain applications always need to be considered.

The methods discussed here may use arbitrarily irregular dx_{Ω_i}. Here we consider the case that the functions fit together continuously at x_i. The points at interval ends are called *principal nodes* or *corner nodes*. The latter expression comes from the fact that these points are at the corners of cells and that the interior grids of grid intervals are on the edges with two and more dimensions. For the methods considered here, principal nodes carry amplitudes and are dynamic points. Together with the interior (1D)/edge (2D and 3D) points, the corner points are collocation points. For functional representations of higher than first order, we will use more collocation points per interval Ω_i. Equations (2) and (3) define these interior points, which are also called *high-order collocation points*. The number of points needed will depend on the method to be defined, and such points are indicated by broken indices, as indicated above. The methods considered here are from the continuous Galerkin (CG) family. Discontinuous Galerkin (DG) methods are also used for discretization (Durran 2010). This book is about practical modeling, and we choose CG for an elaborate description because some members of this discretization family are near operational use in realistic models (Giraldo 2001; Taylor et al. 1997; Herrington et al. 2019). As the fields are polynomials in the interior of intervals, they are continuous and differentiable for such points. The fields differ in the regularity at principal nodes. In this book, we consider the two cases that fields are continuous functions C^0 or differentiable functions C^1 or even more regular.

As we consider methods with a potential for practical application in realistic models, we consider for the order of representation one, two, and three. We follow arguments given in Herrington et al. (2019), being based on practical experience. For practical modeling, meaning 3D models using realistic atmospheric data, currently there is no reason to use polynomials with degree beyond three. Therefore, in this book, we limit ourselves to polynomial degrees one, two, and three. Very often a polynomial representation of third degree leads to an approximation order of four by super-convergence.

The properties of a discretization depend among others on the regularity of the field representations at the principal nodes. We explore such properties just for schemes using C^0 and C^1. Properties for even more regular field representations are left for the reader to explore.

Different amplitudes may be used to define the mentioned polynomials, such as collocation grid points or derivatives at points. This leads to different forms of the dynamic procedures for forecasting. Some of such procedures may be arithmetically equivalent, as the approximation procedures often refer to the functions, not to grid points or other amplitudes directly. The different arithmetic forms of a scheme, however, may lead to computer programs of different efficiencies.

Functional Representations, Amplitudes, and Basis Functions

In this section, we will define 1D piecewise polynomial functions. Such functions are called polynomial splines.

A third-degree polynomial (n^{ord}=3) is determined by four coefficients. Therefore, there will be a maximum of four amplitudes per grid interval, which reduce to three if the field is continuous at nodes (C^0) and to two if also the derivatives are continuous (C^1). To define field functions, basis functions as well as approximation procedures will be defined. Formula for the definition and properties of basis functions will also be summarized in Table 1, Tables 1, and 2 in Chapter "2D Basis Functions for Triangular and Rectangular Meshes". This summary of basis functions and their properties is useful for readers intending to modify or extend

Table 1 The properties of the basis functions $b^2_{i+\frac{1}{2}}(x')$ and $b^3_{i+\frac{1}{2}}(x')$ defined in the interval $\Omega_i = (x_i, x_{i+1})$ with the width $dx_{\Omega_i} = x_{i+1} - x_i$ ($dx'_{\Omega_i} = \frac{1}{2}dx_{\Omega_i}$). The mid-point is $x_{i+\frac{1}{2}} = \frac{1}{2}(x_{i+1} + x_i)$. The local coordinate is defined as $x' = x - x_{i+\frac{1}{2}}$. The lower index x' in f means spatial differentiation with respect to x'

Function $f(x')$	$b^2_{i+\frac{1}{2}}$	$b^3_{i+\frac{1}{2}}$
Expression	$\frac{1}{2}\left(x'^2 - dx'^2_{\Omega_i}\right)$	$\frac{1}{6}x'\left(x'^2 - dx'^2_{\Omega_i}\right) = \frac{1}{3}b^2_{i+\frac{1}{2}}x'$
$f(-dx'_{\Omega_i})$	0	0
$f(dx'_{\Omega_i})$	0	0
$f_{x'}(x')$	x'	$\frac{1}{2}x'^2 - \frac{1}{6}dx'^2_{\Omega_i}$
$f_{x'}(-dx'_{\Omega_i})$	$-dx'_{\Omega_i}$	$\frac{1}{3}dx'^2_{\Omega_i}$
$f_{x'}(dx'_{\Omega_i})$	dx'_{Ω_i}	$\frac{1}{3}dx'^2_{\Omega_i}$
$f_{x'x'}(x')$	1	x'
$f_{x'x'x'}(x')$	0	1
$\int_{-dx'_{\Omega_i}}^{dx'_{\Omega_i}} f(x')dx'$	$-\frac{2}{3}dx'^3_{\Omega_i}$	0

the procedures presented in this section. If amplitudes at principal nodes are given, we can create a piecewise linear spline using linear hat functions. As for flux derivatives, discontinuous functions may arise, the basis functions for discontinuous representation are given. For the cell interval

$$\Omega_i = (x_i, x_{i+1}), \ \text{for } i = 0, 1, 2, ...,\tag{5}$$

we define the discontinuous hat functions as (see Fig. 1)

$$\begin{cases} e^+_i(x) = \dfrac{x_{i+1} - x}{x_{i+1} - x_i}, & \text{for } x \in \Omega_i, \\ e^-_i(x) = \dfrac{x - x_{i-1}}{x_i - x_{i-1}}, & \text{for } x \in \Omega_{i-1}. \end{cases}\tag{6}$$

The two hat functions are defined to be zero when $x \notin \Omega_i$ for $e^+_i(x)$ and $x \notin \Omega_{i-1}$ for $e^-_i(x)$. Even though this book deals with continuous Galerkin methods only, the representation of discontinuous functions will be treated in the following, as they may occur by taking the derivative of continuous functions and also in directions where there is no differentiation for two dimensions. In the following, $fl(x)$ may be any function, such as a density field $h(x)$ or a flux.

In the following, we treat the linear part and the higher-order part of the field representations separately. The linear part is determined by corner points only. The high-order part of fields is obtained by subtracting the linear part. This means that the collocation points of the high-order part are 0 at the corner points. They are

Table 2 The properties of the basis functions $b_i^-(x')$ and $b_i^+(x')$ defined in the interval Ω_j with the width $dx_{\Omega_j} = x_{j+1} - x_j$ $(dx'_{\Omega_j} = \frac{1}{2}dx_{\Omega_j})$. The mid-point is $x_{j+\frac{1}{2}} = \frac{1}{2}(x_{j+1} + x_j)$. The local coordinate is defined as $x' = x - x_{j+\frac{1}{2}}$, where $j = i - 1$ for $b_i^-(x')$ and $j = i$ for $b_i^+(x')$. The lower index x' in f means spatial differentiation with respect to x'

Function $f(x')$	b_i^+	b_i^-
Expression	$-\dfrac{1}{2dx'_{\Omega_i}}b^2_{i+\frac{1}{2}}(x') + \dfrac{3}{2dx^2_{\Omega_i}}b^3_{i+\frac{1}{2}}(x')$	$\dfrac{1}{2dx'_{\Omega_{i-1}}}b^2_{i-\frac{1}{2}}(x') + \dfrac{3}{2dx^2_{\Omega_{i-1}}}b^3_{i-\frac{1}{2}}(x')$
Value $f(x')$ at left boundary	0	0
Value $f(x')$ at right boundary	0	0
$f_{x'}(x')$	$-\dfrac{1}{2dx'_{\Omega_i}}x' + \dfrac{3}{2dx^2_{\Omega_i}}\left(\dfrac{1}{2}x'^2 - \dfrac{1}{6}dx^2_{\Omega_i}\right)$	$-\dfrac{1}{2dx'_{\Omega_{i-1}}}x' + \dfrac{3}{2dx^2_{\Omega_{i-1}}}\left(\dfrac{1}{2}x'^2 - \dfrac{1}{6}dx^2_{\Omega_{i-1}}\right)$
Value $f_{x'}(x')$ at left boundary	1	0
Value $f_{x'}(x')$ at right boundary	0	1
$f_{x'x'}(x')$	$-\dfrac{1}{2dx'_{\Omega_i}} + \dfrac{3}{2dx^2_{\Omega_i}}x'$	$\dfrac{1}{2dx'_{\Omega_{i-1}}} + \dfrac{3}{2dx^2_{\Omega_{i-1}}}x'$
$f_{x'x'x'}(x')$	$\dfrac{3}{2dx^2_{\Omega_i}}$	$\dfrac{3}{2dx^2_{\Omega_{i-1}}}$
$\int_\Omega f(x')dx'$	$\dfrac{1}{3}dx^2_{\Omega_i}$	$-\dfrac{1}{3}dx^2_{\Omega_{i-1}}$

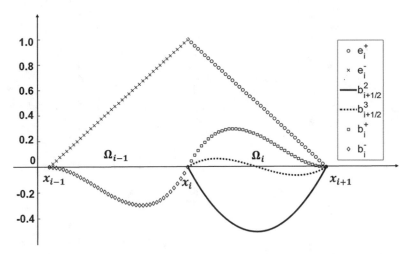

Fig. 1 Images of the basis functions $e_i^+, e_i^-, b_{i+\frac{1}{2}}^2, b_{i+\frac{1}{2}}^2, b_i^+$ in the interval Ω_i and b_i^- in the interval Ω_{i-1}

obtained by subtracting the linear interpolated values of corner amplitudes from the field at edge points.

Using Eq. (6), the piecewise linear basis function representation of a discontinuous function $fl(x)$ has double amplitudes fl_i^+ and fl_i^- at principal nodes. fl is called *a double-valued function* at corner points. Assume now that fl_i is the discontinuous amplitude at principal nodes, the linear part of the field fl_i is approximated by

$$fl^{lin}(x) = \sum_{i=0}^{i_e} \left[fl_i^+ e_i^+(x) + fl_i^- e_i^-(x) \right], \tag{7}$$

where i_e is the number of cells.

Now we define the continuous hat function as

$$e_i(x) = e_i^+(x) + e_i^-(x). \tag{8}$$

Assume that fl_i is the continuous amplitude at principal nodes, we obtain for the representation of the linear part fl_i of continuous fl:

$$fl^{lin}(x) = \sum_{i=0}^{i_e} fl_i e_i(x). \tag{9}$$

Note that Eq. (7) represents discontinuous functions fl^{lin}, while the fields defined in Eq. (9) are continuous. Equation (7) uses double amplitudes at corner points.

Continuous functions of the form Eq. (9) are obtained by making the two amplitudes fl_i^+ and fl_i^- equal. When a discontinuous function is given, the approximation by a continuous function is often obtained by requiring that the two functions have the same mass associated with each grid point.

Consider a discontinuous representation of fl according to Eq. (7); then there is a natural approximation of the discontinuous function by a continuous one, using the principle of mass conservation. Note that only the linear part of fl needs adjustment. We define $m_i^+ = \int_{x_i}^{x_{i+1}} e_i^+(x)dx$, $m_i^- = \int_{x_{i-1}}^{x_i} e_i^-(x)dx$, $m_i = m_i^+ + m_i^-$, and $fl_i = \frac{1}{m_i}(fl_i^+ m_i^+ + fl_i^- m_i^-)$. A mass consistent approximation of the discontinuous function Eq. (7) by the continuous function Eq. (9) is then obtained by

$$\int_{x_0}^{x_{ie}} fl^{lin}(x)dx = \sum_{i=0}^{ie} \int_{x_{i-1}}^{x_{i+1}} \left[fl_i^+ e_i^+(x) + fl_i^- e_i^-(x) \right] dx$$

$$= \sum_{i=0}^{ie} fl_i m_i(x).$$

(10)

Higher-order field representations are obtained by adding a high-order function $h_i^{ho}(x)$ to the piecewise linear representation Eq. (10). $h_i^{ho}(x)$ is continuous with $h_i^{ho}(x) = 0$ for x_i being principal nodes $(i = 0, 1, 2, ...)$. For a third-order representation of $fl(x)$, we obtain

$$fl(x) = fl^{lin}(x) + fl^{ho}(x),$$

(11)

where fl^{lin} is defined in Eq. (10). It will turn out that when $fl(x)$ in Eq. (9) approximates an analytic function, fl^{ho} is a $O(3)$ term. fl^{ho} is defined using a set of basis functions $b_{i+\frac{1}{2}}^2(x)$ and $b_{i+\frac{1}{2}}^3(x)$ (see Fig. 1):

$$\begin{cases} b_{i+\frac{1}{2}}^2(x') = \frac{1}{2}\left(x'^2 - dx_{\Omega_i}'^2\right), \\ b_{i+\frac{1}{2}}^3(x') = \frac{x'}{6}\left(x'^2 - dx_{\Omega_i}'^2\right), \end{cases}$$

(12)

where $x' = x - x_{i+\frac{1}{2}}$ is the local coordinate of the interval Ω_i, $dx_{\Omega_i}' = \frac{1}{2}dx_{\Omega_i}$, $x_{i+\frac{1}{2}} = \frac{1}{2}(x_i + x_{i+1})$. The two basis functions are zero when $x \notin \Omega_i$.

A collection of interpolation formula is given in Table 1. In the following, we will just choose $n^{ord} = 3$. Second-degree polynomial representation is done in an analog manner. It is easily verified that (see Fig. 2)

$$\begin{cases} b_{x',i+\frac{1}{2}}^2(x') = x', \; b_{x'x',i+\frac{1}{2}}^2(x') = 1, \; b_{x'x'x',i+\frac{1}{2}}^2(x') = 0, \\ b_{x',i+\frac{1}{2}}^3(x') = \frac{1}{2}x'^2 - \frac{1}{6}dx'^2, \; b_{x'x',i+\frac{1}{2}}^3(x) = x', \; b_{x'x'x',i+\frac{1}{2}}^3(x) = 1. \end{cases}$$

(13)

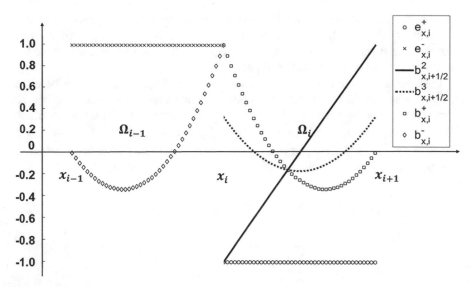

Fig. 2 Image of the spatial first derivatives of basis functions $e_i^+, e_i^-, b_{i+\frac{1}{2}}^2, b_{i+\frac{1}{2}}^2, b_i^+$ in the interval Ω_i and b_i^- in the interval Ω_{i-1}

We obtain for the high-order basis function representation of Eq. (11):

$$fl^{ho}(x) = \sum_{i=0}^{i_e} \left[fl_{xx,i+\frac{1}{2}} b_{i+\frac{1}{2}}^2(x) + fl_{xxx,i+\frac{1}{2}} b_{i+\frac{1}{2}}^3(x) \right]. \tag{14}$$

With Eqs. (10), (11), and (14), we obtain the third-order representation of a field $fl(x)$. In Eqs. (10) and (14), the amplitudes describing the field $fl(x)$ are $fl_i, fl_{xx,i+\frac{1}{2}}$ and $fl_{xxx,i+\frac{1}{2}}$. These amplitudes form the spectral representation of $fl(x)$. They are also called *spectral coefficients* or *spectral amplitudes*, and Eqs. (10) and (14) are called *the spectral representation*. Note that fl_i is the grid point value at a corner point. So corner point amplitudes are both grid point values and spectral coefficients. This is described for the third-order representation. The second-order representation of a field is obtained from Eq. (14) by dropping the second term in the sum that can be treated analogously.

Besides the spectral representation, *a grid point representation* will be used. The grid point representation uses the collocation points $x_i, x_{i+\frac{1}{3}}, x_{i+\frac{2}{3}}, x_{i+1}$. Grid point and spectral representations are equivalent. Collocation points can be distributed arbitrarily in a grid cell. Here we treat the case of equally distant collocation points. The two representations can be transformed into each other. As the spectral coefficients in Eqs. (14) and (10) belong to the principal nodes x_i and the corner point amplitude is $fl(x_i)$, no spectral to grid point transformation is needed for this point. Apart from $fl(x_i)$, the grid point space consists of $fl(x_{i+\frac{1}{3}})$ and $fl(x_{i+\frac{2}{3}})$.

For the points $x_{i+\frac{1}{3}}$ and $x_{i+\frac{2}{3}}$, transformations between spectral and grid point spaces are needed:

$$\begin{cases} fl(x_{i+\frac{1}{3}}) = fl^{lin}(x_{i+\frac{1}{3}}) + fl_{xx,i+\frac{1}{2}} b^2_{i+\frac{1}{2}}(x_{i+\frac{1}{3}}) + fl_{xxx,i+\frac{1}{2}} b^3_{i+\frac{1}{2}}(x_{i+\frac{1}{3}}), \\ fl(x_{i+\frac{2}{3}}) = fl^{lin}(x_{i+\frac{2}{3}}) + fl_{xx,i+\frac{1}{2}} b^2_{i+\frac{1}{2}}(x_{i+\frac{2}{3}}) + fl_{xxx,i+\frac{1}{2}} b^3_{i+\frac{1}{2}}(x_{i+\frac{2}{3}}). \end{cases}$$
(15)

When solving Eq. (15) for $fl_{xx,i+\frac{1}{2}}$ and $fl_{xxx,i+\frac{1}{2}}$, we have the grid point to spectral transform. In $\Omega_i = (x_i, x_{i+1})$, the points x_i together with amplitudes fl_i are called *the grid point space*. A standard representation is the representation by grid points.

Following Eqs. (9), (11), and (14), we obtain the following representations of fields:

$$fl(x) = \sum_{i=0}^{ie} \left[fl_i e_i(x) + fl_{xx,i+\frac{1}{2}} b^2_{i+\frac{1}{2}}(x) + fl_{xxx,i+\frac{1}{2}} b^3_{i+\frac{1}{2}}(x) \right].$$
(16)

From Eq. (16), we see that a set of amplitudes describing the function $fl(x)$ is $fl_i, fl_{xx,i+\frac{1}{2}}, fl_{xxx,i+\frac{1}{2}}$. For the schemes described in this book, the amplitude at principal nodes fl_i will always be used as an amplitude. However, for the amplitudes $fl_{xx,i+\frac{1}{2}}, fl_{xxx,i+\frac{1}{2}}$ describing the high-order part of $fl(x)$, there are alternative possibilities to choose then the second and third derivatives of $fl(x)$ at the mid-point. Amplitudes $fl_{xx,i+\frac{1}{3}}$ and $fl_{xxx,i+\frac{2}{3}}$ computed according to Eq. (15) can be used to derive a basis function representation analog to Eq. (16):

$$fl(x_{i'}) = \sum_{i=0}^{ie} \sum_{i'=0,1,2} fl_i e^{col}_{i,i'}(x).$$
(17)

For the definition of $e^{col}_{i,i'}(x)$, see Durran (2010). The basis function $e^{col}_{i,i'}$ is called *characteristic polynomials*. They are associated with each computational cell. For the definition of the characteristic polynomials, the collocation points of a cell must not necessarily be chosen regular as we mostly do in this book. For SE method, they are chosen as Gauss–Lobatto points. Gauss–Lobatto points are regular in a cell when using second-degree polynomial representation only. An advantage of representation Eq. (16) over Eq. (17) is that the different terms in Eq. (16) have different orders of magnitude. This fact can be used to construct efficient approximations, in particular sparse grids, when we will go to more than one dimension.

The representations Eq. (15) or (16) can be done in the same way for time derivatives as a function of x. Obtaining a representation of time derivatives from field and flux representations is called numerical discretization, and examples for classic schemes were given in Chapter "Simple Finite Difference Procedures". Discretizations based on polynomial field representations will be given in the following. For example, Eq. (15), applied for the time derivatives, gives a grid point representation for the time derivatives. When a discretization has delivered the time derivative and it is transformed into grid point space, the time discretization by RK4 or other time methods can be done as for the classic methods described in Chapter "Local-Galerkin Schemes in 1D". It is also possible to use RK4 with the spectral representation. The algebraic form used does not have impact on the result without physical parameterization. However, in a realistic application with physical parameterizations, there will be a difference (see Herrington et al. 2019 and Durran 2010). Physical parameterizations are done in grid point space only. Therefore, the time derivatives of such models, when using all corner and edge points, must be transformed into grid point space for applying RK4. In Section "The Interface to Physics in High-Order L-Galerkin Schemes" in Chapter "Simple Finite Difference Procedures", special physics interfaces are defined, which use only corner points for the physics part of a model. In this case, it is possible to do the model dynamics entirely in spectral space.

Another set of high-order basis functions $b_i^-(x), b_i^+(x)$ is used to have a set of amplitudes defined at corner points for the high-order part of field fl. These functions are defined as (see Fig. 1)

$$
\begin{cases}
b_i^-(x') = \dfrac{1}{2dx'_{\Omega_{i-1}}} b_{i-\frac{1}{2}}^2(x') + \dfrac{3}{2dx'^2_{\Omega_{i-1}}} b_{i-\frac{1}{2}}^3(x'), & \text{for } x' = x - x_{i-\frac{1}{2}} \text{ and } x \in \Omega_{i-1}, \\
b_i^+(x) = -\dfrac{1}{2dx'_{\Omega_i}} b_{i+\frac{1}{2}}^2(x') + \dfrac{3}{2dx'^2_{\Omega_i}} b_{i+\frac{1}{2}}^3(x'), & \text{for } x' = x - x_{i+\frac{1}{2}} \text{ and } x \in \Omega_i,
\end{cases}
$$

$$(18)$$

where $dx'_{\Omega_{i-1}} = \frac{1}{2} dx_{\Omega_{i-1}}$, $dx'_{\Omega_i} = \frac{1}{2} dx_{\Omega_i}$, $x_{i-\frac{1}{2}} = \frac{1}{2}(x_i + x_{i-1})$, and $x_{i+\frac{1}{2}} = \frac{1}{2}(x_i + x_{i+1})$. The basis functions $b_i^-(x)$ and $b_i^+(x)$ and their one-sided derivatives satisfy the properties (see Fig. 2):

$$
\begin{cases}
b_i^-(\pm dx'_{\Omega_{i-1}}) = 0.0, & \text{for } x \notin \Omega_{i-1}, \\
b_i^+(\pm dx'_{\Omega_i}) = 0.0, & \text{for } x \notin \Omega_i, \\
b_{x',i}^-(-dx'_{\Omega_{i-1}}) = 0, & \\
b_{x',i}^-(dx'_{\Omega_{i-1}}) = 1, & \\
b_{x',i}^+(-dx'_{\Omega_i}) = 1, & \\
b_{x',i}^+(dx'_{\Omega_i}) = 0. &
\end{cases}
$$

$$(19)$$

The two basis functions are zero when $x \notin \Omega_i$ or Ω_{i-1}. In Eq. (19), when we replace the variable x' by x, the basis functions are characterized by $b_i^-(x_{i-1}) = b_i^-(x_i) = b_i^+(x_i) = b_i^+(x_{i+1}) = 0, b_{x,i}^-(x_{i-1}) = 0, b_{x,i}^-(x_i) = 1, b_{x,i}^+(x_i) = 1, b_{x,i}^+(x_{i+1}) = 0$.

To define L-Galerkin methods later, we state the integral of the derivatives of basis functions:

$$
\begin{cases}
\int_{-dx'_{\Omega_i}}^{dx'_{\Omega_i}} b_i^+(x')dx' = \frac{1}{3}dx'^3_{\Omega_i}, \\
\int_{-dx'_{\Omega_{i-1}}}^{dx'_{\Omega_{i-1}}} b_i^-(x')dx' = -\frac{1}{3}dx'^3_{\Omega_{i-1}}.
\end{cases}
\tag{20}
$$

For a third-order representation of $fl(x)$, we obtain from Eq. (11) in analogy to Eq. (16):

$$
fl(x) = \sum_{i=0}^{i_e} fl_i e_i(x) + \sum_{i=0}^{i_e} \left[fl'^+_{x,i} b_i^+(x) + fl'^-_{x,i} b_i^-(x) \right],
\tag{21}
$$

with $fl'^+_{x,i} = fl^+_{x,i} - \frac{1}{dx_{\Omega_i}}[f(x_{i+1}) - f(x_i)]$ and $fl'^+_{x,i} = fl^-_{x,i} - \frac{1}{dx_{\Omega_{i-1}}}[f(x_i) - f(x_{i-1})]$. In Eq. (21), the dynamic amplitudes are $fl^+_{x,i}$ and $fl^-_{x,i}$, and the derivatives are considered to be one-sided. $fl^+_{x,i}$ and $fl^-_{x,i}$ are the left and right derivatives of $fl(x)$. Then the transformation between the amplitudes at the collocation points and the one-sided derivatives is

$$
\begin{cases}
fl(x_{i+\frac{1}{3}}) = fl^{lin}(x_{i+\frac{1}{3}}) + fl^+_{x,i}b_i^+(x_{i+\frac{1}{3}}) + fl^-_{x,i}b_i^-(x_{i+\frac{1}{3}}), \\
fl(x_{i+\frac{2}{3}}) = fl^{lin}(x_{i+\frac{2}{3}}) + fl^+_{x,i}b_i^+(x_{i+\frac{2}{3}}) + fl^-_{x,i}b_i^-(x_{i+\frac{2}{3}}).
\end{cases}
\tag{22}
$$

Equation (22) is the transformation between one-sided principal derivative space and collocation space. This can easily be solved for $fl^+_{x,i}$ and $fl^-_{x,i}$ to obtain the transformation from collocation space and one-sided derivative space. It is left to the reader to work out other transformations between the second and third derivative representations and the principal node derivative representation. The representations by collocation points ($fl(x_{i+\frac{1}{3}})$ and $fl(x_{i+\frac{2}{3}})$), by the second and third derivatives ($fl_{xx,i+\frac{1}{2}}$ and $fl_{xx,i+\frac{1}{2}}$) and by left and right derivatives ($fl^+_{x,i}$, and $fl^-_{x,i}$) are equivalent. In all these 3 representations, the corner point amplitudes are also dynamic variables.

The different descriptions of $h(x)$ or $fl(x)$ given above are equivalent as they lead to the same functional representation. Therefore, any L-Galerkin method working on the basis functions only is independent of the representation used. For example, the collocation point or the second–third derivative representation of a field will lead to the same numerical scheme when this scheme is defined working on the basis functions. Some methods, such as SEs, do not work in functional space but rather on the collocation points. So the SE method results into different versions depending on the choice of the collocation points. According to Durran (2010), only the version using Gauss–Lobatto points is stable.

Based on the above description, we summarize the transformations from grid point space to spectral space as follows.

First, we use Eq. (10) to define $fl^{lin}(x)$ using the two corner points belonging to the cell $\Omega_i = (x_i, x_{i+1})$ and form the linear part for grid amplitudes at $x_{i+\frac{1}{3}}$ and $x_{i+\frac{2}{3}}$:

$$\begin{cases} fl^{lin}_{i+\frac{1}{3}} = fl^{lin}(x_{i+\frac{1}{3}}), \\ fl^{lin}_{i+\frac{2}{3}} = fl^{lin}(x_{i+\frac{2}{3}}). \end{cases} \tag{23}$$

Second, we use these values to compute the high-order deviations at inner collocation points:

$$\begin{cases} fl^{ho}_{i+\frac{1}{3}} = fl_{i+\frac{1}{3}} - fl^{lin}_{i+\frac{1}{3}}, \\ fl^{ho}_{i+\frac{2}{3}} = fl_{i+\frac{2}{3}} - fl^{lin}_{i+\frac{2}{3}}. \end{cases} \tag{24}$$

Third, we can obtain the relation between the spectral and grid point amplitudes at the inner collocation points:

$$\begin{cases} fl_{xx,i+\frac{1}{2}} = -\dfrac{9}{8dx'^{2}_{\Omega_i}}(fl^{ho}_{i+\frac{2}{3}} + fl^{ho}_{i+\frac{1}{3}}), \\ fl_{xxx,i+\frac{1}{2}} = -\dfrac{81}{8dx'^{3}_{\Omega_i}}(fl^{ho}_{i+\frac{2}{3}} - fl^{ho}_{i+\frac{1}{3}}). \end{cases} \tag{25}$$

Eqution (25) provides the spectral amplitudes in the second/third derivative representation. From these, we obtain the amplitudes $fl^{+}_{x,i}$ and $fl^{-}_{x,i}$ for the one-sided derivative representation:

$$\begin{cases} fl_{xx,i+\frac{1}{2}} = -\dfrac{1}{2dx'_{\Omega_i}}(fl^{+}_{x,i} - fl^{-}_{x,i}), \\ fl_{xxx,i+\frac{1}{2}} = \dfrac{3}{2dx'^{2}_{\Omega_i}}(fl^{+}_{x,i} + fl^{-}_{x,i}). \end{cases} \tag{26}$$

Equations (25) and (26) are equivalent representations of third-order piecewise representations of the field $fl(x)$.

$$\begin{cases} fl^{-}_{x,i} = -\dfrac{9}{4dx'_{\Omega_i}}(2fl^{ho}_{i+\frac{2}{3}} - fl^{ho}_{i+\frac{1}{3}}), \\ fl^{+}_{x,i} = -\dfrac{9}{4dx'_{\Omega_i}}(fl^{ho}_{i+\frac{2}{3}} - 2fl^{ho}_{i+\frac{1}{3}}). \end{cases} \tag{27}$$

$fl(x)$ is continuous but has a derivative not necessarily continuous. Note that the representation of piecewise polynomial functions with continuous derivatives is achieved by choosing the fl_i^+ and fl_i^- equal:

$$fl_i^+ = fl_i^-, \tag{28}$$

for differentiable $fl(x)$. Of course, Eq. (28) can be used to approximate fields with discontinuous derivatives by those with continuous derivatives.

The representation Eq. (21) is a continuous function, which is not necessarily differentiable. The fields according to Eq. (16) $fl(x)$ or $h(x)$ are called *a third-order* or *o3 representation*. If in Eq. (16) the third term is omitted, we have *a second-order* or *o2 representation*. The fluxes can be represented as o2 or o3. In this book, we do not consider lower orders than 2 or higher than 3 for fields $h(x)$ and fluxes $fl(x)$. There are two numbers n and m to describe the approximation orders of schemes, according to their order of representation. We name onom a method representing the field $h(x)$ in order n and its flux $fl(x)$ in order m. There is another property of function space that is important for the properties of the resulting discretization, and this is the regularity of the functions at principal nodes. Eq. (16) or (21) implies that $fl(x)$ is continuous at principal nodes. We define a C^0-space that is the set of continuous functions fl according to Eq. (16) or (21). Again, we can demand these properties for the fields $h(x)$ and the fluxes $fl(x)$. Therefore, a scheme using third-order fluxes $fl(x)$ and fields $h(x)$ being continuous at principal nodes may be classified as o3o3C^0C^0. We can also define C^1-space the space of functions that are differentiable at principal nodes. In this book (except for Sections "The L-Galerkin scheme: o3o5C^1C^2 and The L-Galerkin Scheme: o4o5C^1C^2" in Chapter "Simple Finite Difference Procedures"), we do not consider higher regularity than C^1 and no higher order than o3 for fields or fluxes may be represented to a higher order.

We have seen that when we have the time derivative as a function of x, we can use RK4 or any other time differencing method to predict the dynamic field in time. We assume that the fields in basis function formulation are given in the form Eqs. (7) and (14) as continuous piecewise polynomial functions. By applying directly the RHS of the dynamic equations to the functional representation Eqs. (7) and (14), we obtain a discontinuous representation of the time derivative. Any approximation of this discontinuous function by a function of standard form Eqs. (7) and (14) is called *a Galerkin operation*, and this will allow the time translation. Galerkin operations are at the heart of all numerical integration procedures on basis function–generated function spaces. Such methods will be described in the following. The classic Galerkin method (Durran 2010) uses a least square method to achieve this approximation. This is often called the use of the weak formulation of the problem, as the least square method turns out to be equivalent to the scalar multiplication of the functional representations Eqs. (7) and (14). With local basis functions (see Section "The Classic Galerkin Procedure" in Chapter "Simple Finite Difference Procedures"), classic Galerkin methods need the solution of a band diagonal matrix. Even though the basis functions are local, the classic Galerkin method will turn out to be non-local. This means that for the computation of the time derivative for an index i all amplitudes h_i for the whole area are needed. While it is feasible for small problems, the classic Galerkin method poses a major problem with very large problems to be solved with multiprocessor computers where the efficiency may break down due to the necessity of much processor communication. For this reason, local versions of the Galerkin method have been developed, such as the third-degree method, SEs, and o*nom*. Some of such methods will be described in the next sections.

The Classic Galerkin Procedure

The aim of this book is to solve systems of conservation equations such as given in Eq. (1). In this section, we consider the simple test problem Eq. (2) in Chapter "Simple Finite Difference Procedures". Let the field $h(x)$ be given as a representation Eq. (16) or (21). Using the transformations Eq. (22) to collocation point space, the time marching procedure can be done in this space by any time schemes given in (Durran 2010; Ahlberg and Nilson 1967; Szabo and Babuska 1991). In this book, we will use the RK4 scheme for tests (Section "The Runge–Kutta and Other Time Discretization Schemes" in Chapter "Introduction"). The time schemes can also be done in spectral space for the case without physical parameterizations.

Classic Galerkin schemes (Durran 2010; Honnert et al. 2020) can be applied to compute the time derivatives. When computed using a differentiation of the representation Eq. (16), the RHS of the equation of motion Eq. (2) in Chapter "Simple Finite Difference Procedures" will result in a discontinuous function. In order to obtain a system of ordinary FD equations to be solved by a time integration scheme, this must be approximated by a continuous function system of the form Eq. (16) for the time derivatives. The classic Galerkin scheme achieves the continuity property by a least square approximation. This is equivalent to the weak formulation, where RHS and h_t are multiplied with the basis functions, using the scalar product in the Hilbert space of fields (see Section "Linear Algebra" in Chapter "Summary and Outlook"). This leads to a matrix equation for the amplitudes of the time derivative amplitudes, the mass matrix equation. When applied with the function systems Eq. (16) or (21), the classic Galerkin schemes are called finite elements (FE).

The basis functions given in Eq. (16) or (21) are piecewise linear or piecewise cubic. This allows the approximation of any smooth function of up to third order in the parameter dx, representing the cell size. In other words, with decreasing dx, the representation error using the function representations Eq. (16) or (21) goes to 0 in the designed order 3+1. The basis functions defined in Eq. (16) or (21) are called *test functions*, and the set of test functions forms *the spectral space*. In order to approximate a given function by test functions, the amplitudes in Eq. (16) must be chosen appropriately. These amplitudes are called *spectral coefficients*. The Galerkin procedure is mainly concerned with the approximation of the time derivatives of dynamic fields. The spectral coefficients determined in this way will then be amplitudes of the time derivatives of fields, also called the spectral representation of the time derivatives.

The spectral space is a linear space because the dependence of the test functions on the amplitudes is linear. This terminology is explained in Section "Linear Algebra" in Chapter "Summary and Outlook". The emphasis of this book is to approximate the dynamic equation (1) in Chapter "Simple Finite Difference Procedures" by linear function spaces, such as given in Section "Functional Representations, Amplitudes, and Basis Functions" in Chapter "Simple Finite Difference Procedures", and we go up to third order for the piecewise polynomial representations. Such higher-order dependence of the basis functions on x simplifies the construction of high-order approximations. The Galerkin procedure cannot only handle cases of approximations by linear spaces. An example of a powerful approximation based on a nonlinear dependence of the test function on approximation parameters will be given at the end of this section.

However, the classic Galerkin procedure is often applied with linear spaces. Let an equation of the form Eq. (1) in Chapter "Simple Finite Difference Procedures" be given. For simplicity, we consider one equation for a field $h(x)$:

$$h_t = RS(h, h_x, h_{xx}). \tag{29}$$

The classic Galerkin approximation is applicable to all spatial linear basis function (b_i) representations of a field $h(x)$:

$$h(x) = \sum_i h_i b_i(x). \tag{30}$$

In Eq. (30), $b_i(x)$ may be piecewise polynomial functions of any order 1, 2, or 3. Classic Galerkin always concerns the approximation in linear spaces as the dependence in Eq. (30) on h_i is linear. The definitions Eqs. (8), (12), and (16) are examples of a third-order field representation, where the basis functions are $e_i(x), b^2_{i+\frac{1}{2}}(x)$ and $b^3_{i+\frac{1}{2}}(x)$. Another possibility is to use just the linear approximation Eq. (10). The amplitudes for the third-order case are $h_i, h_{xx,i+\frac{1}{2}}$ and $h_{xxx,i+\frac{1}{2}}$. In most applications, the function spaces defined in Eq. (30) are rather high dimensional, as the dimension of the function space corresponds to the number of node points of the grid. As shown in Section "The Boussinesq Model of Convection Between Heated Plates" in Chapter "Introduction", even a dimension as low as 4 can lead to fairly accurate approximations for heat generated convection. A further simplification using 3 amplitudes is the paradigmatic Lorenz model that was introduced in Sections "The Boussinesq Model of Convection Between Heated Plates and The Lorenz Paradigmatic Model" in Chapter "Introduction". For this model, 3 amplitudes are sufficient with the ansatz Eq. (30). High dimensional approximation spaces Eq. (30) allow to approximate any solution of Eq. (29) with a given scale that is defined by the grid length. A low dimensional approximation with Eq. (30) uses knowledge of the solution, which then becomes very efficient and even could be used for parameterization of convection for the case that conventional cells have about the same size as the grid. This problem is called the gray area of convection parameterization in climate models (Moncrieff et al. 2017).

By inserting Eq. (29) into Eq. (30), we obtain an approximation $RS_a(x)$ of the right-hand side. As done often, we will suppress the index a in the following. For both $RS(x)$ and $h_t(x)$, we assume approximations of the form Eq. (30). To obtain the classic Galerkin approximation to Eq. (29), we form the scalar products of the right- and left-hand sides of Eq. (29) with the basis functions $b_i(x)$. The scalar product of two functions $f(x)$ and $g(x)$ is defined as $(f, g) = \int f(x)g(x)dx$. For the (linear) Galerkin approximation, we then obtain the classic Galerkin approximation:

$$(h_t, b_i) = (RS(x), b_i(x)).\tag{31}$$

Note that Eq. (31) is a linear equation for the spectral amplitudes $h_{t,i}$. To see this, insert Eq. (30) into Eq. (31), and the Galerkin equation (31) becomes

$$\left(\sum_{i'} h_{t,i'}b_{i'}, b_i\right) = (RS, b_i),\tag{32}$$

in which we have dropped the argument x of the functions to indicate that we consider them as elements of the Hilbert space of meteorological states. Equation (32) can be transformed into

$$\sum_{i'} h_{t,i'}(b_{i'}, b_i) = RS_i,\tag{33}$$

where $RS_i = (RS, b_i)$. When introducing the vectors $\mathbf{RS} = \{RS_i\}$ and $\mathbf{h}_t = \{h_{t,i}\}$ and the matrix $\mathbf{M} = \{m_{i',i}\} = (b_{i'}, b_i)$, we can write Eq. (33) as

$$\mathbf{M} \cdot \mathbf{h}_t = \mathbf{RS},\tag{34}$$

where $\mathbf{M}$ is called the mass matrix. In Eq. (34), when using grid related approximations, the dimension in phase space of the vectors is rather high and corresponds to the number of node points of the grid.

When the basis functions are local, Eq. (34) is a local set of equations as only few side diagonals are different from 0 for the central element $m_{i',i}$ of $\mathbf{M}$. An efficient solution procedure for such sparse matrix equations is Gaussian elimination (Durran 2010). It turns out that even though we have a local equation, the solution is not local. The computed time derivative depends on all amplitudes of the grid. This is rather efficient on computers with few processors (Steppeler et al. 1990). Gaussian elimination is defined for fields in one spatial dimension. By repeating the operation, it can be applied in 2 or 3 spatial dimensions as well (Steppeler et al. 1990). On few processor computers, the classic Galerkin method is still rather popular to compute flows in complicated geometries. On massively parallel machines, however, the classic Galerkin method can encounter problems of scalability. This means that the computation time is not proportional to the number n of processors, which is the reason for the introduction of the L-Galerkin methods. Further variants of the classic Galerkin method, such as Petrov–Galerkin, have been investigated (Steppeler et al. 1990). Such alternatives can for example be obtained by using alternative weights for the integrals used above. Practical advantages in realistic models of such variants of the classic approach could not be shown.

The classic Galerkin method is limited to linear spaces Eq. (30) of approximation. Classic Galerkin CG can be accurate up to fourth order, at least for regular cells. CG is not local, and SE and onom use high-order basis functions. We have shown that SE2 is second order. SE3 and other L-Galerkin methods are suitable for irregular grids. onom is also very efficient due to sparseness. However, the limitation to linear spaces Eq. (30) is not necessary. In all variational approximation methods, test functions are selected by the algorithm, the Galerkin method. In Eq. (30), b_i is the test function for the approximation in linear spaces, and h_i are the approximation amplitudes.

In the following, we generalize the concept of Galerkin approximations in linear spaces to the concept of a nonlinear dependence of the test function on the amplitude parameter λ. The field h may depend on the test function assumption where an approximation is made for the approximation to h by a parameter λ:

$$h(x) = h'(x, \lambda), \tag{35}$$

where h' is the test function for the more general case that we no longer approximate in linear spaces. In Eq. (35), the dependence of h on λ may be nonlinear. Equation (30) is a special case of Eq. (35). Similar as with the linear Galerkin procedure, a dynamic equation for the amplitudes in Eq. (30) is provided, the nonlinear generalization of the Galerkin equation will give a dynamic equation for λ. The example to be given in the following is intended to exemplify a possible application of the nonlinear Galerkin procedure. The example is described for the case of just one approximation amplitude λ. This already means that the concepts of resolution, approximation order, etc., are no longer valid. In this example, we have previous knowledge of the shape of the solution, which is then used to obtain the solution very efficiently and avoid the high dimensions in phase space. In principle, such nonlinear Galerkin procedures can be combined with the polynomial approximation spaces, such as Eq. (16), and then the approximation would be a difference scheme of a designed order. Having the corner point positions as dynamic variables would lead to an adaptive grid. The application given here is designed to show that with just one dynamic variable, meaningful applications are possible. For this special application, the term "order of approximation" is not applicable. When solving this initial value problem and putting $\lambda(t)$ into Eq. (35), the required approximation to $h(x, t)$ is obtained. The equation of motion for λ is obtained by using a least square minimalization of the functional Q. We assume that the field h is defined for $x > 0$:

$$Q = \int_t^{t+\Delta t} dt' \int_0^\infty \left[h_t'(x, \lambda) - RS(h', h_x', h_{xx}', \lambda) \right]^2 dx. \tag{36}$$

Equation (36) is minimized for $\Delta t ==> 0$ by doing the differentiation to λ and to obtain the Galerkin approximation for approximations not necessarily in linear spaces:

$$\lambda_t \int_0^\infty \frac{\partial}{\partial \lambda} \left[h'(x, \lambda) - h'(x, \lambda) RS(h', h_x', h_{xx}', \lambda) \right] dx. \tag{37}$$

Note that Eq. (37) has the form:

$$\lambda_t = RS'(\lambda), \tag{38}$$

with $RS'(\lambda) = \frac{1}{\int_0^\infty \frac{\partial}{\partial \lambda} h'(x, \lambda) dx} \int_0^\infty RS(h', h_x', h_{xx}' \lambda) h'(x, \lambda) dx$.

The form Eq. (38) is a prognostic equation for the discretizing parameter λ, and it is generalized to the case of more than one discretizing parameter by replacing the derivative to λ by a partial derivative.

Let us consider a very simple example. Assume the equation of motion:

$$\frac{\partial u}{\partial t} = -v \frac{\partial^2 u}{\partial x^2}. \tag{39}$$

The boundary conditions for Eq. (39) are that the atmosphere is at rest for large x, and for $x = 0$, we have a moving plate with velocity $u_0 = $ const. The physical situation described by Eq. (39) is a viscous flow in the x-direction depending on the y-coordinate, where the wall at $x = 0$ moves with velocity u_0. For $u(x)$, we make the test function u' assumption involving the parameter y_0:

$$u'(x, y_0) = \left(u_0 - u_0 \frac{y}{y_0} \right) \theta(x_0 - x), \tag{40}$$

with

$$\theta = \begin{cases} 1, & \text{for } x > 0 \\ 0, & \text{for } x < 0. \end{cases} \tag{41}$$

When computing the RHS of Eq. (38), differentiation through the jump of the derivative of u is done using the Dirac delta function. The formalism of distributions and its application to flow problems are treated in Section "Polygonal Spline Solutions Using Distributions and Discontinuities". From Eq. (38), we obtain the

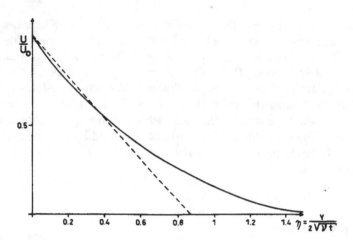

Fig. 3 Comparison of the approximate (dashed) and exact (solid) velocity profile (Reproduced from Steppeler 1978)

equation of motion for y_0 is

$$y_{0,t} = \frac{3}{2} \frac{\nu}{y_0}. \tag{42}$$

The solution of Eq. (42) is $y_0(t) = \sqrt{3\nu t}$. The approximate solution is then

$$u = o_0 \frac{f_1(y)}{2\sqrt{\nu t}}, \tag{43}$$

with $f_1(y) = 1 - y\frac{2}{\sqrt{3}}\theta\left(\frac{\sqrt{3}}{2} - y\right)$. According to Braun (1958), using Eqs. (42) and (43), the approximate and analytic solutions are compared in Fig. 3.

The Galerkin equation (38) generalizes the Galerkin procedure on linear spaces Eqs. (30)–(34), and it is mostly used for approximations in grid related spaces. The linear approximation spaces to be applied in 1D given in Eq. (30) and the special form used in this chapter are described in Section "Functional Representations, Amplitudes, and Basis Functions". Most Galerkin approximations on linear spaces are done in general approximative spaces, meaning spaces being able to approximate up to the scale given by the grid length of the discretization, without prior knowledge of how this solution looks. Figure 3 shows that a few basis functions of a linear space may describe a physical situation well. This was the case with the Lorenz paradigmatic model, which used only 3 amplitudes to describe a conventional cell (described in Section "The Boussinesq Model of Convection Between Heated Plates" in Chapter "Introduction"). For the rest of the book, however, we will concentrate on the general approximation spaces introduced in Section "Functional Representations, Amplitudes, and Basis Functions". These are able to approximate any solution down to a certain scale.

We have seen in Section "Functional Representations, Amplitudes, and Basis Functions" that the amplitudes may be grid point values of h at corner nodes, second derivatives at interior collocation points, or other. To apply the Galerkin formula Eq. (38), we need to form $RS(h)$. For the simple test problem of homogeneous advection Eq. (2) in Chapter "Simple Finite Difference Procedures", $RS(h)$ is obtained by differentiation of the basis functions. For the more general case, more elaborate procedures are applied. Note that according to Eq. (38), certain weighted integrals of $RS(h)$ are needed. Often these are obtained by numerical integration using the collocation points. The integrals occurring in Eqs. (30)–(33) are called Galerkin integrals.

Spectral Elements

Spectral elements (SE, Giraldo 2001; Taylor et al. 1997) are a variant of the classical Galerkin FE method avoiding the problem of the mass matrix solution. Such variants of the original Galerkin method are called L-Galerkin methods. Like the other methods described in this chapter, SE uses the piecewise polynomial spaces of degree n^{ord}. n^{ord} can have any positive value, and it has been tested for values of n^{ord} up to 8 and more. SEs using polynomials of order 2 are called SE2, etc. In Section "Functional Representations, Amplitudes, and Basis Functions" in Chapter "Simple Finite Difference Procedures", polynomial spaces were described up to order 3, and the description of SE in this book is limited to the rather low-order cases SE2 and SE3. So SE uses the same piecewise polynomial function spaces as the other L-Galerkin methods to be discussed in this chapter. Note that recent practical experience from using SE for real atmosphere modeling (Herrington et al. 2019) indicated that it makes no sense to go beyond order 3 for polynomial spaces. However, the collocation grid differs. While in all other L-Galerkin methods discussed in this chapter the collocation grid is regular within an interval $\Omega_i = (x_i, x_{i+1})$, SE uses the irregular Gauss–Lobatto grid for values of n^{ord} greater than 2. The reader should keep in mind that the cell size $dx_{\Omega_i} = x_{i+1} - x_i$ of the interval Ω_i can vary arbitrarily. Therefore, the global collocation grid taken over all cells can be irregular. Just the part of the grid inside a cell can be regular for some L-Galerkin methods.

For SE2, the collocation grid has 3 points, two of them being shared between two cells. Therefore, the average number of collocation points is 2 times the number of cells. For the cell Ω_i, the collocation points $X^c_{i,i'}(i' = 0, 1, 2)$ are

$$\begin{cases} X^c_{i,0} = x_i, \\ X^c_{i,1} = x_{i+\frac{1}{2}}, \\ X^c_{i,2} = x_{i+1}, \end{cases} \qquad (44)$$

where $x_{i+\frac{1}{2}} = \frac{1}{2}(x_i + x_{i+1})$.

SE3 has four collocation points $X^c_{i,i'}(i' = 0, 1, 2, 3)$ per grid cell, and these are

$$\begin{cases} X^c_{i,0} = x_i, \\ X^c_{i,1} = x_{i+\frac{1}{3}} = x_i + \dfrac{1}{2}\left(1 - \dfrac{1}{\sqrt{5}}\right) dx_{\Omega_i}, \\ X^c_{i,2} = x_{i+\frac{2}{3}} = x_i + \dfrac{1}{2}\left(1 + \dfrac{1}{\sqrt{5}}\right) dx_{\Omega_i}, \\ X^c_{i,3} = x_{i+1}. \end{cases} \tag{45}$$

The time-stepping procedure for the simple test equation (2) in Chapter "Simple Finite Difference Procedures" follows to be a large part of the SE method when done in collocation space. The spectral space consists of all polynomials of degree up to three $p_i(x) = a_{i,0} + a_{i,1}x + a_{i,2}x^2 + a_{i,3}x^3$ in Ω_i. With SE3, the polynomial coefficients a_0, a_1, a_2, a_3 belonging to cell Ω_i are the spectral amplitudes, and they form an equivalent description to the four grid point values.

In Section "Functional Representations, Amplitudes, and Basis Functions" some options for polynomial basis functions are given (such as Eqs. (6), (12), and (18)). The approximation space is independent of the basis functions, and this means that the classic Galerkin method and FE and also SE of a given order are independent of the basis functions. So when using the high- and low-order basis functions defined in Section "The Classic Galerkin Procedure", we should get the same results as when using the Legendre functions, which are normally used with SE. The L-Galerkin methods to be introduced in Sections "The L-Galerkin Scheme: o3o3–The L-Galerkin Scheme: o2o3" will use different operations for different basis functions and therefore are dependent on the choice of the basis. For SE, it is customary to use a different basis function consisting of Legendre polynomials. The polynomial basis function used does not have an impact on the method. It has just an impact on the performance of SE on a computer. The result of a time step is the same whatever basis function is used. A version of SE3 using the polynomial basis defined in Section "Functional Representations, Amplitudes, and Basis Functions" has been tried in Steppeler et al. (2019).

The test equation (2) requires the formation of the derivative of the dynamic field $h(x)$. We can achieve this in spectral space. For the derivative polynomial, $p_{x',i}(x') = a_{i,1} + 2a_{i,2}x' + 3a_{i,3}x'^2$, where x' may be a global coordinate or a local coordinate specific to the cell Ω_i, such as $x' = x - x_{i+\frac{1}{2}}$. This derivative is taken at the collocation points, and the grid point values of the derivative are $p_{x'}(X^c_{i,i'}), i \in 0, 1, 2, ..., i_e, i' \in 0, 1, 2, 3$. Sometimes it is considered convenient to use the notation introduced in Eq. (45) : $X^c_{i,1} = x_{i+\frac{1}{3}}$ (Fig. 4).

There is one difficulty: at the corner points, two values of the derivatives are given, as the corner point $X^c_{i,0}$ is the same as $X^c_{i-1,3}$. So we have two values for the time derivative of h, $h^+_{t,i}$ and $h^-_{t,i}$ at corner points. An algorithm must be created to produce a unique value, which is done for regular grid by forming an average of the two values. For an equally spaced cell structure, $dx_{\Omega_i} = \text{const}$, an equally weighted average is a good option: $h_t = \frac{1}{2}(h^+_t + h^-_t)$.

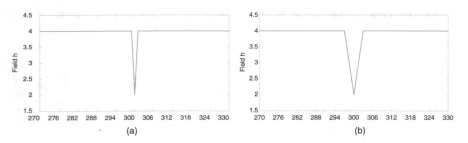

Fig. 4 Condensation after one time step for $h = 4.0$ in grid point space where the state before the condensation step is the constant h field. (**a**) Condensation in a standard way and (**b**) condensation by Eq. (119) (Reproduced from Steppeler et al. 2019)

The SE3 procedure is analog to the performance of classical global spectral method with exponential basis functions in grid point space and consists of the following steps:

- Divide the computational domain and obtain the collocation points $X^c_{i,i'}$ and the amplitudes of field h_i at the corresponding collocation points.
- Define the basis function for SE3 and represent the field h_i for all cells.
- Transform the amplitudes of field h_i at the grid point space into the spectral space.
- Obtain the corresponding amplitudes of $h_{xx,i+\frac{1}{2}}$ and $h_{xxx,i+\frac{1}{2}}$ at the spectral space and transform into grid point space and average at corner points.
- Compute h for the time $t + dt$ using RK4 or another time marching method.

This method is applied to the homogeneous advection equation (2) in Chapter "Simple Finite Difference Procedures" with $dx = 3, dt = 1.5$ and using RK4 to time marching on a grid x_i with $i_e = 200$ and periodic boundary conditions $h(i_e) = h(0)$. The initial values are given by $h_i = 4e^{\frac{(x_i - x_{50})^2}{4}}$. With these parameters, the field h moves by dx in three time steps. Due to the periodic boundary condition at $t = 400dt$, the initial value is reproduced by the analytic solution. Typical for high-order approximations is the appearance of positive and negative field values. The appearance of the negative values is sometimes called *"spectral ringing."* Figure 5 shows the result of advection with SE3 together with a simple parameterization of precipitation.

The L-Galerkin Scheme: o3o3

The L-Galerkin method o3o3 to be introduced in this section uses a third-order piecewise polynomial representation for the dynamic field $h(x)$ according to Eqs. (6), (8), (12), and (16). As explained in Section "Functional Representations,

Amplitudes, and Basis Functions" in Chapter "Simple Finite Difference Procedures", this means that the dynamic amplitudes in spectral space are $h_i, h_{xx,i+\frac{1}{2}}$ and $h_{xxx,i+\frac{1}{2}}(i = 1, 2, 3, ...)$. Basis functions belonging to these amplitudes are defined in Section "Functional Representations, Amplitudes, and Basis Functions" in Chapter "Simple Finite Difference Procedures". The fluxes $fl(x)$ have the same representation as h by amplitudes $fl_i, fl_{xx,i+\frac{1}{2}}$ and $fl_{xxx,i+\frac{1}{2}}(i = 1, 2, 3, ...)$.

The space of all functions defined by Eq. (9) is called $C3^0$. The superscript "0" indicates that we are dealing with continuous functions. If the superscript is "1," it means the functions are differentiable, while C^{-1} means the set of discontinuous functions, etc. The "3" in $C3^0$ indicates the degree of polynomials used is three. For example, SE2 is defined in $C2^0$. It is important to see what makes SE3 and o3o3 different. SE3 and o3o3 use the same function space $C3^0$ for discretization, and the basis functions used conventionally are different. The use of different basis functions does not make two methods different unless the amplitudes belonging to different basis functions are treated differently. For SE3, Legendre polynomials are traditionally used as basis functions, but there are other options. In Steppeler et al. (2019), SE3 was performed using the basis function representation Eqs. (6), (8), (12), and (16). Using such different basis functions for SE3 results in arithmetically equivalent formulations, and the interest in such differences is only the potentially different numerical efficiency of such different formulations.

A feature, such as the choice of grid points or basis functions of the spectral space, is called essential if the alternative would lead to a different result. If an alternative feature does lead to an arithmetically equivalent approximation, the difference is called superficial. For o3o3, the division of the dynamic fields $h(x)$ into a first-order part Eq. (9) and a high-order part Eq. (11) is essential for the numerical procedure. An example for an artificial feature of o3o3 is the choice of collocation points except for the corner points. Any choice of the collocation points for non-corner points will lead to the same scheme within round-off errors. On the other hand, the choice of collocation points for SEs in Section "Spectral Elements" in Chapter "Simple Finite Difference Procedures" is essential because any other choice than Gauss–Lobatto points can lead to unstable SE schemes.

The first-order part of o3o3 is determined by the corner point amplitudes, and these are both spectral and grid point amplitudes. For these points, to compute the time derivative, any high-order FD scheme is admitted. For the example presented here, we use the fourth-order formula Eq. (14) or (15) in Chapter "Simple Finite Difference Procedures". Note that Eq. (14) in Chapter "Simple Finite Difference Procedures" is fourth order for regular grids only, but Eq. (15) in Chapter "Simple Finite Difference Procedures" may be used for any irregular grid if the weights w in Eq. (15) in Chapter "Simple Finite Difference Procedures" are computed correctly. At the corner point, the FD treatment can be done without a consideration of conservation. The FD treatment of the high-order part, in particular the determination of the amplitudes $h_{xx,i+\frac{1}{2}}$ and $h_{xxx,i+\frac{1}{2}}$, is then done such to make the whole scheme conserving. So o3o3 represents a whole family of schemes. There are still different possibilities for the high-order part $h^{ho}(x)$, which are also essential for the version of o3o3 presented in the following. The example given in the following is just an example of possible schemes. Every FD scheme will induce a conserving scheme by using the correct definition of the time derivatives of $h_{xx,i+\frac{1}{2}}$ and $h_{xxx,i+\frac{1}{2}}$.

The FD scheme used at corner points can be done in spectral or grid point space. For the former, there must be a transformation to grid point space if the time integration is done in spectral space. In the example to be presented here, the field representation and time integration will be done in grid point space where we need to transform to spectral space in order to compute the time derivatives of $h_{xx,i+\frac{1}{2}}$ and $h_{xxx,i+\frac{1}{2}}$. Our test example will be homogeneous advection Eq. (1) in Chapter "Simple Finite Difference Procedures". Obviously, the transition from time translation (in this book by RK4) in grid point space to time translation in spectral space would be superficial.

The standard collocation grid is given in Eq. (2) for o3o3, which is the natural definition of regular collocation points within the cell interval. The global collocation grid may be irregular when the size $dx_{\Omega_i} = x_{i+1} - x_i$ of grid cells is irregular. Also irregular distributions of collocation points within a cell are possible. The choice of the collocation points X^c, being either regular or irregular with o3o3, is not essential. The result of a time step does not depend on the choice of points, as long as collocation points will not be used for differentiation. The time step will, however, depend on the differencing stencil grid points $x_{i,i'}^d, (i' = -2, -1, 0, 1, 2)$. We here assume $x_{i,i'}^d$ to be equally distributed on the two intervals neighboring the corner point x_i. If this is not identical to the choice of the collocation points, the stencil points must be computed using the spectral representation. Of course, other choices for the stencil points are possible. An alternative to the standard collocation points used here for differentiation and its effect on CFL will be explored in Section "CPU Time Used with a 3D Version of o3o3 Scheme" in Chapter "Numerical Tests". This will lead to alternative finite difference forms for the derivatives at corner points that certainly have an impact on the properties of the resulting schemes. Such alternative stencil points have not yet been widely explored. At corner points for o3o3, any third-order FD scheme can be used. At this point, no conservation property is observed. The conservation will come with the definition of time derivatives at edge points in Eqs. (49)–(53). The derivatives at corner points are obtained using the set $x_{i,i'}^d$. This is a set of grid points surrounding the target point x_i and will be collected for every x_i from more than one cell. In our example, we use the two neighboring cells of x_i. The set of points $x_{i,i'}^d$ used for differencing may be totally different from the collocation grid points. Different choices of the differencing grid $x_{i,i'}^d$ are essential, resulting into different versions of o3o3, not just different arithmetic formulations. The large number of different versions of o3o3 resulting from different choices of the differencing grid is widely unexplored.

Here we define the differencing grid to be part of the collocation grid:

$$\begin{cases} X^c_{i,-2} = x_{i-\frac{2}{3}}, \\ X^c_{i,-1} = x_{i-\frac{1}{3}}, \\ X^c_{i,0} = x_i, \\ X^c_{i,+1} = x_{i-\frac{1}{3}}, \\ X^c_{i,+2} = x_{i+\frac{2}{3}}. \end{cases} \tag{46}$$

Grid point amplitudes are those of the collocation grid. Once the differencing points $x^d_{i,i'}$ are defined and the flux amplitudes fl_i are formed for these points, the derivatives can be obtained by fourth-order FD scheme Eq. (15) in Chapter "Simple Finite Difference Procedures". Note that the $x^d_{i,i'}$ definition is just one example. Different choices for $x^d_{i,i'}$ and differentiations in spectral space will lead to an infinity of essentially different schemes that are unexplored. There is no reason to expect that the example chosen here is the best.

Here we describe the scheme with time-stepping in grid point space. We assume the initial values describing $h(x)$ for $t = 0$ as the values of h on the collocation grid: as in Eq. (46) the differencing points are chosen to be collocation points. This means that the amplitudes of h at the differencing points are known as well:

$$h_i = h(x_i), i = 0, \frac{1}{3}, \frac{2}{3}, 1, ... \tag{47}$$

Note that for $i = 0, 1, 2, 3, ...$, the grid point values h_i are also amplitudes in spectral space. To compute h in spectral space, the spectral amplitudes $h_{xx,i+\frac{1}{2}}$ and $h_{xxx,i+\frac{1}{2}}$ need to be computed. This is done using Eqs. (24) and (25).

The performance of a time step does not need the transformation into spectral space when differentiation at corner nodes is chosen. At corner nodes $x_i (i = 0, 1, 2, 3, ...)$, differentiation using the differencing points Eq. (46) is done by Eq. (15) in Chapter "Simple Finite Difference Procedures" if the cell structure is irregular. For regular grids, the special form Eq. (14) in Chapter "Simple Finite Difference Procedures" may be used. For grids involving a sudden refinement from dx_{Ω_i} to $\frac{1}{2}dx_{\Omega_i}$, the weights are given in Table 1 in Chapter "Simple Finite Difference Procedures". When observing this, o3o3 will be suitable for irregular grids (Kalnay et al. 1977). Note that by the outlined procedure, the o3o3 scheme has uniform approximation order 3 when the resolution jumps. In practical tests, often order 4 is encountered, as outlined by an example in Section "Test of the o3o3 Scheme on the Cubed Sphere Grid Using the Shallow Water Version of the HOMME Model" in Chapter "Numerical Tests".

Equation (15) in Chapter "Simple Finite Difference Procedures" gives the time derivatives $h_{t,i} (i = 0, 1, 2, ...)$ at corner points. We want to find a representation of the time derivative $h_t(x)$ analog to that of $h(x)$ given in Eqs. (9), (11), and (14):

$$h_t(x) = \sum_i h_{t,i} e_i^-(x) + h_{t,i} e_{i+1}^+(x) + h_{t,xx,i+\frac{1}{2}} b_{i+\frac{1}{2}}^2(x)$$

$$+ h_{t,xxx,i+\frac{1}{2}} b_{i+\frac{1}{2}}^3(x), \text{ for } i = 0, 1, 2, 3, ..., x \in \Omega_i. \tag{48}$$

After the time derivatives at corner points are now known, we must compute time derivatives $h_{t,xx,i+\frac{1}{2}}$ and $h_{t,xxx,i+\frac{1}{2}}$ in Eq. (48) of the spectral coefficients $h_{xx,i+\frac{1}{2}}$ and $h_{xxx,i+\frac{1}{2}}$.

We consider the mass balance in the cell interval Ω_i. Let $m_i(t)$ be the mass contained in Ω_i:

$$m_i(t) = \int_{\Omega_i} h(x', t) dx' = \int_{-dx'_{\Omega_i}}^{dx'_{\Omega_i}} h(x', t) dx', \tag{49}$$

where $dx'_{\Omega_i} = \frac{1}{2} dx_{\Omega_i} = \frac{1}{2}(x_{i+1} - x_i), x' = x - x_{i+\frac{1}{2}}, x \in \Omega_i$. Then, we assume $m_{t,i}$ to be its time derivative. Using Eq. (48), we obtain the temporal variation of the mass balance in Ω_i:

$$m_{t,i} = \frac{\partial}{\partial t} \int_{\Omega_i} h(x', t) dx' = \int_{\Omega_i} \frac{\partial h(x', t)}{\partial t} dx'$$

$$= h_{t,i} \alpha^- + h_{t,i+1} \alpha^+ + h_{t,xx,i+\frac{1}{2}} \alpha^{xx} + h_{t,xxx,i+\frac{1}{2}} \alpha^{xxx}, \tag{50}$$

with $\alpha^- = \int_{\Omega_i} e_i^-(x') dx', \alpha^+ = \int_{\Omega_i} e_{i+1}^+(x') dx', \alpha^{xx} = \int_{\Omega_i} b_{i+\frac{1}{2}}^2(x') dx', \alpha^{xxx} = \int_{\Omega_i} b_{i+\frac{1}{2}}^3(x') dx'$. To the derivation of Eqs. (50), (6) and (12) have been used:

$$\begin{cases} \alpha^- = \int_{\Omega_i} e_i^-(x') dx' = dx'_{\Omega_i}, \\ \alpha^+ = \int_{\Omega_i} e_{i+1}^+(x') dx' = dx'_{\Omega_i}, \\ \alpha^{xx} = \int_{\Omega_i} b_{i+\frac{1}{2}}^2(x') dx' = -\frac{2}{3} dx'^3_{\Omega_i}, \\ \alpha^{xxx} = \int_{\Omega_i} b_{i+\frac{1}{2}}^3(x') dx' = 0. \end{cases} \tag{51}$$

Note that the mass of basis function $\int_{\Omega_i} b_{i+\frac{1}{2}}^3(x') dx'$ belonging to $h_{xxx,i+\frac{1}{2}}$ is 0 because $b^3(x')$ is anti-symmetric with respect to the mid-point $x_{i+\frac{1}{2}}$ in Ω_i.

Using Stokes Theorem, we can obtain another formula for $m_{t,i}$:

$$m_{t,i} = -u_0(h_{i+1} - h_i). \tag{52}$$

Based on Eqs. (50)–(52), we get an equation where all quantities except $h_{t,xx,i+\frac{1}{2}}$ are known. When solving for $h_{xx,t,i+\frac{1}{2}}$, we obtain

$$
\begin{aligned}
h_{t,xx,i+\frac{1}{2}} &= \frac{1}{\alpha^{xx}}[-u_0(h_{i+1}-h_i)-(h_{t,i}\alpha^-+h_{t,i+1}\alpha^+)] \\
&= \frac{2}{3dx_{\Omega_i}'^3}[u_0(h_{i+1}-h_i)+(h_{t,i}+h_{t,i+1})dx_{\Omega_i}'].
\end{aligned}
\tag{53}
$$

Equation (48) can be used to predict h in spectral space, where $h_{t,xxx,i+\frac{1}{2}}$ is still missing so far. Equation (48) can be transformed into a grid point value at $x_{i+\frac{1}{2}}$, which is plausible, but it must be observed that $x_{i+\frac{1}{2}}$ is not one of the collocation points for o3o3. Using $h_{t,i+\frac{1}{2}} = h_{t,i}e_i^+(x_{i+\frac{1}{2}}) + h_{t,i+1}e_i^-(x_{i+\frac{1}{2}}) + h_{t,xx,i+\frac{1}{2}}b^2(x_{i+\frac{1}{2}}) = \frac{1}{2}(h_{t,i}+h_{t,i+1})+h_{t,xx,i+\frac{1}{2}}b^2(x_{i+\frac{1}{2}})$ and $b^2(x_{i+\frac{1}{2}}) = -\frac{1}{2}dx_{\Omega_i}'^2$, we obtain

$$
\begin{aligned}
h_{t,i+\frac{1}{2}} &= h_{t,i}e_i^+(x_{i+\frac{1}{2}}) + h_{t,i+1}e_i^-(x_{i+\frac{1}{2}}) + h_{t,xx,i+\frac{1}{2}}b^2(x_{i+\frac{1}{2}}) \\
&= \frac{1}{2}(h_{t,i}+h_{t,i+1}) + \frac{2}{3dx_{\Omega_i}'^3}[u_0(h_{i+1}-h_i)+(h_{t,i}+h_{t,i+1})dx_{\Omega_i}']b^2(x_{i+\frac{1}{2}}) \\
&= \frac{1}{2}(h_{t,i}+h_{t,i+1}) - \frac{1}{3dx_{\Omega_i}'}[u_0(h_{i+1}-h_i)+(h_{t,i}+h_{t,i+1})dx_{\Omega_i}'] \\
&= \frac{1}{6}(h_{t,i}+h_{t,i+1}) - \frac{u_0}{3dx_{\Omega_i}'}(h_{i+1}-h_i).
\end{aligned}
\tag{54}
$$

The temporal integration will be based on Eq. (54).

When the domain is divided into a regular grid that means $dx_{\Omega_i} = dx = \text{const}$, for $h_{t,xxx,i+\frac{1}{2}}$ we define

$$
h_{t,xxx,i+\frac{1}{2}} = \frac{1}{2dx}(h_{t,xx,i+\frac{3}{2}} - h_{t,xx,i-\frac{1}{2}}).
\tag{55}
$$

The transformation of time derivatives from spectral space Eqs. (54)–(55) to grid point space is done analog to Eq. (15):

$$
\begin{cases}
h_{t,i+\frac{1}{3}} = h_{t,i}e_i(x_{i+\frac{1}{3}}) + h_{t,i+1}e_{i+1}(x_{i+\frac{1}{3}}) + h_{t,xx,i+\frac{1}{2}}b_{i+\frac{1}{2}}^2(x_{i+\frac{1}{3}}) + h_{t,xxx,i+\frac{1}{2}}b_{i+\frac{1}{2}}^3(x_{i+\frac{1}{3}}), \\
h_{t,i+\frac{2}{3}} = h_{t,i}e_i(x_{i+\frac{2}{3}}) + h_{t,i+1}e_{i+1}(x_{i+\frac{2}{3}}) + h_{t,xx,i+\frac{1}{2}}b_{i+\frac{1}{2}}^2(x_{i+\frac{2}{3}}) + h_{t,xxx,i+\frac{1}{2}}b_{i+\frac{1}{2}}^3(x_{i+\frac{2}{3}}).
\end{cases}
\tag{56}
$$

Note that while o3o3 can be used for irregular grids, Eq. (55) is for regular grid only.

For the corner points, no transformation from spectral to grid point space is necessary, as corner point amplitudes at x_i are also spectral amplitudes meaning that the spectral to grid point transformation is the identity transformation at these points. Using Eqs. (48), (52), (55), (56), the temporal integration is performed in the following way:

- Step 1: Assume the grid point values of $h_i (i = 0, \frac{1}{3}, \frac{2}{3}, 1, \frac{4}{3}, ...)$ for the initialization.
- Step 2: Compute the time derivatives $h_{t,i} (i = 0, 1, 2, 3, ...)$ at corner points according to Eq. (48).
- Step 3: Compute the time derivatives of the spectral amplitude $h_{t,xx,i+\frac{1}{2}}$ according to Eqs. (52) and (53).
- Step 4: Compute the time derivatives of the spectral amplitude $h_{t,xxx,i+\frac{1}{2}}$ according to Eq. (55).
- Step 5: Compute the time derivatives at collocation points $x_{i+\frac{1}{2}}, x_{i+\frac{2}{3}}$ by transforming to grid point space according to Eq. (56).
- Step 6: Compute h in grid point space for the new time level $t + dt$ with RK4 method.

Note that all steps except for the fourth one are valid for irregular grids. The fourth step, meaning the computation of $h_{t,xxx,i+\frac{1}{2}}$, is valid for a regular grid only. This is done for the simplicity of the representation. Using methods described in Chapter "Simple Finite Difference Procedures", the fourth step can be generalized to be valid for irregular grids.

Some numerical experiments were done in the same way as for the homogeneous classical FD schemes. The results are presented in Figs. 3, 4, and 5 in Chapter "Simple Finite Difference Procedures". Figure 3 in Chapter "Simple Finite Difference Procedures" shows advection experiments, and the quality of simulation is similar as for the classical fourth-order FD scheme. However, o3o3 is conserving as is SE3, which has the same order of approximation as o3o3. In two dimensions, another difference between SE3 and o3o3 is that o3o3 similar as classic Galerkin allows for a sparse dynamic amplitude grid, where SE3 used the full grid. L-Galerkin schemes, such as SE3 or o3o3, combine high order of approximation with conservation (see Section "Some Further Properties of Finite Difference Schemes" in Chapter "Introduction").

The order of approximation of the space derivative by o3o3 can be analyzed as done in Fig. 2 in Chapter "Simple Finite Difference Procedures" for the classical fourth-order FD differencing. We leave this to the reader. It turns out that o3o3 shows super-convergence to fourth order for both regular and irregular grids. By construction, o3o3 should be third-order converging. This is the same for the spectral elements SE3. In both cases, super-convergence brings the approximation from third to fourth order.

It was already indicated that the computation of time derivatives at the corner points can be done in many ways, and Eq. (48) is just one special example. By the spectral representation (see Eqs. (10) and (14) for the example of o3o3), we know the field for every point in the vicinity of the target point x_i. So the collocation points can be chosen in different ways to produce different results. There are more options for the computation of derivatives at corner points. We mention a few of such options, not going deeply into the description of their properties.

The derivative can be done in spectral space. From Eq. (16), we obtain two one-sided derivatives h_i^+ and h_i^- at corner points, which are not equal, as with o3o3 the test functions are assumed to be continuous but not differentiable. h_i^+ is derived in Ω_i:

$$
\begin{aligned}
h_{x,i}^+ &= h_i e_{x,i}^-(-dx'_{\Omega_i}) + h_{i+1} e_{x,i}^+(-dx'_{\Omega_i}) \\
&\quad + h_{xx,i+\frac{1}{2}} b_{x,i+\frac{1}{2}}^2(-dx'_{\Omega_i}) + h_{xxx,i+\frac{1}{2}} b_{x,i+\frac{1}{2}}^3(-dx'_{\Omega_i}), \\
&= \frac{1}{2dx'_{\Omega_i}}(h_{i+1} - h_i) - h_{xx,i+\frac{1}{2}} dx'_{\Omega_i} + \frac{1}{3} h_{xxx,i+\frac{1}{2}} dx'^2_{\Omega_i},
\end{aligned}
\tag{57}
$$

while h_i^- is derived in Ω_{i-1}:

$$
\begin{aligned}
h_{x,i}^- &= h_{i-1} e_{x,i-1}^-(+dx'_{\Omega_{i-1}}) + h_i e_{x,i-1}^+(+dx'_{\Omega_{i-1}}) \\
&\quad + h_{xx,i-\frac{1}{2}} b_{x,i-\frac{1}{2}}^2(+dx'_{\Omega_{i-1}}) + h_{xxx,i-\frac{1}{2}} b_{x,i-\frac{1}{2}}^3(+dx'_{\Omega_{i-1}}), \\
&= \frac{1}{2dx'_{\Omega_{i-1}}}(h_i - h_{i-1}) + h_{xx,i-\frac{1}{2}} dx'_{\Omega_{i-1}} + \frac{1}{3} h_{xxx,i-\frac{1}{2}} dx'^2_{\Omega_{i-1}}.
\end{aligned}
\tag{58}
$$

For regular grid, we can define the spectral derivative at corner points:

$$
h_{x,i}^{sp} = \frac{1}{2}(h_{x,i}^+ + h_{x,i}^-).
\tag{59}
$$

This version is called o3o3 with spectral differencing. Advection results for this version are included in Figs. 3 and 4 in Chapter "Simple Finite Difference Procedures".

Apart from different versions concerning the corner point derivatives, the computation of the high-order part (amplitudes $h_{t,xx,i+\frac{1}{2}}$ and $h_{t,xxx,i+\frac{1}{2}}$) can be done in alternative ways. As an alternative to Eq. (55), we can use the following definition for $h_{t,xxx,i+\frac{1}{2}}$:

$$
h_{t,xxx,i+\frac{1}{2}} = \frac{h_{xx,i+\frac{3}{2}} - 2h_{xx,i+\frac{1}{2}} + h_{xx,i-\frac{1}{2}}}{4dx'^2_{\Omega_i}}.
\tag{60}
$$

There are also alternatives to the spectral difference at corner points. We can compute a fourth derivative of $h_t(x_{i+\frac{1}{2}})$, based on the fact that $h_{t,xx,i+\frac{1}{2}}$ is already known:

$$h_{t,xxxx,i+\frac{1}{2}} = \frac{h_{t,xx,i+\frac{3}{2}} - 2h_{t,xx,i+\frac{1}{2}} + h_{t,xx,i-\frac{1}{2}}}{4dx_{\Omega_i}'^2}, \tag{61}$$

and $h_{t,xxxx,i+\frac{1}{2}}$ can be used as amplitude for basis function:

$$b^4(x') = \frac{x'^4 - dx_{\Omega_i}'^4}{24}. \tag{62}$$

The term $h_{xxxx,i+\frac{1}{2}} b^4(x_{i+\frac{1}{2}})$ can then be used as a fourth-order correction to the one-sided derivatives Eq. (54). Equation (61) thus defines an alternative to the spectral differencing at corner points.

One feature of spectral differencing at corner points is the possibility to compute the time development just in spectral space. If RK4 is applied to the spectral coefficients, there is no need to transform to grid point space. When all amplitudes stay in spectral space during time-stepping, this may enhance the efficiency of computer implementations of this method. This feature even transfers to realistic models with physics when the physical processes are performed at corner points only (see Section "A Numerical Test for Irregular Resolution" in Chapter "Simple Finite Difference Procedures").

The L-Galerkin Scheme: o2o3

There exist few reports about local basis function systems using function representations being more regular than C^0, the space of continuous test functions. Steppeler (1988) used the classical Galerkin method for vertical discretization in a 3D Galerkin model, using tension splines as basis functions. Tension splines result into ten side diagonals in the mass matrix equation, which made the execution of this scheme rather expensive. L-Galerkin schemes use computationally less expensive ways to introduce differentiable test functions.

We want to solve the dynamic equation (2) in Chapter "Simple Finite Difference Procedures" by methods using differentiable function space representations. The dynamic field $h(x)$ is to be approximated in C^0 function spaces. We call C^0 the set of functions being continuous at corner points. C^1 are the functions being one time differentiable, etc. o2o3 is an example where the flux $fl(x)$ is approximated by differentiable functions, while $h(x)$ is approximated by a function being just continuous.

The function spaces introduced in Section "Functional Representations, Amplitudes, and Basis Functions" by piecewise polynomials are differentiable in the interior of the intervals. So we must only be concerned with the regularity at corner points. o2o3 is explained for the test example of homogeneous advection Eq. (2) in Chapter "Simple Finite Difference Procedures".

For $h(x)$ in Eq. (2) in Chapter "Simple Finite Difference Procedures", we assume a second-order representation. This is the basis function representation Eqs. (13)–(14) with the definition that $h_{xxx,i+\frac{1}{2}} = 0$ in Eq. (13). The collocation grid belonging to this representation of $h(x)$ is

$$x_i \left(i = 0, \frac{1}{2}, 1, \frac{3}{2}, 2, \frac{5}{2}, ... \right),$$
(63)

and the corresponding amplitudes are h_i. Note that the collocation grid within the cell Ω_i is assumed to be regular. In addition to the corner points x_i and x_{i+1} ($i = 0, 1, 2, 3, ...$), there is just one interior collocation point $x_{i+\frac{1}{2}}$. The regularity of the collocation grid within a cell means that $x_{i+\frac{1}{2}} = \frac{1}{2}(x_i + x_{i+1})$. As opposed to this situation, the cell structure is allowed to be irregular. This means that x_i ($i = 0, 1, 2, 3, ...$) can be chosen in an arbitrary way, and $dx_{\Omega_i} = x_{i+1} - x_i$ may be arbitrarily irregular.

Defining the flux $fl(x)$ in Eq. (2) in Chapter "Simple Finite Difference Procedures" as $fl(x) = -u_0 h(x)$, the dynamic equation (2) in Chapter "Simple Finite Difference Procedures" becomes

$$h_t(x) = fl_x(x).$$
(64)

Both h and fl are represented by Eqs. (10), (11) and (14). For $h(x)$, a second-order field representation means $h_{xxx,i+\frac{1}{2}} = 0$. For $fl(x)$, we search a third-order representation ($fl_{xxx,i+\frac{1}{2}} \neq 0$), such that $fl(x)$ is differentiable at corner points. Note that at the collocation points x_i for h: , $fl_i = fl(x_i)(i = 0, \frac{1}{2}, 1, \frac{3}{2}, ...)$ is defined as

$$fl_i = -u_0 h_i.$$
(65)

Third-order piecewise polynomial functions have one-sided derivatives at corner points x_i ($i = 0, 1, 2, 3, ...$) that are right- and left-sided derivatives. We must define $fl_{xx,i+\frac{1}{2}}$ and $fl_{xxx,i+\frac{1}{2}}$ ($i = 0, 1, 2, 3, ...$) in such a way that right- and left-sided derivatives are equal. This definition will be given below. It should be noted that $fl_{xx,i+\frac{1}{2}}$ will not be determined by the collocation point at $x_{i+\frac{1}{2}}$. This means that we will not define $fl_{xx,i+\frac{1}{2}}$, even though this would be an intuitive definition. We define the derivative of fl at corner points that has to be equal to the right- and left-sided derivatives of fl. The derivative $fl_{x,i}$ of fl at corner point is defined using Eq. (15) in Chapter "Simple Finite Difference Procedures" using the collocation point values $fl_i (i = 0, \frac{1}{2}, 1, \frac{3}{2}, ...)$:

$$fl_{x,i} = w^i_{-1} fl_{i-1} + w^i_{-\frac{1}{2}} fl_{i-\frac{1}{2}} + w^i_0 fl_i + w^i_{\frac{1}{2}} fl_{i+\frac{1}{2}} + w^i_1 fl_{i+1}$$

$$= \sum_{j=-2}^{2} w^i_{\frac{j}{2}} fl_{i-\frac{j}{2}}, \text{ for } i = 0, 1, 2, 3, ... \tag{66}$$

It was shown in Section "The Runge–Kutta and Other Time Discretization Schemes" in Chapter "Introduction" that Eq. (66) gives the derivative of fl in fourth-order accuracy. For the definition of the weights $w^i_{i'} (i' \in -1, -\frac{1}{2}, 0, \frac{1}{2}, 1)$, see Chapter Simple Finite Difference Procedures.

From Eqs. (10), (11), and (14), we see that for the definition of $fl(x)$ in the third-order representation, the corner amplitudes $fl_i (i = 0, 1, 2, 3, ...)$ and second and third derivatives at the inner collocation point $x_{i+\frac{1}{2}}$ must be defined. For the corner point derivatives, we use Eq. (66). To define second and third derivatives, we define the high-order part $fl^{ho+}_{x,i}$ and $fl^{ho-}_{x,i}$ of derivatives at corner points. The high-order part of the derivative is discontinuous at point x_i and must be given as left- and right-hand derivatives. For the cell $\Omega_i = (x_i, x_{i+1})$, we get

$$\begin{cases} fl^{ho+}_{x,i} = fl_{x,i} - h_i e^-_{x,i}(-dx'_{\Omega_i}) - h_{i+1} e^+_{x,i}(-dx'_{\Omega_i}) = fl_{x,i} - \dfrac{fl_{i+1} - fl_i}{2dx'_{\Omega_i}}, \\ fl^{ho-}_{x,i+1} = fl_{x,i+1} - h_i e^-_{x,i}(dx'_{\Omega_i}) - h_{i+1} e^+_{x,i}(dx'_{\Omega_i}) = fl_{x,i+1} - \dfrac{fl_{i+1} - fl_i}{2dx'_{\Omega_i}}, \end{cases} \tag{67}$$

in which $fl_{x,i}$ is defined in Eq. (66). Using the high-order basis representation Eq. (14), we obtain the following equations for $fl_{xx,i+\frac{1}{2}}$ and $fl_{xxx,i+\frac{1}{2}}$:

$$\begin{cases} fl^{ho+}_{x,i} = fl_{xx,i+\frac{1}{2}} b^2_{x,i+\frac{1}{2}}(-dx'_{\Omega_i}) + fl_{xxx,i+\frac{1}{2}} b^3_{x,i+\frac{1}{2}}(-dx'_{\Omega_i}), \\ fl^{ho-}_{x,i+1} = fl_{xx,i+\frac{1}{2}} b^2_{x,i+\frac{1}{2}}(dx'_{\Omega_i}) + fl_{xxx,i+\frac{1}{2}} b^3_{x,i+\frac{1}{2}}(dx'_{\Omega_i}), \end{cases} \tag{68}$$

which includes two linear equations for $fl_{xx,i+\frac{1}{2}}$ and $fl_{xxx,i+\frac{1}{2}}$. With Eq. (13), we obtain the higher derivatives of the flux at cell centers:

$$\begin{cases} fl_{xx,i+\frac{1}{2}} = -\dfrac{fl^{ho+}_{x,i} - fl^{ho-}_{x,i+1}}{2dx'_{\Omega_i}}, \\ fl_{xxx,i+\frac{1}{2}} = \dfrac{3}{2} \dfrac{fl^{ho+}_{x,i} + fl^{ho-}_{x,i+1}}{dx'^2_{\Omega_i}}. \end{cases} \tag{69}$$

The flux amplitudes in Eqs. (68) and (69) inserted into the equation of motion Eq. (64) give a form of the time derivative of h as a x-differentiated third-order piecewise polynomial spline Eqs. (10)–(14). The differentiation of a spline Eqs. (10)–(14) results into a (potentially discontinuous) spline of second order. As the spline, Eq. (67) is constructed to be differentiable; the result is a continuous second-order spline for the time derivative of h.

This spectral representation Eqs. (13) and (14) for $h_t(x)$ can be transformed into grid point space and the time step be done by RK4. This time-stepping is analog to what was done for o3o3, which is in detail described in Section "The L-Galerkin Scheme: o3o3" in grid point space. Alternatively, time-stepping can also be done in spectral space. Such change from doing the time step in grid point to spectral space is a superficial change. The result does not change, just the arithmetic form. If no physics parameterizations are called at corner points, the time translation can alternatively be done entirely in the spectral space.

Some results of o2o3 are included in Fig. 5 in Chapter "Simple Finite Difference Procedures". The convergence shown in Fig. 2 in Chapter "Simple Finite Difference Procedures" comes out in fourth order even though by construction only third order may be expected. This is the case for regular and irregular resolution. The occurrence of super-convergence to fourth order is common to the schemes o3o3, SE3, and o2o3. The von Neumann analysis of o2o3 (diagram not shown) shows that o2o3 has no large 0-space, which some versions of o3o3 have.

Splines of High Smoothness

The global spectral method (Durran 2010) uses the classic Galerkin method (see Section "The Classic Galerkin Procedure") with exponential basis functions or spherical harmonics on the sphere. According to the definition of regular function spaces, the global spectral method is classified as $C^\infty C^\infty$ as the discretization spaces for the fields and the fluxes use functions being differentiable arbitrarily often. The global spectral method uses the classic Galerkin method. The existing Galerkin methods with local basis functions use much less regular basis functions. Most applications use C^0 functional approximations for both fields and fluxes. Discontinuous basis functions using spaces C^{-1} are in use (Durran 2010), though not treated in this book. CG methods with continuous basis functions in C^0-space are very common. For vertical discretization in classic FE Galerkin method, tension spline basis functions (Steppeler 1988) have been used. This method is classified as o3o3$C^2 C^2$ (see Section "The L-Galerkin Scheme: o3o3"), as tension splines are two times differentiable. The o2o3 method treated in Section "The L-Galerkin Scheme: o2o3" is classified as o2o3$C^0 C^1$. The fluxes are approximated by differentiable splines. It is fair to say that methods used in atmospheric models to a large majority use approximations in C^0-spaces. L-Galerkin methods do not go beyond C^1 for fluxes, and for the approximation of the fields, C^1 regularity is the maximum used. So there is a large difference in regularity of basis functions between global spectral and L-Galerkin methods.

In order to facilitate the development of methods of higher regularity, a few properties of higher-order splines will be given. The presentation will be limited to spaces C^1 and C^2. Tension splines (Steppeler 1988) will not be described, as they are numerically rather expensive, and this book concentrates on methods with a potential for applications in realistic models. These methods are largely unexplored.

The linear part of the field representation $fl(x)$ is the same for splines of all orders and regularities. It is determined by the amplitudes at corner points as given by Eq. (9). The representation of the field $fl(x)$ is given in Eq. (16), and the high-order part $fl^{ho}(x)$, as given in Eq. (14), must be generalized. We assume basis function representations by splines up to order 6. Using the definitions of Section "Functional Representations, Amplitudes, and Basis Functions", we generalize Eq. (14) to

$$fl^{ho}(x') = \sum_0^{i_e} [fl_{xx,i+\frac{1}{2}} b^2_{i+\frac{1}{2}}(x') + fl_{xxx,i+\frac{1}{2}} b^3_{i+\frac{1}{2}}(x')$$
$$+ fl_{xxxx,i+\frac{1}{2}} b^4_{i+\frac{1}{2}}(x') + fl_{xxxxx,i+\frac{1}{2}} b^5_{i+\frac{1}{2}}(x')].$$

(70)

As in Eq. (14), we have introduced the local variable $x' = x - x_{i+\frac{1}{2}}$, $x_{i+\frac{1}{2}} = \frac{1}{2}(x_i + x_{i+1})$, $dx_{\Omega_i} = x_{i+1} - x_i$. Basis functions $b^2_{i+\frac{1}{2}}(x)$ and $b^3_{i+\frac{1}{2}}(x)$ are defined in Eq. (12). Their derivatives are given in Eq. (13) and Table 1. For the other two basis functions occurring in Eq. (70), we define

$$\begin{cases} b^4_{i+\frac{1}{2}}(x') = \dfrac{1}{4!}(x'^4 - dx'^4), \\ b^5_{i+\frac{1}{2}}(x') = \dfrac{1}{5!}x'b^4(x'), \end{cases}$$

(71)

where $dx'_{\Omega_i} = \frac{1}{2}dx_{\Omega_i}$. The basis functions have the property:

$$\begin{cases} b^4_{4x',i+\frac{1}{2}} = b^4_{x'x'x'x',i+\frac{1}{2}} = 1, \\ b^5_{5x',i+\frac{1}{2}} = b^5_{x'x'x'x'x',i+\frac{1}{2}} = 1. \end{cases}$$

(72)

Following the notation in Eqs. (72) and (70) can be rewritten as $fl^{ho}(x') = \sum_0^{i_e} \sum_{j=2}^5 fl_{jx,i+\frac{1}{2}} b^j_{i+\frac{1}{2}}(x')$. There are many ways to approximate a spline by a more regular one. Section "The L-Galerkin Scheme: o2o3" gives an example of approximation of a continuous spline by a C^1 spline and bases the numerical scheme o2o3 on it.

Using definitions Eq. (71), we define the functions $b'^4_{i+\frac{1}{2}}$ and $b'^5_{i+\frac{1}{2}}$, requiring in addition to Eq. (71) that the derivatives at corner points are 0:

$$\begin{cases} b'^4_{i+\frac{1}{2}}(x') = b^4_{i+\frac{1}{2}}(x') + \alpha b^2_{i+\frac{1}{2}}(x'), \\ b'^5_{i+\frac{1}{2}}(x') = b^5_{i+\frac{1}{2}}(x') + \beta b^3_{i+\frac{1}{2}}(x'), \end{cases} \tag{73}$$

with defining properties:

$$\begin{cases} b'^4_{x',i+\frac{1}{2}}(\pm dx'_{\Omega_i}) = 0, \\ b'^5_{x',i+\frac{1}{2}}(\pm dx'_{\Omega_i}) = 0. \end{cases} \tag{74}$$

From Eq. (74), we obtain

$$\begin{cases} \alpha = -\frac{1}{6} dx'^2_{\Omega_i}, \\ \beta = -\frac{1}{5} dx'^2_{\Omega_i}. \end{cases} \tag{75}$$

Equation (18) gives the oriented basis functions b_i^+ and b_i^- at corner point x_i. In an analog way, we define the second derivative of basis function $b_i'^+$ and $b_i'^-$ at corner points:

$$\begin{cases} b'^+_{xx,i}(x_i) = 1, \\ b'^+_{xx,i}(x_{i+1}) = 0, \\ b'^+_{x,i}(x_i) = b'^+_{x,i}(x_{i+1}) = 0. \end{cases} \tag{76}$$

Analog to Eq. (76), we define the basis function with index $-$:

$$\begin{cases} b'^-_{xx,i}(x_i) = 1, \\ b'^-_{xx,i}(x_{i-1}) = 0, \\ b'^-_{xx,i}(x_i) = b'^-_{xx,i}(x_{i-1}) = 0. \end{cases} \tag{77}$$

Note that the functions $b^2_{i+\frac{1}{2}}$ and $b^3_{i+\frac{1}{2}}$ defined in Section "Functional Representations, Amplitudes, and Basis Functions" have the second derivatives: $b^2_{xx,i+\frac{1}{2}}(x) = 1$ and $b^3_{i+\frac{1}{2}}(x) = x'$. Therefore, the b'^+ and b'^- can be created from $b^2_{i+\frac{1}{2}}(x)$ and $b^3_{i+\frac{1}{2}}(x)$ as third-degree polynomial splines. Basis functions Eqs. (71)–(77) can be used to approximate a continuous piecewise polynomial function of order 5 by a function with a continuous fourth-order derivative. Using Eqs. (73)–(75), we are able to generalize Eq. (11) and Eq. (14) to piecewise fifth-order polynomials. As

with the third-order case, the functions are continuous, and we can request the field
to be also differentiable. For the fifth-order case using Eqs. (73)–(75), we now can
ask in addition for the continuity of the second derivative. Note that a continuous
second-order derivative does not mean that this function is two times differentiable
in C^2-space. This is in complete analogy to the approximation of a continuous
function by a differentiable function, as was done for the performance of a time
step with the o2o3 method (see Section "The L-Galerkin Scheme: o2o3"). Note that
a piecewise polynomial function of the form Eqs. (11) and (14) is differentiable
if it is continuous and the derivative of order 1 is continuous. Also in analogy to
the o2o3 scheme, a scheme having continuous second derivatives but discontinuous
derivatives is possible. In this book, however, we demand that for a continuous
second derivative also the first derivative is continuous and therefore the second
derivative exists. We consider schemes that use differentiable field or flux functions
and do not go beyond this regularity. Two times differentiable polynomial spline
functions are possible by going to degree 5 for function representations.

We can then write Eq. (70) in an equivalent form using one-sided first- and
second-order derivatives at corner points:

$$h(x') = h^{lin}(x') + \sum_i h^+_{x,i} b^+_i(x') + h^-_{x,i} b^-_i(x') + h^+_{xx,i} b'^+_i(x') + h^-_{xx,i} b'^-_i(x'),$$

$$(78)$$

with the corner point amplitudes are $h_i, h^+_{x,i}, h^-_{x,i}, h^+_{xx,i}, h^-_{xx,i}$. Equation (78) is the
representation of a continuous fifth-order spline. The following requests will make
the field representation Eq. (78) C^2 regular.

$$\begin{cases} h^+_{x,i} = h^-_{x,i} = h_{x,i}, \\ h^+_{xx,i} = h^-_{xx,i} = h_{xx,i}. \end{cases} \quad (79)$$

Many numerical approximations including those used in this book are based on
Taylor expansions and require smooth fields in order to be accurate. Eq. (79)
reduces the number of dynamic amplitudes used to describe the field h from 4 to
2, and we may assume that we take away degrees of freedom where the numerical
treatment is not accurate. So we may assume that the accuracy of the representation
does not suffer by reducing the degrees of freedom. Such reduction of dynamic
amplitudes without that the accuracy suffers is called grid sparseness. This, as
described in Chapter "2D Basis Functions for Triangular and Rectangular Meshes",
is rather common in 2 and 3 dimensions. We have here with the one-sided derivative
representation an example of grid sparseness even in one space dimension.

Note that in this book we consider only CG or even higher regularity, such as
differentiable basis functions, represented as polynomial splines. So they can be
irregular just at the boundary of the neighboring cells, meaning at the corner points.
Higher regularity compared to CG has been very little investigated for local basis
functions. For the global spectral method, we have infinite regularity, meaning that

the field and flux functions are analytic and arbitrarily differentiable. So global spectral method is classified as $C^\infty C^\infty$, while the most popular Galerkin approach (Griffiths 1986) for Galerkin methods using local cells belongs to $C^0 C^0$. This section gave the foundation for more regularity with L-Galerkin methods going up to C^2. For polynomial splines, we only discuss the regularity at corner points, as the fields are polynomials and infinitely differentiable in the interior of cells. At corner points, the regularity is determined by the continuity of the fields and the continuity of their higher derivatives.

The basis functions associated with corner points allow polynomial degree up to 5. The use of corner based amplitudes $h_i, h^+ x, i, h^-_{x,i}$ is equivalent to the use of the function value h_i, the derivative values $h_{xx,i+\frac{1}{2}}, h_{xxx,i+\frac{1}{2}}, h_{xxxx,i+\frac{1}{2}}$, and $h_{xxxxx,i+\frac{1}{2}}$. If we admit the representation of discontinuous functions or those with discontinuous first or second derivatives, we may have two amplitudes at a corner point: $h_i^+, h_i^-, h^+_{x,i+\frac{1}{2}}, h^-_{x,i+\frac{1}{2}}, h^+_{xx,i+\frac{1}{2}}, h^-_{xx,i+\frac{1}{2}}$. Regular functions are obtained when the two amplitudes indicated by $+$ or $-$ are equal. Obviously, the options using discontinuities in fields are defined in C^{-1}, the space of discontinuous functions.

A Conserving Second-Order Scheme Using a Homogeneous FD Scheme

Conserving high-order schemes (e.g., third order) can be constructed by SE3, o3o3, and o2o3. This is probably an efficient solution, regarding numerical costs. However, many existing models are based on the neighbor point FD approximation Eq. (9) in Chapter "Simple Finite Difference Procedures". For regular resolution, this is called the centered FD approximation and also finite volume approximation. This FV approximation has the disadvantage that it is second order only for regular resolution and often drops to first-order approximation when the grid becomes irregular. In particular, many model codes are written for homogeneous FD schemes and may be difficult to rewrite for inhomogeneous FD schemes that are used by all SE and the high-order versions of FE, o3o3, and o2o3.

For a non-conserving FD scheme, Eq. (31) in Chapter "Simple Finite Difference Procedures" gives an alternative second-order scheme, generalizing the scheme to second order. This section describes a conserving homogeneous second-order FD scheme for both regular and irregular grids. Potentially L-Galerkin schemes o*nom* (for example, o2o3) are numerically much more efficient due to their use of sparse grids. However, conserving centered FD schemes are often used in models formulated for homogeneous FD schemes. Centered FD schemes in their simplest version are second order for regular grids only. As some of these models may not be suitable for inhomogeneous differencing, it may be worthwhile to define a generalization of second-order centered FD scheme for irregular grid using homogeneous differencing. The reader may be warned of the complexity of these schemes. For the same computational effort, it could be possible to run the o2o3 scheme including the intrinsic double resolution. Another aim to be followed by low-order homogeneous schemes is grid staggering meaning that the linear function systems for different fields, such as velocity or density, are not defined in the same grid.

Let us assume an irregular division for the domain $\Omega = \cup_i \Omega_i$, where the interval width is $dx_{\Omega_i} = x_{i+1} - x_i$ and $dx_{\Omega_i} \neq dx_{\Omega_{i-1}}$. The amplitudes of the density field h_i are given at the corner points x_i. The second derivative of the field h is computed as

$$
\begin{cases}
h_{x,i+\frac{1}{2}} = \dfrac{1}{dx_{\Omega_i}} (h_{i+1} - h_i), \\
h_{xx,i} = \dfrac{1}{\frac{1}{2}\left(dx_{\Omega_i} + dx_{\Omega_{i-1}}\right)} \left(h_{x,i+\frac{1}{2}} - h_{x,i-\frac{1}{2}}\right), \\
h_{xx,i+\frac{1}{2}} = \dfrac{1}{2} \left(h_{xx,i} + h_{xx,i+1}\right).
\end{cases}
\tag{80}
$$

Note that Eq. (80) is a second-order accurate representation of $h_{xx,i+\frac{1}{2}}$.

Now assume that the cell boundary grid $x'_{i+\frac{1}{2}}$ is given and x_i is generated from

$x'_{i+\frac{1}{2}}$: $x_i = \frac{1}{2}\left(x'_{i+\frac{1}{2}} + x'_{i-\frac{1}{2}}\right)$. With h_i and $h_{xx,i+\frac{1}{2}}$, we can use the second-order basis function representation to compute h at points $x'_{i+\frac{1}{2}}$:

$$
h\left(x'_{i+\frac{1}{2}}\right) = G^4(h_{i-1}, h_i, h_{i+1}),
\tag{81}
$$

where G^4 is a second-order interpolation operator. With these values for fluxes: $fl\left(x'_{i+\frac{1}{2}}\right) = -u_0 h\left(x'_{i+\frac{1}{2}}\right)$, we can obtain the centered FD formula:

$$
h_{x,i} = \frac{1}{x'_{i+\frac{1}{2}} - x'_{i-\frac{1}{2}}} \left[fl\left(x'_{i+\frac{1}{2}}\right) - fl\left(x'_{i-\frac{1}{2}}\right)\right],
\tag{82}
$$

which is a second-order version of centered FD for irregular grids. Note that for a regular grid $(x'_{i+\frac{1}{2}} = x_{i+\frac{1}{2}})$, the classical centered FD formula is not recovered. The linear averaging implicit in Arakawa schemes is replaced by a second-order term.

Equation (82) has the same form as Eq. (29) in Chapter "Simple Finite Difference Procedures" . However, the methods are different. Equation (82) uses a second-order interpolation at half-level points, which Eq. (29) in Chapter "Simple Finite Difference Procedures" does not use. There is an obvious alternative to Eq. (82). It is possible to use directly the functional form Eq. (29) in Chapter "Simple Finite Difference Procedures" and differentiate the flux. This will result into a discontinuous linear representation. It can be approximated by a regular function. These methods have not been tested. It is not known if the difference of the two versions is superficial or they produce different results.

Boundaries and Diffusion

This section bases diffusion on the polynomial spline representation. As this is based on cells within the computational domain, the resulting diffusion approximations take automatically account of boundaries, and it is possible to create diffusion schemes that are conserving and do not have mass going through the boundaries. Only the basic formula will be given. There is very little work available on this subject.

There is a huge number of possibilities to formulate fourth-order diffusion operators in polygonal spline space for o*nom* schemes. Here we give one example in 1D to show the principle. As the spectral space is the polygonal spline space equivalent to the collocation grid space, diffusions can also be formulated in grid point space, at least for regular or quasi-regular grids. Many existing models (see Table 1 in Chapter "Introduction") use more or less regular grids and the formulation of diffusion in the collocation grid as described in Section "Diffusion" in Chapter "Introduction" in Chapter "Simple Finite Difference Procedures" .

Potential advantages of a formulation in spectral space are mass conservation and the suitability for totally irregular cell structure. The example given here is a field h in 1D second-order spectral representation:

$$h(x) = \sum_{i=0}^{i_e} \left[h_i e_i(x) + h_{xx,i+\frac{1}{2}} b^2_{i+\frac{1}{2}}(x) \right].$$ (83)

We want to obtain discretization formula for diffusion within the function system of piecewise second derivatives. The first derivative can be directly obtained by differentiating Eq. (83). This will result into a discontinuous function, as in Eq. (83); both $e_i(x)$ and $b^2_{i+\frac{1}{2}}$ have discontinuous derivatives. At corner point x_i, the derivative of h is discontinuous. Let h_x^+ and h_x^- be the two one-sided derivatives at point x_i. We assume a regular resolution with constant cell interval. Then by directly differentiating the basis function representation, we have the discontinuous approximation $h_{xx}^d(x)$ Eq. (83) for h_{xx}:

$$h_{xx}^d(x) = \sum_i (h_{x,i}^+ - h_{x,i}^-)\delta(x - x_i) + h_{xx,i+\frac{1}{2}}\chi(x).$$ (84)

The upper index d means diffusion, and δ is the Dirac delta function. $\chi(x)$ is the characteristic function of the interval Ω_i:

$$\chi(x) = \begin{cases} 1, & \text{for } x \in \Omega_i, \\ 0, & \text{otherwise} . \end{cases}$$ (85)

Note that $h_{xx,i+\frac{1}{2}}$ can be interpreted as a point value of the second derivative at $x_{i+\frac{1}{2}}$. We do the second derivative by two times performing the first derivative. So the representation Eq. (84) is a representation as a distribution. Distributions are a generalization of the concept of functions. A simple introduction to distributions will be given in Section "Polygonal Spline Solutions Using Distributions and Discontinuities". The approximation of $h_{xx}(x)$ within the original basis function representation is obtained from Eq. (84) through approximating $\delta(x - x_i)$ by $\frac{1}{2}\frac{e_i(x)}{2dx}$ and approximating the characteristic function $\chi(x - x_{i+\frac{1}{2}})$ by $\alpha b^2(x)$:

$$h_{xx}^{da}(x) = \sum_i (h_i^+ + h_i^-)\frac{e_i(x)}{4dx} + h_{xx,i+\frac{1}{2}}\alpha b_{i+\frac{1}{2}}^2. \tag{86}$$

The upper index da in Eq. (86) means "approximation to the second derivative," where Eq. (84) is the second derivative obtained by direct differentiation. We obtain α from the condition:

$$\alpha \int \chi(x)dx = \alpha dx = \int b^2(x)dx, \tag{87}$$

which is $\alpha = -\frac{dx^2}{6}$.

Fourth-order diffusion, approximating the operator $\frac{\partial^4}{\partial x^4}$, can be obtained by iterating the formation of the second derivative. This means that the second derivative applied to Eq. (86) is applied in the same way using steps Eqs. (84)–(86). Similar as with the collocation form of diffusion Eq. (86), Euler forward time-stepping rather than RK4 should be used. This was done in a similar way in Section "Diffusion" in Chapter "Simple Finite Difference Procedures". The time step and coefficient of diffusion are limited by the request that the smallest wave becomes 0 after one diffusion step. For higher values, the diffusion step remains stable, but such values do not make sense.

Consider the special case that the corner point amplitudes and the second derivatives at edge points are constant: $h_i = c_1 = $ const, $h_{xx,i+\frac{1}{2}} = $ const. When choosing $c_1 = -2b_{i+\frac{1}{2}}^2(x_{i+\frac{1}{2}})$, the collocation grid points are $\pm b_{i+\frac{1}{2}}^2(x_{i+\frac{1}{2}})$, which is the $2dx$ wave. The approximation Eq. (86) applied to this again results into a $2dx$ wave, with a reduced amplitude. This means that applying two times the approximation for second derivative in spectral space will reduce the amplitude of the $2dx$ wave. This wave therefore is filtered by the application of the product of two second-order derivatives, which is the fourth-order diffusion operator. The application of the function space fourth-order diffusion Eq. (86) filters the $2dx$ wave. This means that the second-order diffusion operator smoothes quite well. The time step and the coefficient of numerical diffusion can be chosen such that the $2dx$ wave is removed in one time step. The concentration of diffusion on the small scales is investigated by spectral diagrams, which for the case of diffusion in collocation point space was done in Chapter "Simple Finite Difference Procedures".

For the diffusion in spectral space as described, qualitatively similar results can be expected. If we form the fourth-order operator by iteration, this will damp the $2dx$ wave.

When there is a boundary at one of the corner points, there is no regularization necessary at this corner point for the second-order diffusion. So the second derivative in Eq. (84) can be used as corner point value, which means that differences are taken as one-sided. Generally, basis function method allows to correctly implement boundary conditions. We did not point this out in detail, and this whole area would benefit from more research.

Transfer Function Analysis

So far we have considered transport by advection. For short wavelength, negative group velocities may appear near $2dx$, and this part of the spectrum is not useful for practical applications. For advection, no schemes without computational modes for wavelengths near $2dx$ are known. In addition to the advection waves, atmospheric models contain fast waves that are important for the geostrophic adjustment process. This is described in books about theoretical meteorology (Pedlosky 1987). Many models artificially change the speed of the fast waves in order to achieve numerical efficiency. Atmospheric models are known to give good predictions if the fast waves do not have the correct speeds. In fact, semi-implicit time integration methods slow down the simulations of the fast waves in order to obtain more efficient numerical schemes. However, for the fast waves as well as for advection, it is important that we do not create negative group velocities by the numeric procedure. For the fast waves, as opposed to advection, numerical schemes are known, which maintain positive group velocities down to the $2dx$ wave (Durran 2010). Such schemes with uniformly positive group velocities use staggered grids, which means grids where density and momentum collocation points are defined not at the same places within a cell (Durran 2010). These schemes are often in a low order. Many models in practical use staggered grids, as the useful resolution of fast waves is increased.

There will always be negative group velocities due to advection. These produce non-realistic irregular motions in addition to the meteorological signal, and if these non-realistic features accumulate, even instability may occur. To remove such features, diffusive filters are used, such as described in Section "Diffusion" in Chapter "Simple Finite Difference Procedures" and Section ". This is called numerical diffusion to distinguish it from the physical diffusion of second order, which represents molecular or turbulent friction as indicated in the Navier Stokes equation. As the computational modes are concentrated near the $2dx$ wave, for numerical diffusion we need schemes that filter only small waves. It is customary to use fourth or even higher diffusion for the numerical diffusion. This implies that this purely numerical effect does not change the physical diffusion for the larger scales. The analysis of fast waves and the effect on the physical solution are in analogy to the situation with telephone communication where it is necessary to filter the shortest wavelengths in order to achieve noise-free telephone connections (Honnert et al. 2020; Haltiner and Williams 1983; Schoenstadt 1979).

The von Neumann method as introduced in Section "The Von Neumann Method of Stability Analysis" in Chapter "Introduction" delivers wave velocities as a function of wavenumber k. It is considered of particular importance to consider fast waves of the meteorological system. Let us look at the fast waves of the shallow water equation, which we neglect the Coriolis and advection terms in one dimension:

$$\begin{cases} U_t = -H_x, \\ H_t = -(HU)_x. \end{cases} \tag{88}$$

Equation (88) is nonlinear, as the product of the fields $H(x, t)$ and $U(x, t)$ appears. We use the linearized version of Eq. (88) that is assuming the form $U(x, t) = U_0(x) + u(x, t)$, $H(x, t) = H_0(x) + h(x, t)$ for the fields. u and h are called the perturbation variables. It is customary to choose $U_0 = 0$, H_0 = constant, so we have $U = u$. When after inserting we neglect the product hu_x, we get the linearized form of Eq. (88):

$$\begin{cases} u_t = -h_x, \\ h_t = -H_0 u_x. \end{cases} \tag{89}$$

The solution space of the linear equation is easy to obtain and can be used for test purposes. For wave solutions, we suppose $u = A^u e^{-I\omega t} e^{Ikx}$ and $h = A^h e^{-I\omega t} e^{Ikx}$, where $I = \sqrt{-1}$, k is the wavenumber, and ω is the angular frequency. Inserting this wave solution into Eq. (89), we obtain a 2×2 linear equation for A^u and A^h. This equation has a solution under the condition that ω is a function of k. The function $\omega(k)$ is called the (analytic) dispersion relation of the system. For Eq. (89), the analytical dispersion relation is

$$\omega = \pm\sqrt{H^0}k. \tag{90}$$

Equation (90) means that there are two solutions, representing waves moving to the right and to the left. For Eq. (89), the range of k is to infinity. Arbitrarily small waves are possible in the undiscretized system Eq. (89).

For regular discretized systems with constant grid length dx, no wave corresponding to a wavelength smaller than $2dx$ can be used. So dispersion diagrams of approximated systems on a limited number of grid points cover a finite range of k. For the example of homogeneous advection Eq. (2), in Chapter "Simple Finite Difference Procedures" dispersion relations have been computed in Section "The Von Neumann Method of Stability Analysis" in Chapter "Simple Finite Difference Procedures". For the fast waves described by Eq. (88), dispersion relations have been obtained in (Steppeler 1989).

For the fast waves represented in Eq. (89), it would be desirable to deviate from the dispersion $\omega = \pm\sqrt{H^0}k$ from the analytic system as little as possible. For advection, one might disregard all waves where the deviation from the analytic solution is more than 10%. The discarded wavenumbers should be filtered using numerical diffusion (see Section "Diffusion" in Chapter "Simple Finite Difference Procedures"). The diffusion operator should be adjusted such that it affects mostly the discarded part of the spectrum. For the fast waves, a systematic deceleration of waves is not detrimental to the meteorological waves (Robert 1982).

Apart from the phase velocity $u_{ph} = \frac{\omega}{k}$ (Eq. (50)), the group velocity of a system $u_{gr} = \frac{\partial \omega}{\partial k}$ (Eq. (51) in Chapter "Simple Finite Difference Procedures") is important, especially for the speed of transporting information. So the group velocity of all waves is supposed to be smaller than the velocity of light, and negative group velocities are ruled out by principles of causality. As this book is not about philosophy, we do not go deeper into this. It is sufficient to say that the part of the spectrum showing negative group velocities needs to be filtered, as is standard in the information transporting industry.

The group velocity is an important measure for a numerical scheme. The group velocity is negative in a small range of k. In a dispersion diagram, negative group velocities are easy to spot as they occur when the dispersion diagram has a maximum. When there is a maximum, determined by $\frac{\partial \omega(k)}{\partial k} = 0$ for $k \neq k_0$, the largest possible k value, an area of negative group velocities is implied. Looking at the dispersion diagrams for advection obtained in Fig. 7, we can identify that the area of negative group velocities is even large, indicating some room for improvements. Negative group velocities H for fast waves are a reason for the use of grid staggering in models using second-order approximations (Steppeler 1990).

For simplicity, we present a simple example based on centered differences. Equation (89) is approximated by u_i and $h_i (i = 0, 1, 2, 3, ...)$. Both fields are represented in the same grid, which is called the A-grid representation. For the A-grid, the spatial discretization is

$$
\begin{cases}
u_{t,i} = -\dfrac{1}{dx_{\Omega_i} + dx_{\Omega_{i-1}}}(h_{i+1} - h_{i-1}), \\[2mm]
h_{t,i} = -\dfrac{H_0}{dx_{\Omega_i} + dx_{\Omega_{i-1}}}(u_{i+1} - u_{i-1}).
\end{cases}
\tag{91}
$$

As an alternative, we use the C-grid representation where u_i is defined at half levels: $u_{i+\frac{1}{2}} (i = 1, 2, 3, ...)$. For the C-grid, we obtain

$$
\begin{cases}
u_{t,i+\frac{1}{2}} = -\dfrac{1}{dx_{\Omega_i}}(h_{i+1} - h_i), \\[2mm]
h_{t,i} = -\dfrac{H_0}{dx_{\Omega_{i-\frac{1}{2}}}}(u_{i+\frac{1}{2}} - u_{i-\frac{1}{2}}).
\end{cases}
\tag{92}
$$

For the staggered scheme, the region of negative group velocities is absent. Negative group velocities are limited to a very narrow region near $\lambda = 0$. Note that for the advection process, no schemes with exclusively positive group velocities are known. Also the classic fourth-order FD scheme Eq. (14) in Chapter "Simple Finite Difference Procedures" has a small area of negative group velocity. This area of negative group velocities is smaller than that in the second-order FD case. So the conventional wisdom (Schoenstadt 1979) is that a high-order approximation eliminates the need for grid staggering.

It may be strange to provide a priority to the simulation of fast waves that have a small amplitude after an adjustment of the initial state. However, this is important as fast waves are essential for adjusting the atmosphere (e.g., induced by a mountain). In theoretical physics, negative group velocities are ruled out. However, in the world of numerics, they exist, and one could ask how they act on a system. In the system Eq. (88), it is possible to introduce a force by a mountain $m(x)$.

$$\begin{cases} U_t = -(H_x - m_x), \\ H_t = -(H - m)U_x. \end{cases} \tag{93}$$

In Eq. (93), $m = m(x)$ is the mountain. The stationary solution with this mountain is a flow with velocity U at the outflow, and $U(x, t)$ has a maximum over the mountain. The analytic solution is stationary. As we want to see the effect of negative group velocities, Eq. (93) is solved by the A-grid Eq. (91), and the mountain $m(x)$ is chosen to be different from 0 in one point only. The stationary solution is obtained by averaging the numerical solution over many time steps. Because the scale of the mountain is very small, the solution is not very accurate, while a nearly stationary solution is achieved. However, the stationary solution is overlayed by small scales, which appear to be very irregular and not stationary. Such small scales are the computational modes. As modelers try to use as little diffusion as possible, computational modes are rather common in models. They can be the result of programming errors or special meteorological events. Staggered grids, having no negative group velocities, do not have this particular error. With the unstaggered grid, non-stationary features in the form of numerical noise will appear. This undesirable effect is avoided if schemes with positive group velocities are used or the negative part of the spectrum is filtered.

A Numerical Test for Irregular Resolution

Simple tests, such as homogeneous advection or fast waves, are a good start before applying a scheme in realistic mode. We have seen in Section "A Conserving Second-Order Scheme Using a Homogeneous FD Scheme" in Chapter "Simple Finite Difference Procedures" that relevant information can be obtained using the dispersion diagram. However, the dispersion diagram and the von Neumann analysis are limited to a regular grid. As in this book we want to investigate irregular grids, we need simple tests for this. In particular, information on stability and accuracy is interesting and not obtainable for irregular grids by the von Neumann method.

Let an irregular grid x_i be given on which advection Eq. (1) in Chapter "Simple Finite Difference Procedures" is to be tested. We use the initial values for advection already used in Fig. 2 in Chapter "Simple Finite Difference Procedures" but use it for a range of scales.

$$h_i = h(x_i) = e^{-\frac{(x_i-x_0)^2}{l^2}}. \tag{94}$$

Periodic boundary conditions $h_{i+i_e} = h_i$ are applied, with i_e being the total number of points. When the time for integration is large, the value of x_0 should have only a minor impact on the outcome of tests. Let dx be an average value of dx_i, $dx^{min} = min_i dx_i$. The range of l is from dx^{min} to l^{max}. l^{max} is to be chosen in such a way that the function on a periodic boundary does not overlap with itself. For the example presented here, we use 600 points, $i_e = 600$. If information on stability is required, the integration time must be such that many rotations are achieved. Otherwise one may miss an instability where the eigenvalue has an absolute value > 1 but rather near to 1.

We can for example use this method to compare the nearest neighbor approximation (centered differences for regular grid) with the generalization Eq. (44) in Chapter "Simple Finite Difference Procedures" for uniform second-order approximation and the 2 versions of the classic fourth-order method in comparison to the fourth-order uniform-order method. Unfortunately, very little of this kind of research has been done.

Internal Boundaries for Vertical Discretization

As opposed to the horizontal resolution in NWP models, the vertical discretization is always not uniform. For example, clouds should be represented by a few vertical layers with short dz. Near the Earth surface, the vertical resolution should be increased to resolve the planetary boundary layer well. However, in the stratosphere and the upper atmosphere, there are no frequent activities of the synoptic system. That means in this vertical area, the resolution can change smoothly. In Chapter "Simple Finite Difference Procedures", we demonstrated that for the advection process sudden changes of resolution can cause accuracy problems with systems showing a non-uniform order of approximation.

In the vertical, the fast waves are considered most problematical. Sudden changes of resolution are known to create noisy solutions. Here we use the linearized 1D shallow water equation 89). Shallow water modeling is not done in the vertical, but the character of the fast waves is the same. We use the irregular grid defined in Eq. (24) in Chapter "Simple Finite Difference Procedures" using $i_e = 600$ points.

Again, periodic boundary conditions are not applicable in the vertical. However, we want to investigate the internal boundary without being disturbed by boundary problems. Periodic boundary conditions $u(i_e) = u(0), h(i_e) = h(0)$ are considered to be un-problematical.

The solutions are obtained for classical o4 (see Section "The Interface to Physics in High-Order L-Galerkin Schemes"), the modified o4 with uniform order of approximation and o2o3. The initial values are as

$$\begin{cases} h_i(t) = h(x_i) = e^{-\frac{(x_i - x_0')^2}{8^2}}, \\ u_{in} = 0. \end{cases} \tag{95}$$

For advection, the behavior of such jumps in resolution was tested in Chapter "2D Basis Functions for Triangular and Rectangular Meshes". In particular in Fig. 5 in Chapter "Simple Finite Difference Procedures", the results are shown, which indicate that a uniform order of approximation prevents the artificial reflection and noise generation at such internal boundaries.

The solution for fast gravitational waves can be done in a similar way as for advection. For the fast wave, we have 2 waves moving in the opposite directions. We expect that o4, which has the order dropping to one at the point with change of resolution, shows in Fig. 5 in Chapter "Simple Finite Difference Procedures" a strong development of noise. The modified o4 and o2o3 have a uniform approximation order. So far the experiments indicate that the generation of noise and inaccuracies is prevented by using schemes with uniform approximation order. 2D examples of the solution of the shallow water equations on an irregular mesh on the sphere are presented in Section "Shallow Water Tests on the Sphere: Solid Body Rotation, Solid Body Flow, Advection, and Williamson Test No. 6" in Chapter "Numerical Tests". Section "Shallow Water Tests on the Sphere: Solid Body Rotation, Solid Body Flow, Advection, and Williamson Test No. 6" in Chapter "Numerical Tests" shows for a 2D solution on the sphere that a model of uniform approximation order 3 is able to prevent artificial reflections and grid imprinting at points, where the resolution changes.

Open Boundary Condition

Open boundary conditions are those where a wave package leaves the model area without being reflected. Durran (2010) gave a thorough discussion of this problem. A remarkable result of open boundary problem is that there is no analytic theory that after discretization makes the boundary open. A heuristic argument in this respect is that the boundary is just one point, and discretizations need several points using a smooth solution to give good results. Open boundaries occur with the advection problem and with fast waves. Fast waves also have the reflecting boundary, and both can be implemented without problem. For advection, only the open boundary exists, and any attempt to introduce another boundary with advection will lead to problems. For advection, the open boundary is connected to the cut-cell boundaries, and an example is described in Sections "A Simple Cut-Cell System Based on the Staggered Low-Order Basis Functions and A Conserving Version of the Cut-Cell Scheme" in Chapter "Finite Difference Schemes on Sparse and Full Grids".

In Section "Internal Boundaries for Vertical Discretization", we used the linearized shallow water equation as a test example. We refer to Section "A Numerical Test for Irregular Resolution" for the discussion of the applicability of shallow water to this problem that concerns mainly the vertical. Initial values are the same as in Section "A Numerical Test for Irregular Resolution". We use o3o3 discretization for both fields h and u for a test. o3o3 allows for an irregular resolution. We use the regular resolution dx for all grid intervals, except for the last. The boundary element $dx^b = x_{i_e} - x_{i_e-1}$, where the superscript b represents boundary. For the boundary point x_{i_e}, we pose the boundary condition $h_{i_e} = h_{i_e-1}$ and $u_{i_e} = u_{i_e-1}$. For the boundary interval, we pose the condition $h_{xx,i_e-\frac{1}{2}} = u_{xx,i+\frac{1}{2}} = h_{xxx,i_e-\frac{1}{2}} = u_{xxx,i+\frac{1}{2}} = 0$.

A numerical example for an open upper boundary for fast waves is given in Section "A Numerical Example of Open Boundary Condition for a Fast Wave" in Chapter "Numerical Tests".

The L-Galerkin scheme: o3o5C^1C^2

Many methods are possible to define highly regular schemes. We try to increase the regularity by defining the schemes o4o5C^1C^2 and o3o5C^1C^1 in analogy to the o2o3C^0C^1 and o3o3C^0C^0 schemes presented in Sections "The L-Galerkin Scheme: o3o3–The L-Galerkin Scheme: o2o3". Both schemes use C^1 functions for the fields, and o4o5C^1C^2 uses a fourth-order component to define the high-order part. The flux is then defined to be two times differentiable, and its differentiation directly gives the form assumed for the field, meaning that after appropriate formulation of the flux, no further approximation to the time derivative of the field is necessary.

For o3o5C^1C^1, the fields and fluxes are differentiable C^1 functions. The representation of h can be done in standard form Eqs. (9) and (11), where h is determined by $h_i, h_{xx,i+\frac{1}{2}}$ and $h_{xxx,i+\frac{1}{2}}$. However, these amplitudes cannot be chosen independently, as the continuous requirement of the derivative must be satisfied. An equivalent polynomial spline representation of $h(x)$ by $h_{xx,i}^+$ and $h_{xx,i}^-$ ($i = 0, 1, 2, 3, ...$) is

$$\begin{cases} h_{xx,i+\frac{1}{2}} = \frac{1}{2}(h_{xx,i}^+ + h_{xx,i+1}^-), \\ h_{xxx,i+\frac{1}{2}} = \frac{1}{dx_{\Omega_i}}(h_{xx,i}^+ - h_{xx,i+1}^-), \end{cases} \tag{96}$$

which can be solved for $h_{xx,i}^+$ and $h_{xx,i+1}^-$ to obtain the transformation from the amplitudes defined at $x_{i+\frac{1}{2}}$ to those defined at x_i. The analog can be done for $h_{x,i}^+$ and $h_{x,i}^-$.

While the set of amplitudes h_i^+, h_i^-, $h_{xx,i}^+$ and $h_{xx,i+1}^-$ defines the general continuous polynomial spline, one of these amplitudes, for example $h_{xx,i}^+$, is sufficient to define the general differentiable polynomial spline. When $h_{xx,i}^+$ is given, the condition of second differentiability implies the amplitude $h_{xx,i}^-$, and the value of the first derivative of h at x_i is subject to the condition that the one-sided derivatives are equal. In order to obtain the derivative at corner points, use the transformations and the general spline representation Eqs. (9) and (11) to obtain the right-sided derivative $h_{x,i}^+$ and use the field representation:

$$\begin{cases} h(x) = h^{lin}(x) + h^{ho}(x), \\ h^{ho}(x) = \sum_i h_{x,i}^+ b_i^+(x) + h_{x,i}^- b_i^-(x). \end{cases} \qquad (97)$$

The condition of differentiability is

$$h_{x,i}^+ = h_{x,i}^-. \qquad (98)$$

In case that Eq. (98) is satisfied, define

$$h_{x,i} = h_{x,i}^+ = h_{x,i}^-. \qquad (99)$$

Equation (97) under condition Eq. (99) gives a representation of $h(x)$ as a one time differentiable function. We can in particular take the first derivative of Eq. (97) and obtain $h_{x,i}$ and $h_{x,i+\frac{1}{2}}$. Because of Eq. (99), $h_{x,i}$ is uniquely determined, as the two one-sided derivatives are equal. Using approximation Eq. (66) applied to grid points $h_{x,i+i'}$, $i' = 0, \frac{1}{2}, 1, \frac{3}{2}, ...$, we obtain an approximation for $h_{x,i}$. We apply this to the flux $fl(x) = -u_0 h(x)$ and now have the grid point values fl_i, $fl_{x,i}$ and $fl_{xx,i}$, $i = 1, 2, 3, ...$ So we can have a representation of $fl(x)$ in the space of piecewise fifth-degree polynomials with continuous second and first derivatives using Eq. (70). This means that the flux divergence is a piecewise fourth-degree polynomial with continuous derivatives. Then $h_t^a = \frac{\partial fl}{\partial x}$ is obtained by a direct differentiation that is continuous C^0. As a differentiation of an o3 field h_t^a comes out as an o2 spline, which has the basis function representation:

$$h_t(x) = h_t^{lin}(x) + \sum_i h_{t,x,i} \left[b_i^+(x) + b_i^-(x) \right] + h_{t,xxxx,i+\frac{1}{2}} b_{i+\frac{1}{2}}^4(x). \qquad (100)$$

For h_t using Eq. (100), we can obtain the corner point representation. To achieve this, note that according to Eq. (18) we have

$$b_{i+\frac{1}{2}}^2(x) = [b_{i+1}^-(x) - b_i^+(x)] dx_{\Omega_i}. \qquad (101)$$

We get b^3 as a linear combination of b^- and b^+. Insert Eq. (101) into Eq. (100) and obtain

$$h_t(x) = h_t^{lin}(x) + \sum_i h_{t,xx,i+\frac{1}{2}} b^2_{i+\frac{1}{2}}(x) + h_{t,xxx,i+\frac{1}{2}} b^3_{i+\frac{1}{2}}(x) + h_{t,xxxx,i+\frac{1}{2}} b^4_{i+\frac{1}{2}}(x).$$

$$(102)$$

Using Eq. (100) and piecewise fourth-degree polynomials as representation for h, we immediately have a C^1C^2 scheme. If an o3 representation is desired for h, we must approximate the o4 function by o3 functions in Eq. (100). Using Eq. (100), we can obtain the mass associated with point x_i. m_i^+ is the mass right of x_i, and m_i^- is the mass left of x_i. We define $m_i = m_i^0 + m_i^+ - m_i^-$, where m_i^0 is the mass associated with the linear part of the field at point x_i and m_i^+, m_i^- are the masses associated with the amplitudes for b_i^+ and b_i^-.

Now we consider the collocation grid on dx_{Ω_i}: $h_{t,i}, h_{t,i+\frac{1}{2}}, h_{t,i+1}, ..., (i = 0, 1, 2, 3, ...)$. $h_{t,i}$ is already known. The mid cell values are obtained from the polynomial spline representation for h_t:

$$h_{t,i+\frac{1}{2}} = \frac{1}{2}(h_{t,i} + h_{t,i+1}) - \frac{1}{2}h_{t,xx,i+\frac{1}{2}} dx_{i+\frac{1}{2}}'^2 + \frac{1}{24}h_{t,xxxx,i+\frac{1}{2}} dx_{i+\frac{1}{2}}'^4, \qquad (103)$$

with $dx_i' = \frac{1}{2}(x_{i+\frac{1}{2}} - x_i)$.

To perform a time step, we must approximate the continuous polynomial spline h_t^a by a differentiable spline h_t. Let us first obtain a fourth-order approximation for the derivative of h_t. This is done by the fourth-order FD formula for irregular grids Eq. (15) in Chapter "Simple Finite Difference Procedures" using the collocation points x_i and $x_{i+\frac{1}{2}}$:

$$h_{t,x,i} = -u_0(w_i^{-1} h_{t,i-1} + w_i^{-\frac{1}{2}} h_{t,i-\frac{1}{2}} + w_i^0 h_{t,i} + w_i^{\frac{1}{2}} h_{t,i+\frac{1}{2}} + w_i^1 h_{t,i+1}). \qquad (104)$$

Equation (104) is obtained from Eq. (15) in Chapter "Simple Finite Difference Procedures" for the case of the function h_t. Using the mass associated with the point x_i and the representation Eq. (103) and its analog for the field h_t, we obtain an equation for the corner point value and its associated high-order amplitudes $h_{t,i}^+$ and $h_{t,i}^-$. We create a set of 3 test functions associated with x_i and choose the corner point amplitude for h_t such that the local mass balance is fulfilled.

$$\begin{cases} h_{t,i}^+ = m_i^+ (h_{t,x,i} - \dfrac{1}{dx_{\Omega_i}}(h_{t,i+1} - h_{t,i})), \\[2mm] h_{t,i}^- = m_i^- (h_{t,x,i+1} - \dfrac{1}{dx_{\Omega_{i-1}}}(h_{t,i} - h_{t,i-1})), \end{cases} \qquad (105)$$

where m_i^+ and m_i^- are the masses contained in the basis functions $b_i^+(x)$ and $b_i^-(x)$.

The mass balance equation for the determination of $h_{t,x,i}$ is

$$fl_{i+1} - fl_i = \frac{1}{2}h_{t,i+\frac{1}{2}}dx_{\Omega_i} + h_{t,i}^+ m_i^+ + h_{t,i}^- m_i^-. \tag{106}$$

In Eq. (106), $h_{t,i}$ is the only unknown quantity and can be solved for Eq. (105), which gives the associated high-order amplitudes for corner point x_i. Eq. (106) is solved recursively. In the $h_{t,i}^a$ field, the value x_i is replaced by the $h_{t,i}$ computed according to Eq. (106). This is done for all even x_i and then repeated for the uneven x_i. The whole procedure is similar to the o2o3 scheme with added complications.

In order to generalize the o2o3 and o3o3 schemes to a higher order, we must assume a polynomial spline representation for h of fourth order. The flux will be prepared as a fifth-order polynomial spline that is 2 times differentiable.

The L-Galerkin Scheme: $o4o5C^1C^2$

The standard polynomial spline representation of h with test equation (2) in Chapter "Simple Finite Difference Procedures" is an extension of Eq. (14) to fourth or fifth order:

$$h(x) = h^{lin}(x) + \sum_i h_{xx,i+\frac{1}{2}}b_{i+\frac{1}{2}}^2(x) + h_{xxx,i+\frac{1}{2}}b_{i+\frac{1}{2}}^3(x)$$
$$+ h_{xxxx,i+\frac{1}{2}}b_{i+\frac{1}{2}}^4(x) + \varepsilon h_{xxxxx,i+\frac{1}{2}}b_{i+\frac{1}{2}}^5(x). \tag{107}$$

The first three terms of RHS of Eq. (107) define the third polynomial order representation according to Eq. (16) in Section "Functional Representations, Amplitudes, and Basis Functions" in Chapter "Simple Finite Difference Procedures". The term with $h_{xxxx,i+\frac{1}{2}}$ is a fourth-order correction. For the representation of h in Eq. (107), the fifth-order term is not used for $\varepsilon = 0$. $\varepsilon = 1$ is used when representing the flux $fl(x)$.

In Eq. (107), each of the high-order amplitudes $h_{xx,i+\frac{1}{2}}, h_{xxx,i+\frac{1}{2}}, h_{xxxx,i+\frac{1}{2}}$, $h_{xxxxx,i+\frac{1}{2}}$ contributes to the derivative of h at corner points x_i. We define an alternative representation to Eq. (107) such that the fourth- and fifth-order terms proportional to $h_{xxxx,i+\frac{1}{2}}, h_{xxxxx,i+\frac{1}{2}}$ do not contribute to the derivatives of h at corner points. We define alternative fourth- and fifth-order basis functions $b_{i+\frac{1}{2}}^{4a}(x)$ and $b_{i+\frac{1}{2}}^{5a}(x)$ as fourth- or fifth-degree polynomial spline basis functions characterized by the properties:

$$\begin{cases} b^{4a}_{xxxx,i+\frac{1}{2}}(x_{i+\frac{1}{2}}) = 1, \\ b^{5a}_{xxxxx,i+\frac{1}{2}}(x_{i+\frac{1}{2}}) = 1, \\ b^{4a}_{x,i+\frac{1}{2}}(x_i) = 0, \\ b^{5a}_{x,i+\frac{1}{2}}(x_i)) = 0, \\ b^{4a}_{x,i+\frac{1}{2}}(x_{i+1}) = 0, \\ b^{5a}_{x,i+\frac{1}{2}}(x_{i+1}) = 0. \end{cases} \tag{108}$$

Equation (108) is solved by the ansatz:

$$\begin{cases} b^{4a}_{i+\frac{1}{2}}(x) = b^4_{i+\frac{1}{2}}(x) + \alpha^4 b^2_{i+\frac{1}{2}}(x), \\ b^{5a}_{i+\frac{1}{2}}(x) = b^5_{i+\frac{1}{2}}(x) + \alpha^5 b^3_{i+\frac{1}{2}}(x). \end{cases} \tag{109}$$

b^4 and b^5 are defined in Eq. (71).

From Eqs. (108) and (109), we obtain

$$\begin{cases} \alpha^4 = -\dfrac{1}{6}dx'^2, \\ \alpha^5 = -\dfrac{1}{10}dx'^2. \end{cases} \tag{110}$$

Note that $b^{4a}_{xxxx,i+\frac{1}{2}}(x_{i+\frac{1}{2}}) = 1$ and $b^{5a}_{xxxxx,i+\frac{1}{2}}(x_{i+\frac{1}{2}}) = 1$; alternative to Eq. (107), we obtain the representation:

$$h(x) = h^{lin}(x) + \sum_i h_{xx,i+\frac{1}{2}} b^2_{i+\frac{1}{2}}(x) + h_{xxx,i+\frac{1}{2}} b^3_{i+\frac{1}{2}}(x)$$
$$+ h_{xxxx,i+\frac{1}{2}} b^{4a}_{i+\frac{1}{2}}(x) + \varepsilon h_{xxxxx,i+\frac{1}{2}} b^{5a}_{i+\frac{1}{2}}(x). \tag{111}$$

Equation (111) allows to separate the high-order part of h into two parts, being in third and fifth order, respectively:

$$\begin{cases} h(x) = h^{lin}(x) + h^{ho,2,3}(x) + h^{ho,4,5}(x), \\ h^{ho,2,3}(x) = \sum_i h_{xx,i+\frac{1}{2}} b^2_{i+\frac{1}{2}}(x) + h_{xxx,i+\frac{1}{2}} b^3_{i+\frac{1}{2}}(x), \\ h^{ho,4,5}(x) = \sum_i h_{xxxx,i+\frac{1}{2}} b^{4a}_{i+\frac{1}{2}}(x) + \varepsilon h_{xxxxx,i+\frac{1}{2}} b^{5a}_{i+\frac{1}{2}}(x). \end{cases} \tag{112}$$

For $h^{ho,2,3}(x)$, we have the property:

$$h^{ho2,3}_x(x_i) = h^{ho,2,3}_x(x_{i+1}) = 0. \tag{113}$$

In Section "Functional Representations, Amplitudes, and Basis Functions", it was shown that the one-sided derivatives at corner points x_i can be used as amplitudes for a third-order representation. Equation (113) means that the derivatives determine only h^{lin} and $h^{ho,2,3}$. Our goal is to define the flux as a polynomial spline such that it approximates $fl(x) = -u_0 h(x)$ and is two times differentiable. This means it is differentiable at corner points x_i. As we assume h to be differentiable and basis function b^{4a} does not contribute to derivatives at corner points, we have the trivial solution:

$$fl(x) = fl^{1,2,3}(x) + fl^{4,5}(x), \tag{114}$$

with $fl^{1,2,3}(x) = -u_0[h^{lin}(x) + h^{ho,2,3}(x)]$.

From Eqs. (111)–(113), it follows that $fl^{1,2,3}(x)$ is differentiable, and when introducing corner point based amplitudes fl_i^+, fl_i^-, these amplitudes determine $fl^{1,2,3}(x)$ such that the one-sided amplitudes $fl_x^{+,1,2,3}$ and $fl_x^{-,1,2,3}$ are equivalent to the first derivatives of the part $fl^{+,1,2,3}$. $fl^{4,5}$ has 0 derivatives at corner points and may be defined by second derivatives at corner points. Note that h is a continuous and differentiable polynomial spline leading to $fl_x^{+,1,2,3} = fl_x^{-,1,2,3}$. This means that one of the one-sided corner point amplitudes are convenient as amplitudes. They can be chosen independently. In contrast, the mid-point amplitudes $fl_{xx,i+\frac{1}{2}}^{1,2,3}$ and $fl_{xxx,i+\frac{1}{2}}^{1,2,3}$ must be chosen subject to the conditions that the derivatives at corner points exist.

While $fl^{1,2,3}(x)$ has continuous derivatives, the second derivatives are not continuous. The task is to define $fl^{4,5}(x)$ in such a way that $fl_{xx}(x)$ is continuous. So fl is two times differentiable, and the flux divergence being $h_t(x)$ is continuous with the discretization form Eq. (108) of $\varepsilon = 0$ and satisfying the differentiability at corner points. This means that after forming the flux derivative no further approximation is necessary.

As we have seen in Section "Splines of High Smoothness", the second derivatives of a function, as obtained from fourth- and fifth-order basis, can be obtained by corner amplitudes. This makes it easy to implement the condition that second derivatives are continuous. This is done by obtaining high-order approximations for second derivatives of the flux at corner points and using the obtained value for both one-sided second derivatives at this corner point.

The approximation of the second derivative at corner and center points is done in a similar way as with the o2o3 scheme it was done for first derivatives. To obtain the second derivatives, the values of h and fl at collocation points are formed. For collocation points, we choose: fl_i as given and

$$fl_{i+\frac{1}{2}} = \frac{1}{2}(fl_i + fl_{i+1}) - u_0 h_{xx,i+\frac{1}{2}} dx'^2_{i+\frac{1}{2}}. \tag{115}$$

Using these basis functions, we use the fourth-order second derivative formula with target point x_i, $i = 0, 1, 2, 3, ...$, being the corner points:

$$fl_{xx,i} = w^2_{i,-1}fl_{i-1} + w^2_{i,-\frac{1}{2}}fl_{i-\frac{1}{2}} + w^2_{i,0}fl_i + w^2_{i,+\frac{1}{2}}fl_{i+\frac{1}{2}} + w^2_{i,+1}fl_{i+1}. \quad (116)$$

Equation (116) is rather similar to the formula for the computation of the first derivatives Eq. (15) in Chapter "Simple Finite Difference Procedures". It differs by the weights only. The upper index 2 indicates that Eq. (116) is for the computation of the second derivatives. As we did for the weights $w_{i,i'}$ used in (**), the $w^2_{i,i'}$ are computed numerically by the Galerkin compiler MOW_GC. No regularity of the corner grid x_i, $i = 0, 1, 2, 3, ...$ is required.

We define $fl^{2,3+}_{xx,i}$ and $fl^{2,3,-}_{xx,i}$ to be the one-sided second derivatives of $fl^{2,3}$ to the right and the left of x_i. They can be computed using Eq. (114). Now define the second derivative deficiency fl^d_i, $i = 0, 1, 2, 3, ...$ as the value to be added to $fl^{2,3+}_i$ or $fl^{2,3-}_i$ to obtain the high-order approximated values Eq. (116):

$$\begin{cases} fl^{d+}_i = fl_{xx,i} - fl^{2,3+}_{xx,i}, \\ fl^{d-}_i = fl_{xx,i} - fl^{2,3-}_{xx,i}. \end{cases} \quad (117)$$

Now consider target point $x_{i+\frac{1}{2}}$ and use Eq. (112) to define $fl_{xxxx,\frac{1}{2}}$ and $fl_{xxxxx,i+\frac{1}{2}}$:

$$\begin{cases} fl_{xxxx,i+\frac{1}{2}} = \dfrac{1}{b^{4a}_{i+\frac{1}{2}}(x_i) + b^{4a}_{i+\frac{1}{2}}(x_{i+1})}(fl^{d+}_i + fl^-_i), \\ fl_{xxxxx,i+\frac{1}{2}} = \dfrac{1}{b^{4a}_{i+\frac{1}{2}}(x_i) - b^{4a}_{i+\frac{1}{2}}(x_{i+1})}(fl^{d+}_i - fl^-_i). \end{cases} \quad (118)$$

Note that the basis function $b^{4a}_{i+\frac{1}{2}}(x)$ is symmetric with respect to the cell center $x_{i+\frac{1}{2}}$. Therefore, the two values of b in the bracket in Eq. (118) are identical. Similarly, the function $b^{4a}_{i+\frac{1}{2}}(x)$ is anti-symmetric. With Eq. (118), the flux $fl(x)$ is constructed as being continuous and two times differentiable. This means that the differentiation will provide a finite difference scheme that is directly suitable for time discretization, for example, by RK4.

The reader will notice that many of the discretization choices are arbitrary. No claim is made that the choices we made are optimal, and the reader is encouraged to explore alternative choices.

The intention of Sections "The L-Galerkin scheme: o3o5C^1C^2–The L-Galerkin Scheme: o4o5C^1C^2" is to give a first idea of the construction of L-Galerkin schemes with field representations of higher regularity than just being continuous. Few investigations of such methods have been done. So the reader wanting to explore these methods must be aware that he/she enters uncharted terrain.

The Interface to Physics in High-Order L-Galerkin Schemes

For all L-Galerkin schemes o*nom*, the field representation for the field $h(x)$ can be divided into a linear part $h^{lin}(x)$ and the high-order part $h^{ho}(x)$ Eq. (112). The high-order part of a field is essential for accurate transport. In particular, all methods envisaged need a minimum degree of smoothness in order to achieve a good advection. Many physics routines, however, do not deliver smooth fields at all. It is typical that physics routines are independent for different grid points. For example, the albedo in radiation transfer schemes may be randomly different between grid points. This means that an accurate transport of the heat produced by radiation is not possible. For the general philosophy of physical parameterization, we refer to the introduction where it was pointed out that physical processes do not belong to the field of numerics and therefore are not the subject of this book. However, the interface between the physical parameterizations and the numerical routines in a model belongs to the subject of numerics and should be mentioned here.

It seems reasonable to call the physics routines at the corner points where they determine the linear part h^{lin} of a field. The amplitudes produced by physics are interpolated linearly to the interior points. This results in more than one grid point to describe the amplitude of a physics routine. In other words, the corner point physics interface produces smoother fields than the condensation at each corner and edge point. This means that for corner point physics the advection of the amplitude created by physics is more accurate. For the example of condensation, the difference of corner point physics and physics called at every grid point is shown in Fig. 5a, b. As this means that physics routines are called on a subset of grid points, there is also a significant saving in computer time associated. However, here we are mainly concerned with the potential increase of forecast accuracy (Fig. 6).

The concept of calling parameterizations at corner points is explored for an example of moisture advection with condensation by SE3. The test will be performed on a grid with 600 points (200 intervals) with $dx = 1$ and periodic boundary conditions for Eq. (2) in Chapter "Simple Finite Difference Procedures". The initial moisture field is the bell-shaped curve $h(x) = 4e^{\frac{9}{4}(x_i - x_{150})}$, and the advection part is already described in Section "Spectral Elements". The result is shown in Fig. 5. As this is a rather small-scale structure, there is a small deformation of the solution by the advection process seen in Fig. 5. Due to this numerical error of advection, the maximum of the moisture field drops by about 10% during the transport of 600 dx.

We use this solution to test two interfaces to the physics routines. We choose a simple condensation routine. Assume that due to condensation at point x_{350} the moisture condenses, and for this grid point only, we change the value of h_{350}:

$$h_{350} = min(2, h_{350}), \tag{119}$$

and the moisture loss may appear as precipitation. Equation (119) is a moisture condensation parameterization that could be found in realistic models some time

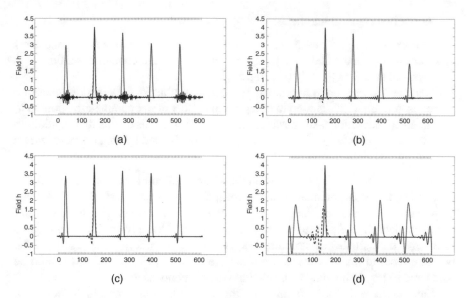

Fig. 5 The condensation of moisture as computed by (**a**) interface 1, (**b**) interface 2, (**c**) advection as in (**a**) and (**b**) with no condensation, (**d**) same interface as used in (**b**) for a second-order accurate advection scheme (Reproduced from Steppeler et al. 2019)

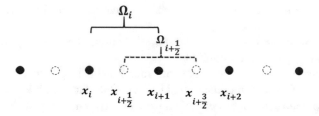

Fig. 6 The definition of Ω_i and $\Omega_{i+\frac{1}{2}}$

ago. Recently, moisture condensation in realistic models is parameterized in a much more refined way using cloud modeling techniques. For a test of interfaces to physics, the simple routine can still be useful.

Note that the SE3 method used for prediction in Fig. 5 is done in a form where a linear part of h is determined by corner points and the deviation is the high-order part $h^{ho}(x)$ of $h(x)$:

$$h^{ho}(x) = h(x) - \sum_{i=0,3,6,\dots} h_i e_i(x). \tag{120}$$

The linear hat function $e_i(x)$ is defined in Section "Functional Representations, Amplitudes, and Basis Functions".

We test two physical interfaces:

- Interface 1 (full grid interface): Eq. (119) is applied in collocation space. This means that from the 600 points in the model domain the condensation Eq. (119) is applied at point x_{350} only (see Fig. 5a).
- Interface 2 (corner point interface): The condensation Eq. (119) is done at corner points, and Eq. (120) is used to compute the effect of the parameterization on h, and no condensation takes place on the part $h^{ho}(x)$. This means that in addition to the corner point x_{350} 4 neighboring points are affected by the parameterization (see Fig. 5b).

As the condensation is performed on the 200 corner point only and not on all 600 grid points, there is a potential of saving computer time with Interface 2. The results of the 1D test also indicate that Interface 2 leads to better results. The analytic result of the condensation Eq. (119) is that the amplitude of the advected moisture drops to 2, after the bell-shaped structure has passed the point x_{350}. The results of the two interfaces to the condensation routine are shown in Fig. 5, and this should be compared to Fig. 5 showing the advection without condensation.

With Interface 1, the maximum of the moisture field drops to 3, not to 4. After in this case just the advection errors reduce the amplitude by 10%, the condensation drops it by another 10%, where the analytic solution requests a decrease of amplitude to 50 %. Also, the condensation with Interface 1 causes a computational mode in a similar way, as we have seen in Fig. 5 with low-order schemes and irregular resolution.

With Interface 2, the computational model is absent, and the amplitude drops to 2, as requested by the analytic solution.

Our simple example suggests that Interface 2 leads to parameterizations being less noisy and more accurate than obtained with Interface 1. Note that in realistic models the potential savings of computer time due to computing physics on a subset of points can be very substantial. In realistic models, such as those in Table 1 in chapter "Introduction", almost universally Interface 1 is used.

Polygonal Spline Solutions Using Distributions and Discontinuities

We consider here discontinuities in the definition of fluxes. It remains true that we deal with CG only. Our treatment is quite different from DG as reported by Durran (2010). Considered as distributions, the fluxes are differentiable at points of discontinuity. The concept of distributions was introduced for the computation of diffusion of C^1 functions in Section "Boundaries and Diffusion", and here we apply this formalism for the computation of divergence for the case that the functional representation of fields is low order. The treatment is analog to that given in Section "Boundaries and Diffusion". In fact, the derivatives of discontinuous polygonal splines exist as distributions (Schwarz 1950). Here we do not use the strict mathematical definition of the term "distribution" and rather take a heuristic approach. Let a function $h(x)$ have a jump at point x_i, meaning that we have different right and left limit values h_i^+ and h_i^- at these points. Now we approximate $h(x)$ by a sequence of functions $h_k(x)$ differing from h only for an x-interval $|x - x_i| < \epsilon_k$, where $\epsilon_k \to 0$ for $k \to \infty$. The functions $h_k(x)$ do not converge in the classical sense. However, the integrals $\int_{-\epsilon_k}^{\epsilon_k} h_k(x)dx$ converge. While the limit of $h_k(x)$ in the classical sense does not exist, the limit of the integrals exists. The limit is the distribution called Dirac delta function $\delta(x)$, $h_x(x) = (h_i^+ - h_i^-)\delta$. The distribution δ to be defined below is not a function, and taking the functional value of $\delta(x - x_i)$ at $x = x_i$ does not make sense. Integrals of distributions are defined. Sometimes we see the definition that $\delta(x)$ is 0 everywhere except at x_i, where it is ∞. However, this definition is not correct, as ∞ is not a real number. When treating discontinuities with distribution derivatives, intrinsic conservation laws are present.

Discontinuous basis function representation can be useful when treating low-order and staggered systems. So this section corresponds to Section "A Conserving Second-Order Scheme Using a Homogeneous FD Scheme" in Chapter "Simple Finite Difference Procedures" as such low-order representation exists in many realistic models. Section "A Conserving Second-Order Scheme Using a Homogeneous FD Scheme" in Chapter "Simple Finite Difference Procedures" and distributions can be used to understand low-order staggered systems as distributions. The polygonal spline versions of such low-order systems may be useful when going to irregular grids and maintaining second-order approximation conservation in such systems (Schwarz 1950). The full mathematical theory of distributions is complicated, and in the following, we limit ourselves to the simple application of the δ function. In this book, only very elementary features are described, which already lead to interesting applications.

The Dirac delta δ function occurs by differentiation of a function with a jump. The integral over $\delta(x)$ is 1. The reader should not be confused by the fact that δ is normally written as a function $\delta(x)$, while it is defined only as the integral $\int g(x)\delta(x)dx = g(0)$, with $g(x)$ being an arbitrary sufficiently regular function. This integral is a mapping forming the space of functions $g(x)$ to real numbers. This mapping is the definition of δ. Higher-order differentiations are possible within distributions, but this book does not go beyond the δ function.

The exact mathematical theory of distributions is given in Schwarz (1950). However, for this book, it is sufficient to have a simple understanding of it. If a function $f(x)$ has a discontinuity at $x = x_1$, we can choose an interval $(x_1 - \varepsilon, x_1 + \varepsilon)$ to modify $f(x)$ in this interval such that it becomes differentiable. We are interested in properties of f, which in the limit $\varepsilon \to 0$ do not depend on the function chosen for interpolation in the small interval.

We define the function

$$\theta(x) = \begin{cases} 0, & \text{for } x < 0, \\ 1, & \text{for } x \geq 0, \end{cases} \tag{121}$$

and then

$$\frac{\partial \theta}{\partial x} = \delta(x), \tag{122}$$

with $\delta(x)$ being the Dirac delta function. Let now $x_1 < 0$ and $x_2 > 0$; then we have

$$\int_{x_1}^{x_2} \delta(x) = 1 = \theta(x_2) - \theta(x_1). \tag{123}$$

If now θ is a flux, then Eq. (123) means that the integral over the flux divergence is the difference of boundary values. For continuous and differentiable functions, the determination of integrals of flux divergences by boundary values is essential for establishing conservation laws. According to Eq. (123), this is true if fluxes have points where they are not differentiable and the derivatives are distributions.

The most obvious application is the advection of a field $h(x)$ being defined at half-level points with amplitudes $h_{i+\frac{1}{2}}$. Such fields occur in staggered grids where some fields of an equation system are defined at full-level points x_i and others at half-level points $x_{i+\frac{1}{2}}$. For a field defined at half-level points, we have the representation by piecewise constant basis functions $\chi_{i+\frac{1}{2}}$ that is given for regular grids: $dx_i = dx$.

$$\chi_{i+\frac{1}{2}}(x) = \begin{cases} 1, & \text{if } x \in \Omega_i, \\ 0, & \text{otherwise.} \end{cases} \tag{124}$$

$$\chi_i'(x) = \begin{cases} 1, & \text{if } x \in \Omega_{i+\frac{1}{2}}, \\ 0, & \text{otherwise.} \end{cases} \tag{125}$$

χ was used in the same way in Section "Boundaries and Diffusion" in Chapter "Simple Finite Difference Procedures", and χ' is the same definition for the interval $\Omega_{i+\frac{1}{2}}$, being the interval (x_i, x_{i+1}).

The following formula can be done for irregular grid, but for simplicity, here we stick to the regular case. As analog explanations were given in Section "Boundaries and Diffusion", they are not repeated here. For the field h, we assume the half-level representation:

$$h(x) = \sum_i h_{i+\frac{1}{2}} \chi_{i+\frac{1}{2}}(x). \tag{126}$$

Then we define the full-level value of h and its flux as

$$\begin{cases} h_i = \frac{1}{2}(h_{i+\frac{1}{2}} + h_{i-\frac{1}{2}}), \\ fl(x) = -u_0 \sum_i h_i \chi_i'(x). \end{cases} \tag{127}$$

We define

$$fl_i = -u_0 h_i.$$ (128)

From flux differentiation, we obtain

$$h_t(x) = \sum_i \delta(x - x_{i+\frac{1}{2}}) u_0 (h_{i+1} - h_i).$$ (129)

For the derivation of Eq. (129), see Section "Boundaries and Diffusion".

We replace δ by the χ recovery, which is the χ function having the same integral as δ:

$$\delta(x - x_{i+\frac{1}{2}}) = \frac{1}{dx_{\Omega_i}} \chi(x - x_{i+\frac{1}{2}}).$$ (130)

Equation (130) together with Eq. (129) and assuming a χ representation for $h_t(x)$ imply that

$$h_{t,i+\frac{1}{2}} = -\frac{u_0}{2dx_{\Omega_i}} (h_{i+\frac{3}{2}} - h_{i-\frac{1}{2}}),$$ (131)

which is a well-known FD formula for fields defined at half levels, coming out of classical finite difference considerations (see Chapter "Simple Finite Difference Procedures"). Equation (131) shows that using low-order basis function assumptions it is possible to obtain classical FD formulations. Coming out of the low-order Galerkin formulations, conservation laws are implied, and the scheme generalizes to irregular grids. Unfortunately, for a general grid, the approximation order is low, mostly one. In this way, distributions based on L-Galerkin methods allow to transfer schemes to irregular grids, which have the same scope as the FV method (Durran 2010). For regular grids, Eq. (131) achieves second order by super-convergence.

It is unknown for irregular grids whether the parameters defining a scheme can be adjusted leading to an approximation order higher than 1. Approximation order 1 is generally considered unsuitable for realistic modeling. If methods similar to Eq. (131) are applied, the grid is modified in such a way that the resolution is nearly regular. In 1D, the grid can be chosen to be regular. However, we will see in the following chapters that on the sphere a regular grid can be obtained in approximation only.

Von Neumann Analysis of Some o*nom* Schemes

In Section "The Von Neumann Method of Stability Analysis" in Chapter "Simple Finite Difference Procedures", the von Neumann analysis for the case of homogeneous FD schemes was described. In this section, the technical performance of the von Neumann method for the more general inhomogeneous schemes is described. We consider the case that there are sets of three different FD equations as is the case with the scheme o3o3 described above. The example presented here is the analysis for the scheme o3o3 defined in Section "The L-Galerkin Scheme: o3o3" and o2o3 defined in Section "The L-Galerkin Scheme: o2o3".

To derive the numerical dispersion relation for o3o3, we use spectral solutions following Ullrich et al. (2018) and Steppeler et al. (2019). The field h for the example of o3o3 is assumed to be

$$h(x_j) = h_0 e^{Ik(j \cdot dx - ct)} = h_0 e^{Ik(j \cdot dx)} e^{-I\omega t}, \ j = 0, 1, 2, ..., \tag{132}$$

where $x_j = j \cdot dx$, $dx =$ const, $I = \sqrt{-1}$ is the imaginary unit and c, k, and ω are the phase velocity, non-dimensional wavenumber, and the frequency of the wave, respectively. The exact solution of Eq. (2) in Chapter "Simple Finite Difference Procedures" is obtained by Eq. (132). When inserting Eq. (2) in Chapter "Simple Finite Difference Procedures" into Eq. (132), we obtain

$$\begin{cases} \dfrac{\partial h_j}{\partial x(x_j)} = Ik \cdot h_j, \\ \dfrac{\partial h}{\partial t} = -I\omega \cdot h_j, \end{cases} \tag{133}$$

where $\omega = -\frac{k}{c}$.

Then, we define the amplitudes $\mathbf{A}_j = (h_j, h_{xx,j}, h_{xxx,j})$ in the spectral space. For each k, the linear relation between $\mathbf{A}_k$ and $\mathbf{A}_{t,k}$ for advection equation (2) in Chapter "Simple Finite Difference Procedures" is given by

$$\mathbf{A}_{t,k} = \mathbf{M}^k \mathbf{A}_k, \tag{134}$$

where $\mathbf{A}_{t,k}$ is the temporal derivative of $\mathbf{A}_k$ and the 3×3 matrix $\mathbf{M}^k$ depends on the wavenumber k. The evolution matrix M^k is given by

$$\mathbf{M}^k = \sum_{m=-N}^{N+1} \mathbf{M}^m e^{mI\delta} (N = 2)$$
$$= \mathbf{M}^{-2} e^{-2I\delta} + \mathbf{M}^{-1} e^{-I\delta} + \mathbf{M}^0 + \mathbf{M}^1 e^{I\delta} + \mathbf{M}^2 e^{2I\delta} + \mathbf{M}^3 e^{3I\delta}, \tag{135}$$

where $\mathbf{M}^k$ is applied to the amplitudes $\mathbf{A} = (h_j, h_{xx,j}, h_{xxx,j})$ and $\delta = \frac{k}{1000} \cdot 2\pi$, $k = 0, 1, 2, ..., 1000$. The matrices $\mathbf{M}^1, \mathbf{M}^2, \mathbf{M}^3, \mathbf{M}^4, \mathbf{M}^5, \mathbf{M}^6$ are given by

$$\mathbf{M}^1 = \begin{pmatrix} 0 & 0 & 0 \\ 0 & 0 & 0 \\ -\frac{1}{6dx}\mathbf{M}_{2,1}^2 & -\frac{1}{6dx}\mathbf{M}_{2,2}^2 & -\frac{1}{6dx}\mathbf{M}_{2,3}^2 \end{pmatrix} \tag{136}$$

$$\mathbf{M}^2 = \begin{pmatrix} \mathbf{M}^2_{1,1} & \mathbf{M}^2_{1,2} & \mathbf{M}^2_{1,3} \\ \frac{2}{3dx^2}\mathbf{M}^2_{1,1} & \frac{2}{3dx^2}\mathbf{M}^2_{1,2} & \frac{2}{3dx^2}\mathbf{M}^2_{1,3} \\ -\frac{1}{6dx}\mathbf{M}^3_{2,1} & -\frac{1}{6dx}\mathbf{M}^3_{2,2} & -\frac{1}{6dx}\mathbf{M}^3_{2,3} \end{pmatrix} \tag{137}$$

$$\mathbf{M}^3 = \begin{pmatrix} \mathbf{M}^3_{1,1} & \mathbf{M}^3_{1,2} & \mathbf{M}^3_{1,3} \\ \frac{4}{9dx^3} + \frac{2}{3dx^2}\left(\mathbf{M}^3_{1,1} + \mathbf{M}^2_{1,1}\right) & \frac{2}{3dx^2}\left(\mathbf{M}^3_{1,2} + \mathbf{M}^2_{1,2}\right) & \frac{2}{3dx^2}\left(\mathbf{M}^3_{1,3} + \mathbf{M}^2_{1,3}\right) \\ \frac{1}{6dx}\left(\mathbf{M}^2_{2,1} - \mathbf{M}^4_{2,1}\right) & \frac{1}{6dx}\left(\mathbf{M}^2_{2,2} - \mathbf{M}^4_{2,2}\right) & \frac{1}{6dx}\left(\mathbf{M}^2_{2,3} - \mathbf{M}^4_{2,3}\right) \end{pmatrix} \tag{138}$$

$$\mathbf{M}^4 = \begin{pmatrix} \mathbf{M}^4_{1,1} & 0 & 0 \\ -\frac{4}{9dx^3} + \frac{2}{3dx^2}\left(\mathbf{M}^4_{1,1} + \mathbf{M}^3_{1,1}\right) & \frac{2}{3dx^2}\left(\mathbf{M}^4_{1,2} + \mathbf{M}^3_{1,2}\right) & \frac{2}{3dx^2}\left(\mathbf{M}^4_{1,3} + \mathbf{M}^3_{1,3}\right) \\ \frac{1}{6dx}\left(\mathbf{M}^3_{2,1} - \mathbf{M}^5_{2,1}\right) & \frac{1}{6dx}\left(\mathbf{M}^3_{2,2} - \mathbf{M}^5_{2,2}\right) & \frac{1}{6dx}\left(\mathbf{M}^3_{2,3} - \mathbf{M}^5_{2,3}\right) \end{pmatrix} \tag{139}$$

$$\mathbf{M}^5 = \begin{pmatrix} 0 & 0 & 0 \\ \frac{2}{3dx^2}\mathbf{M}^4_{1,1} & \frac{2}{3dx^2}\mathbf{M}^4_{1,2} & \frac{2}{3dx^2}\mathbf{M}^4_{1,3} \\ \frac{1}{6dx}\mathbf{M}^4_{2,1} & \frac{1}{6dx}\mathbf{M}^4_{2,2} & \frac{1}{6dx}\mathbf{M}^4_{2,3} \end{pmatrix} \tag{140}$$

$$\mathbf{M}^6 = \begin{pmatrix} 0 & 0 & 0 \\ 0 & 0 & 0 \\ \frac{1}{6dx}\mathbf{M}^5_{2,1} & \frac{1}{6dx}\mathbf{M}^5_{2,2} & \frac{1}{6dx}\mathbf{M}^5_{2,3} \end{pmatrix}, \tag{141}$$

where $\mathbf{M}^k_{j_1,j_2}$ is the element of Matrix $\mathbf{M}^k$ in Row j_1 and Column j_2 (we assume $dx = \frac{1}{3}$ and $u_0 = 1$ for simplification), $\mathbf{M}^2_{1,:} = [-\frac{1}{2}, \frac{7}{36}, \frac{1}{72}]$, $\mathbf{M}^3_{1,:} = [0, -\frac{7}{36}, \frac{1}{72}]$, and $\mathbf{M}^4_{1,1} = \frac{1}{2}$ for standard o3o3, while $\mathbf{M}^2_{1,:} = [-1, \frac{1}{2}, \frac{1}{12}]$, $\mathbf{M}^3_{1,:} = [0, -\frac{1}{2}, \frac{1}{12}]$, and $\mathbf{M}^4_{1,1} = 1$ for spectral o3o3.

We assume that e^k_1, e^k_2, and e^k_3 are defined as the eigenvalues of the matrix $\mathbf{M}^k$. Therefore, the imaginary components of e^k_1, e^k_2 and e^k_3 represent the frequency $\omega(k)$ of the wavenumber k, while the real components are the diffusivity. The phase velocity of the wavenumber k becomes $c(k) = \frac{\omega(k)}{k}$. A negative derivative to k of the phase velocity means negative group velocities, which is inherently unphysical and should be filtered. A negative real part is indicative of implicit diffusion.

The phase velocities in dependence of $k \cdot dx$ are shown in Fig. 7 for the two versions of o3o3 (standard and spectral o3o3). This plot is generated by appending the eigenvalues e^k_1, e^k_2, e^k_3 as a function of k in sequence (see Ullrich et al. 2018). For o3o3, a large part of the spectrum is stationary (phase velocity is 0). For these states, the corresponding physical states are not changing in time, and the physical waves are not captured (Steppeler et al. 2019). Such spaces are called a *0-space*. These stationary modes in 0-space arise when the discrete fourth derivative at corner points evaluates to zero. If the 0-space occurs for small wavenumbers, they can be filtered by diffusion (see Section "Diffusion" in Chapter "Simple Finite Difference

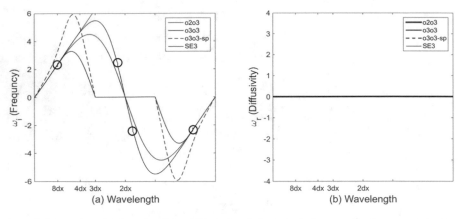

Fig. 7 Dispersion relations for the o2o3, SE3, o3o3, and o3o3 spectral schemes. (**a**) is the imaginary part and (**b**) is the deviation of the eigenvalues from one, which is a measure of the intrinsic diffusivity of the scheme. All four schemes are non-diffusive. The black circles mark the spectral gap for SE3, which produces negative group velocities in an area where the neighboring wavenumber values indicate a well-converged solution. The range of resolutions where the group velocity is positive is called the useful resolution

Procedures"). For another part of the spectrum, the phase velocity is negative. While it clearly indicates that there is room for improvements, the comparison with the two control runs o4 and SE3 shows that o3o3 is competitive in accuracy.

To derive the numerical dispersion relation for o2o3, we use the same theory and equations (132)–(135) where the amplitudes $\mathbf{A} = (h_j, h_{xx,j})$ in Eq. (134), $N = 1$ in Eq. (135) and $\mathbf{M}^k(k = -1, 0, 1, 2)$ are 2×2 matrices. The evolution matrix $\mathbf{M}^k$ is given by

$$\mathbf{M}^1 = \begin{pmatrix} -\frac{u_0}{12dx} + \frac{2u_0}{3dx}\left(\frac{1}{2}\right) & \frac{2u_0}{3dx}\left(-\frac{dx^2}{2}\right) \\ \frac{3}{2dx^2}\mathbf{M}^1_{1,1} & \frac{3}{2dx^2}\mathbf{M}^1_{1,2} \end{pmatrix} = \begin{pmatrix} \frac{1}{4} & -\frac{1}{3} \\ \frac{3}{8} & -\frac{1}{2} \end{pmatrix} \tag{142}$$

$$\mathbf{M}^2 = \begin{pmatrix} \frac{2u_0}{3dx}\left(\frac{1}{2}\right) - \frac{2u_0}{3dx}\left(\frac{1}{2}\right) & -\frac{2u_0}{3dx}\left(-\frac{dx^2}{2}\right) \\ -\frac{3u_0}{2dx^3} + \frac{3}{2dx^2}\left(\mathbf{M}^2_{1,1} + \mathbf{M}^1_{1,1}\right) & \frac{3}{2dx^2}\left(\mathbf{M}^2_{1,2} + \mathbf{M}^1_{1,2}\right) \end{pmatrix} = \begin{pmatrix} 0 & \frac{1}{3} \\ -\frac{9}{8} & 0 \end{pmatrix} \tag{143}$$

$$\mathbf{M}^3 = \begin{pmatrix} -\frac{1}{4} & 0 \\ \frac{3u_0}{2dx^3} + \frac{3}{2dx^2}\left(\mathbf{M}^3_{1,1} + \mathbf{M}^2_{1,1}\right) & \frac{3}{2dx^2}\mathbf{M}^2_{1,2} \end{pmatrix} = \begin{pmatrix} -\frac{1}{4} & 0 \\ \frac{9}{8} & \frac{1}{2} \end{pmatrix} \tag{144}$$

$$\mathbf{M}^4 = \begin{pmatrix} 0 & 0 \\ \frac{3}{2dx^2}\mathbf{M}^3_{1,1} & 0 \end{pmatrix} = \begin{pmatrix} 0 & 0 \\ -\frac{3}{8} & 0 \end{pmatrix}, \tag{145}$$

where $dx = 1$ and $u_0 = 1$ for simplification.

For o2o3, Fig. 7a, b shows the phase velocity and the deviation of the amplification factor from unity, respectively, depending on the non-dimensional wavenumber. For comparison, the corresponding results for SE3 and o3o3 are also given. In Fig. 7a, we focus on the maximum of the frequency curve because the corresponding wavelength is the resolution limit for each scheme. The phase velocity of the analytical solution is a straight line that coincides with the curve for the long wave part in Fig. 7. For the short wave part meaning high wavenumbers, there are large deviations. For o2o3 and SE3, the approximated phase velocities are accurate for wavelengths greater than $3dx$, with o3o3 performing somewhat worse. For smaller wavelengths, the derivative of the frequency curve becomes negative, which results in a negative group velocity. This means that for this range of wavenumbers, the solution is not useful in terms of the propagation of wave packets. Judging from this criterion, the useful wavelength range for o2o3 is larger than that for o3o3 by approximately dx. Finally, as shown in Fig. 7b, the amplification factor is one for all three schemes, and thus, the schemes are all non-diffusive. As FD and L-Galerkin schemes depend on power series developments, they need a certain smooth filter to the solution for high accuracy. It is of interest how fast accuracy is achieved with decreasing wavenumber and what part of the solutions with large wavenumbers, though less accurate, are still useful.

The part of the spectrum with negative group velocities should be filtered. Sometimes, a more elaborate definition of the essential resolution is used (Ullrich 2014). This is based on the realistically approximated part of the dispersion diagram, which is the frequency as a function of the wavenumber. Note that o2o3 does not have the large 0-space of o3o3. Thus, in comparison to o3o3, o2o3 is simpler to perform.

Although SE3 has a marginally larger useful wavelength range than o2o3, we note that the dispersion relation for SE3 shows a spectral gap, as indicated by black circles in Fig. 7a. A spectral gap is a small wiggle on the frequency curve leading to a small area of negative group velocities in an otherwise well-converged k-area. Due to this spectral gap of SE3, Steppeler et al. (2019) concluded that o3o3 has performance advantages over SE3. In contrast, o2o3 and o3o3 have no spectral gap. How to reduce the effect of the spectral gap for SE3 by applying hyper-diffusion has been discussed in the literature (Ullrich et al. 2018). In the absence of a spectral gap, an estimate of the usefully resolved wavenumber k is the range of k up to the maximum.

At first sight, it may appear that o3o3 is at a disadvantage compared to o2o3 because of the large 0-space. However, it is possible to consider the amplitude $h_{xxx,i+\frac{1}{2}}$ as a prognostic variable that is diagnosed from other prognostic variables, namely h_i and $h_{xx,i+\frac{1}{2}}$.

2D Basis Functions for Triangular and Rectangular Meshes

Abstract This chapter defines 2D field representations used for L-Galerkin methods, such as SE, FE, and o*nom*. For convenience and ease of understanding, some formulas are given for regular cell structures. A generalization to irregular cells is possible. More general cell structures, such as quadrilaterals or hexagons, can be derived from the triangular representation by treating some of the amplitudes as diagnostic. Even if applicable to irregular meshes, the meshes will be presented as structured. For research in new numerical methods, structured test models have the advantage of being easy to create. Unstructured programming is conveniently done by using a neighborhood management system, which is present in the MPAS, COSMO, and Fluidity models.

Keywords Basis functions · Sparse grids · Euclid's lemma · Triangular basis functions · Corner representation · Boundaries

Grids in 2D are constructed by using lines of grid points. The grids are generated by two vectors $\mathbf{n}^{x'}$ and $\mathbf{n}^{y'}$:

$$\begin{cases} \mathbf{n}^{x'} = (1, 0), \\ \mathbf{n}^{y'} = (cos\varphi, sin\varphi). \end{cases} \tag{1}$$

We allow for non-rectangular grids determined by the angle φ. The $'$ indicates that we are admitting vectors $\mathbf{n}^{x'}$, $\mathbf{n}^{y'}$ to be non-orthogonal. When $\mathbf{n}^x = (1, 0)$, $\mathbf{n}^y = (0, 1)$ describe the rectangular normalized basis, we have $\mathbf{n}^{y'} = cos\phi\mathbf{n}^x + sin\phi\mathbf{n}^y$. For $\varphi = \frac{\pi}{2}$, we have a rectangular grid and for $\varphi = \frac{\pi}{3}$ the grid is rhomboidal and with this angle it is suitable for the construction of uniform triangular or hexagonal cells (see Chapter "Finite Difference Schemes on Sparse and Full Grids").

For a regular 2D grid with the grid lengths dx and dy, the position of the point at the cell is defined as

$$\mathbf{r}_{i,j} = \left(x_i, y_j\right) = idx\mathbf{n}^{x'} + jdy\mathbf{n}^{y'}. \tag{2}$$

© Springer Nature Switzerland AG 2022

J. Steppeler, J. Li, *Mathematics of the Weather*, Springer Atmospheric Sciences, https://doi.org/10.1007/978-3-031-07238-3_4

With this definition, all points $\mathbf{r}_{i,j}$ $(i, j = 0, \frac{1}{3}, \frac{2}{3}, 1, \ldots)$ are used to define the full grid collocation points of rhomboidal or triangular cells where we have chosen the collocation points for a third-order representation. The grid Eq. (2) is called a structured grid. The L-Galerkin methods to be described are possible also for unstructured and irregular grids. The choice Eq. (2) is done for simplicity. An example of an irregular structured grid is $\mathbf{r}'_{i,j} = \mathbf{r}_{i,j} + \boldsymbol{\delta}_{i,j}$, with the x- and y-components of random variables $\boldsymbol{\delta}_{i,j}$ being smaller than dx and dy. The irregular structured grid is shown in Fig. 1f. Such slightly irregular grids can be used to do tests for the geometry of the sphere, where regular grids do not exist. Many examples and demonstrations of this book will use the regular grid $\mathbf{r}_{i,j}$. The numerical methods, however, will be suitable for irregular and unstructured grids. A particular concern is that for irregular cell structures, the methods maintain a uniform approximation order meaning that at no point the order of approximation falls below the target order. Most methods proposed will have approximation orders greater than or equal to 3.

The grid $\mathbf{r}_{i,j}$ defines a rectangular or rhomboidal cell structure in a natural way. Rectangular cells can be indexed by the lower left corner $\mathbf{r}_{i,j}$. The rectangular cell is defined by the corner points $\mathbf{r}_{i,j}, \mathbf{r}_{i,j+1}, \mathbf{r}_{i+1,j+1}, \mathbf{r}_{i+1,j}$. The sides of cells are called edges. In order to obtain triangular cells, the rectangles can be divided into 2 triangles, as shown in Fig. 1a, b. This division into triangles can be done for any ϕ in Fig. 1. For $\phi = \frac{2}{3}\pi$, we obtain equal-sided triangles. Triangular and rectangular cells can exist in the same grid, as exemplified in Fig. 1a. Figure 1a gives the example of an arbitrary division of squares into triangles. In this case a corner point has between 4 and 8 neighboring cells and neighboring corners if 4 quadrilaterals meet at a corner. For a corner point, a neighboring corner is a corner point on the same edge. As seen from Fig. 1c, for $dx = dy$ and $\varphi = \frac{\pi}{3}$, the cells are regular triangles. Another rather regular division to create regular cells is shown in Fig. 1d. Here corner points have either 4 or 8 neighbors of cells when the lines dividing the rectangles have always the same direction.

As shown in Fig. 1b, 6 triangles can be combined to form a hexagon. The example shown in Fig. 1b creates irregular hexagons, as sides may differ by a factor of $\sqrt{2}$ and the angles also are not equal. Regular hexagons are obtained by choosing $\phi = \frac{2}{3}\pi$. For the case of regular triangles shown in Fig. 1c, the hexagons are regular. These hexagons formed in a triangular grid are not unique. They may overlap and not necessarily induce a hexagonal cell structure. Figure 1h shows a triangular mesh, where two hexagons are highlighted. These two cells do not belong to the same hexagonal cell structure, as they cannot be part of a hexagonal cell system covering the whole area. A hexagonal cell structure is supposed to cover the computational area by non-overlapping cells. Figure 1h indicates that for the same triangular mesh the creation of hexagonal cells is not unique. Several hexagonal cell grids are possible. However, when one of the hexagonal cells in the triangular mesh is defined, the hexagonal grid is then determined. To obtain a hexagonal cell structure meaning a set of non-overlapping hexagons, special hexagons must be selected. In Fig. 1h, each of the highlighted hexagons can induce a hexagonal cell structure and these

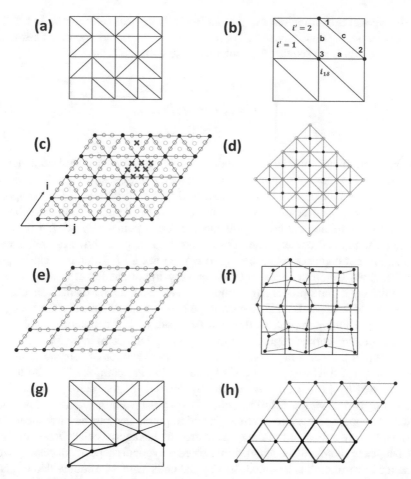

Fig. 1 Rhomboid, triangular, and hexagonal cells. (**a**) Regular quadrilateral grid with subdivision of some cells into triangles. The division into triangles is in irregular directions. Not all cells must be triangular. Corner points have between 4 and 8 corner point neighbors. (**b**) Creating hexagonal cell by combining triangles. Hexagonal cells must be chosen in a special way to obtain a disjunct covering by hexagonal cells. (**c**) Order 3 full grid for rhomboidal/triangular cells. A triangular cell contains 10 points, a rhomboidal 16. Many collocation points are shared between cells. (**d**) Triangular cell division in different directions. Corner points have alternatively 4 or 8 neighbors. (**e**) Serendipity sparse grid for $\varphi = \frac{\pi}{3}$ where corner points is black, and high order or edge points are white. (**f**) An irregular cell structure presented as a structured grid indexed by $i, j = 0, 1, 2, 3, \ldots$. This grid can be used to test irregular cell structures without having access to a management program for unstructured data. (**g**) Triangular cells based on a square grid structure with a lower boundary represented by a linear spline. (**h**) Example of two hexagons in a triangular grid, which cannot be members of the same hexagonal grid cell structure. Groups of 6 triangles form a hexagonal cell. We speak of a hexagonal cell structure only if a set of hexagonal cells covers the whole computational area

two hexagonal cell structures are not identical. Hexagonal cells can be indexed by their center point. Obviously, not all points (i, j) become center points of hexagonal cells. From Fig. 1, we see that the centers of hexagonal cells are

$$\begin{cases} i = 0, j = 1, 4, 7, \ldots, \\ i = 1, j = 2, 5, 8, \ldots, \\ i = 2, j = 0, 3, 6, \ldots. \end{cases} \tag{3}$$

This example shows that the grid notation Eq. (3) does not deliver a compact cell indexing for hexagons. An index system is called compact, if each index points to an object, here a hexagonal cell. A structured grid with indices i, j is compact if all i, j define a point being used. The indices Eq. (3) define a compact subset of Eq. (2) if we have the rhomboidal grid and use corner points only. All i, j in Eq. (3) define a rhomboidal corner point. The centers of hexagons, however, as given in Eq. (3), are not compact, as according to Eq. (3) not all i, j define a center point of a hexagon. Compact index notation is very important for efficient computer programming if multiprocessing computers are used. In particular, an efficient message passing depends on the data storage to be compact. Note that the triangular cells in Fig. 1b are compact, as this is a full grid, both for the corner points and for the combined corner, edge, and interior points. Each combination i, j defines a triangular point. When going to sparse grids, the construction of compact structured grids is not trivial. Unstructured grids by definition are compact. The rhomboidal cells defined by Eq. (2) are determined by index i, j. As each rhomboidal cell consists of two triangles, a further index is necessary. Further amplitudes with 2 values for corner points and 6 values for edge points need to be introduced. It is possible to have only a part of the grid covered by hexagonal cells. Other cells may be triangles or rhomboids. Such grids of mixed polygonal shape are necessary when a change of resolution is intended. Hexagonal cells cannot be subdivided to give a higher resolution hexagonal grid. This is easily seen from Fig. 1b. Triangular and rhomboidal cells may easily be subdivided. The grid refinement of a hexagonal grid can be achieved by bordering the hexagonal grid by triangular cells, which then are subdivided and a higher resolution hexagonal grid is then created.

In this book most examples use structured programming and in particular sparse grids in this environment require some thoughts if the representation is also required to be compact. For introducing numerical methods, structured grids have the advantage that the whole grid construction can be easily handled by the modeler. However, for unstructured grids, it may be easier to obtain a compact representation. With unstructured grids, a storage management system is often available as a program system and the modeler can concentrate on the approximations.

Collocation points are introduced to help defining the field inside a cell. An easy case is the rhomboidal cell. We will use collocation grids with collocation grid length $dx^c = \frac{1}{2}dx, dy^c = \frac{1}{2}dy$ leading to half index values and $dx^c = \frac{1}{3}dx, dy^c = \frac{1}{3}dy$ leading to triple index values. This formalism can be done for regular and irregular grids. In the following, for simplicity, some of the formula are given for the special case of a regular structured grid. The basic collocation grid is the grid generated by dx^c analog to Eq. (1). The indices are either $i, j = 0, \frac{1}{2}, 1, \frac{3}{2}, \ldots$ for $dx^c = \frac{1}{2}dx, dy^c = \frac{1}{2}dy$ or $i, j = 0, \frac{1}{3}, \frac{2}{3}, 1, \frac{4}{3}, \ldots$ for $dx^c = \frac{1}{3}dx, dy^c = \frac{1}{3}dy$. The collocation grid for the regular triangular or rhomboidal cells in the regular case $dx^c = \frac{1}{3}dx, dy^c = \frac{1}{3}dy$ is shown in Fig. 1c.

The basic collocation grid is the full grid. It is used by most of the classic schemes, in particular those described in Chapter "Simple Finite Difference Procedures". Here we are concerned how the L-Galerkin schemes described in Chapter "Local-Galerkin Schemes in 1D" can be performed in two dimensions. The rhomboidal grids are included in our analysis because when projecting the icosahedron of similar semi-Platonic bodies to the sphere, the resulting grid for small areas of the sphere is approximately plane. It is rhomboidal, rather than rectangular (see Chapter "Platonic and Semi-Platonic Solids" and Fig. 1 in Chapter "Introduction").

The discretizations defined in this chapter use the target points and their near neighbors. The advantage of structured grids is that neighboring points are immediately known. For example, (x_{i+1}, y_j) is the neighbor to the right of (x_i, y_j). Unstructured grids are increasingly used in atmospheric modeling. In this case just a list of grid points (x_i, y_j) is given as a table, in order to do a discretization, the neighboring points must be given as a table $(x_{i,k}, y_{j,k}), k = 1, \ldots, n^{neigh}$, where n^{neigh} is the number of neighboring points considered for a discretization. Often, the different kinds of neighboring points need to be tabulated, such as neighboring on an edge or on a corner.

Figure 1c shows the full grid for the example $\varphi = \frac{\pi}{3}$. Figure 1e shows the corresponding sparse collocation grid. As a cell is defined by the corner points with index $i, j = 1, 2, 3, 4$, each rhomboidal cell has 9 points for $dx^c = \frac{1}{2}dx, dy^c = \frac{1}{2}dy$ and 16 points for $dx^c = \frac{1}{3}dx, dy^c = \frac{1}{3}dy$. Some of these points belong to more than one cell. In the following, we treat the case $dx^c = \frac{dx}{3}, dy^c = \frac{1}{3}dy$ and the case $dx^c = \frac{dx}{2}, dy^c = \frac{1}{2}dy$ is analog. Let us first consider Fig. 1c as a rhomboidal cell structure. Corner points belong to 4 cells and edge points belong to 2 cells. Considering the same grid with triangular cells, each corner point belongs to 6 cells and edge points belong to two cells. Inner points, being neither corner nor edge points, belong to one cell only. For efficient computer programming, it is not suitable to store globally all 16 points belonging to a rhomboidal cell. This would be non-compact storage. Rather, it is efficient to store for each cell only a subset of points called the compact points, such that the combination of all points is the whole grid. In Fig. 1c, the compact points belonging to the rhomboidal cell marked i, j are marked as crosses. The combination of all compact points for all cells is the full grid. This means that the points indicated by crosses in Fig. 1c combine to a compact representation of the full grid. In this case the cell i, j has 16 grid points and 9 compact points. The sum of the numbers of compact points over the cells is the number of grid points of the whole grid. From the grid points of a cell, many are shared with other points. Table 1 gives the grid statistics for a number of cell systems. This means that during computations, the points belonging to a cell must be collected from distributed memory.

Table 1 Grid statistics of different cell structures for second- or third-order field representation ($dx^c = \frac{dx}{2}$) and third-order cell structure ($dx^c = \frac{dx}{3}$).

Grid	Grid length	Grid points per cell	Compact points per cell	Standard serendipity	Standard serendipity compact points	Sparseness factor
2nd-order S[a] or R[b]	$\frac{dx}{2}$ (2D)	9	4	8	3	$\frac{3}{4}$
3rd-order S or R	$\frac{dx}{3}$ (2D)	16	9	12	5	$\frac{5}{9}$
2nd-order S or R	$\frac{dx}{2}$ (3D)	27	8	20	4	$\frac{4}{8}$
3rd-order S or R	$\frac{dx}{3}$ (3D)	64	27	32	7	$\frac{7}{27}$
2nd-order triangular	$\frac{dx}{2}$	6	Not the same 2D for different cells	No reduction of grid points for triangle		1

[a] S represents Square
[b] R represents Rhomboidal

In a grid according to Eq. (2), rhomboidal cells $\Omega_{i,j}$ are defined as

$$\Omega_{i,j} = \{c_{i,j}\} = \left[(x, y), x = x_i + k dx^c, y = y_i + k dy^c, dx^c = dy^c \cdot cos\varphi, k = 0, 1, 2, 3 \right]. \tag{4}$$

The index of the cell is the lower left corner point index. The corner points of $\Omega_{i,j}$ are $c_{i,j}$; $c_{i+1,j}$; $c_{i+1,j+1}$; $c_{i,j+1}$. For each $\Omega_{i,j}$, there exist two triangular cells, Ω^{tr_1} and Ω^{tr_2}, with the corners

$$\begin{cases} c_{i,j}, c_{i+1,j}, c_{i,j+1}; & \text{for } \Omega^{tr_1}, \\ c_{i+1,j+1}, c_{i+1,j}, c_{i,j+1}; & \text{for } \Omega^{tr_2}. \end{cases} \tag{5}$$

Both for rhomboidal and for triangular cells, we can define local coordinates x', y'. For the cell $\Omega_{i,j}$ defined in Eq. (4), we have:

$$\mathbf{r} = (x, y) \in \Omega_{i,j} : \mathbf{r} = \mathbf{r}_{i,j} + x'\mathbf{n}^{x'} + y'\mathbf{n}^{y'}, \tag{6}$$

where $\mathbf{n}^{x'}$, $\mathbf{n}^{y'}$ are defined in Eq. (1). For the triangular cells, we get the following representation by local coordinates for the case that the division line runs from the upper left to the lower right corner of the quadrilateral:

$$\begin{cases} \mathbf{r} = (x, y) \in \Omega_{i,j}^{tr_1} : \mathbf{r} = \mathbf{r}_{i,j} + x'\mathbf{n}^{x'} + y'\mathbf{n}^{y'}, \dfrac{x'}{dx} + \dfrac{y'}{dy} \leq 1, \\ \mathbf{r} = (x, y) \in \Omega_{i,j}^{tr_2} : \mathbf{r} = \mathbf{r}_{i,j} + x'\mathbf{n}^{x'} + y'\mathbf{n}^{y'}, \dfrac{x'}{dx} + \dfrac{y'}{dy} \geq 1, \end{cases} \tag{7}$$

where $\mathbf{n}^{x'} = (0, 1)$, $\mathbf{n}^{y'} = (cos\varphi, sin\varphi)$, $x' \in [0, dx]$, $y' \in [0, dy]$.

In the following field, amplitudes will be associated with collocation points and other amplitudes may be associated with a cell. Amplitudes being not collocation point values are called spectral amplitudes. Spectral amplitudes are normally equivalent to the collocation point amplitudes, meaning that transformations between the two representations exist. Dynamic equations will be derived for such amplitudes using L-Galerkin methods. These methods will rely for each dynamic field $h(x, y)$ on an approximation assumption $h^a(x, y)$. $h^a(x, y)$ depends on all dynamic amplitudes, collocation point or other amplitudes and is called the basis function representation. The upper index a stands for approximation. Following a general practice, in the following the distinction between $h(x, y)$ and $h^a(x.y)$ is omitted and the approximated field will be called $h(x, y)$.

Within triangular or rhomboidal cells, fields will be represented as polynomials resulting in a piecewise polynomial representation. At corner and edge points, various continuity requirements are requested.

So within a cell $\Omega_{i,j}$ ($x, y \in \Omega_{i,j}$), the field $h(x, y)$ is represented as a sum of basis functions $b_\nu(x.y)$:

$$h(x, y) = \sum_\nu h_\nu b_\nu(x, y), \tag{8}$$

where the basis functions b_ν are a polynomial in x and y. For second-order representation, the polynomial is of at least second degree, and for third-order representation it is of at least third degree.

For this choice of basis functions, we have the following property for the third-order representation:

Let $h(x, y)$ be a locally analytic function meaning that a power series development exists ($h = \sum_{i,j} a_{i,j} x^i y^j$). Then an approximation $h^a(x, y)$ within the function system exists, such that

$$max_{(x,y)} |h(x, y) - h^a(x.y)| = o(dx^4). \qquad (9)$$

For a third-order representation and dx being the maximum cell size, $o(dx^4)$ is a term converging to 0 with dx in fourth order. Other choices than polynomials are possible. For example, the global spectral method (Durran 2010) uses trigonometric functions as basis. With the global spectral method, all basis functions overlap on the whole computational area. For people interested in using other basis functions than defined by polynomials, the minimum overlap theorem is stated here in its 1D form. The definition of order of functional representation in 1D is given by Eq. (9), where h is assumed to be not dependent on y. The minimum overlap theorem says:

Theorem 4.1 *Let $h(x)$ be represented by the 1D form of Eq. (9), $h(x) = \sum_{\nu \in (0,n^o)} h_\nu b_\nu(x)$, let the basis function be of the form $b_\nu^{dx}(x') = b_\nu^0(\frac{x'}{dx})$, and let the approximation of an analytic function $h(x)$ be of n^{order}: $max_{(x)} |h(x) - h^a(x)| = o(dx^{n^{order}+1})|$, where x' is the local coordinate $x' = x - x^m$, with $x^m = \frac{1}{2}(x_\nu + x_{\nu+1})$. Then the overlap n^o is at least $n^{order} + 1$, and if $n^o = n^{order} + 1$, then all basis functions $b^o(x')$ are piecewise polynomials of at most degree n^{order}.*

The word minimum overlap comes from the fact that for the combination of all cells, n^o is the maximum number of functions overlapping on a cell. In other words, the theorem of minimum overlap says that if a minimum number of basis functions overlap and during the convergence process the shape of basis functions does not change, they have to be polynomials in order to achieve the desired rate of convergence.

In 1D, it is always possible to choose basis functions of exactly degree n^{order}. In 2D, depending on the shape of the cell, it will turn out in the following sections that it is convenient to use a few more than the minimum number of polynomial coefficients according to the minimum overlap theorem. For example, for a square cell and approximation order $n^{order} = 3$, the number of linear independent polynomials is 10, being $1, x, y, x^2, xy, y^2, x^3, x^2 y, xy^2, y^3$. It will turn out in the following that the fourth-order polynomials xy^3 and $x^3 y$ are used with the serendipity representation. This brings the number of linear independent basis functions associated with a square cell from 10 to 12. The increase in the number of basis functions is called *the polynomial order excess*, which is 2 for the case of serendipity interpolation on squares (see Section "Rhomboidal Basis Functions and Sparse Grids for the Regular Grid Case"). For a square cell, it is seen from Fig. 1c that each rhomboidal cell has 16 collocation points. These allow to determine 16 polynomial coefficients. Then the polynomial order excess is 6 when using the full grid. For SE method (see Chapter "Local-Galerkin Schemes in 1D" and Durran 2010), in fact 16 polynomial coefficients are computed for a square cell. This gives the third-order spectral element method SE3, the polynomial order excess of 6.

The presence of a polynomial order excess means that not all 16 collocation points of the rhomboidal cell in Fig. 1c are needed to determine the polynomial basis functions. 12 collocation points are sufficient to determine $h(x, y)$. It is usual to omit the inner points of a square cell and retain the corner and edge points. In this way a sparse grid is obtained as shown in Fig. 1e. *The sparseness factor* is defined as the number of points in the sparse grid divided by the number of points of the full grid. Here the full grid points are the points of the compact grid. As from the 16 collocation points of a square grid 9 are compact points, the sparseness factor is $\frac{5}{9}$ for the case of the square serendipity grid. The sparseness factor indicates a potential of saving computer time from sparse grids. Note that the triangular grid, when using all edge, center, and corner amplitudes, is a full grid representation. We can arrive at sparse triangular representations by treating some of these amplitudes as diagnostic.

For a third-order representation on triangular cells, the number of collocation points is 10. This equals the number of linear independent third-order polynomial coefficients. The same is the case for second-order representations on triangles. So in this case there is no polynomial order excess and no sparseness of the grid. However, triangles are often combined to obtain other cell shapes, such as hexagons, pentagons, or quadrilaterals. If this is the case, there is the possibility of treating some of the collocation points as diagnostic, which then offers the possibility of sparse grids with rather small sparseness factors.

There is one more important point to observe when constructing grids on limited domains. Figure 1g shows a regular triangular grid, where just the lower boundary is irregular to accommodate for an irregular lower boundary. It should be noted that for amplitudes on this lower boundary not just the dynamic equations (see Chapter "Finite Difference Schemes on Sparse and Full Grids") must be solved. On the edge lines of the lower surface, the condition must be satisfied that the velocity must be parallel to the surface meaning the normal component of the velocity at each point of the surface must be 0. When we assume as illustrated in Fig. 1g the lower boundary is a linear spline, the velocity field will be discontinuous at cell boundaries. Then for boundary treatment, we must make an exception to the rule that we use CG methods where fields are continuous at cell boundaries. For grids aligned in x- and z-directions, it will be shown that the discontinuities need not be in the direction of differentiation. Another possibility is to require the boundary condition only as an average over the basis functions. Curved cell boundaries are another possibility.

Rhomboidal Basis Functions and Sparse Grids for the Regular Grid Case

Grids and basis functions are described for cubic basis functions only. The quadratic case can be done in an analog way. In local coordinates x', y' according to Eq. (6), the rhomboidal cell has size dx and dy. The full grid is the grid according to Eq. (2) when i, j are allowed to take all values i, $j = 0, \frac{1}{3}, \frac{2}{3}, 1, \frac{4}{3}, \ldots$. Each rhomboidal cell has $4 \times 4 = 16$ points, some of which are shared between cells. Corner points i, $j = 0, 1, 2, 3, \ldots$ are shared between four cells and each cell has four corner points. Edge points are shared between two cells and each cell has eight edge points. Therefore, there are $4 \cdot \frac{1}{4} + 8 \cdot \frac{1}{2} + 4 = 9$ independent points in a cell (see Fig. 1c). Let i_0, j_0 be the lower left corner of rhomboidal cell Ω_{i_0, j_0}. Let us associate the nine points $i_0 + i'$, $j_0 + j' \left(i', j' = 0, \frac{1}{3}, \frac{2}{3} \right)$ in this cell. Then the combination of all cells associated with grid points is the full grid and each point is counted in one cell only. This means that the indicated points form a compact representation of collocation points. For discretizations on rhomboidal cells, the distinction between inner, corner, and edge points is important which for cell Ω_{i_0, j_0} is defined as follows:

$$\mathbf{r}_{i_0+i', j_0+j'}, \tag{10}$$

where corner points are obtained for $(i', j') = (0, 0)$, edge points are for $(i', j') = (0, \frac{1}{3}), (0, \frac{2}{3}), (\frac{1}{3}, 0), (\frac{2}{3}, 0)$, and inner points are for $(i', j') = (\frac{1}{3}, \frac{1}{3}), (\frac{1}{3}, \frac{2}{3}), (\frac{2}{3}, \frac{1}{3}), (\frac{2}{3}, \frac{2}{3})$. L-Galerkin discretizations on rhomboidal cells are done on the sparse grid which is defined to contain corner and edge points. From Eq. (10), it can be seen that the sparseness factor is 5:9. In 3D, the sparseness factor 7:27 makes the numerical cost of a model faster under the assumption that the numerical effort is proportional to the number of points to be computed. Using corner point derivatives as amplitudes (see Chapter "Simple Finite Difference Procedures"), the sparseness factor in 3D for rhomboids is 4:27. For hexagonal grids in 3D (see Chapter "Full and Sparse Hexagonal Grids in the Plane"), the sparseness factor with derivative amplitudes will turn out to be 8:81. The corner point physics interface (see Section "The Interface to Physics in High-Order L-Galerkin Schemes") can double the corresponding saving of computer time. These are substantial potential savings. An example of a sparse 2D solution with o3o3 is given by Steppeler et al. (2019).

The full or sparse grids define the grid point space. For the spectral representations, the spectral coefficients or amplitudes are distributed in the cell meaning that the spectral amplitudes in the same way as the grid point values have locations within the grid cells. This will be described for third-order representations. Second-order or first-order representations can be obtained by analogy. The corner points, obtained for i', j' being integers $(1, 2, 3, \ldots)$, always carries grid point values. As we have already seen in 1D, the grid point values at corner points are also spectral amplitudes. Note that corner point amplitudes define the linear part of a field representation.

The grid and basis functions will first be explained for a regular rhomboidal cell structure and the spectral amplitudes are second and third derivatives in x and y directions for the spectral representation. $fl_{xx,i+\frac{1}{2},j}$ and $fl_{xxx,i+\frac{1}{2},j}$ are defined at half-level points $\left(x_{i+\frac{1}{2}}, y_j\right)$. $fl_{yy,i,j+\frac{1}{2}}$ and $fl_{yyy,i,j+\frac{1}{2}}$ are defined at half-level points $\left(x_i, y_{j+\frac{1}{2}}\right)$. The spectral representation grid is shown in Fig. 2c, where only the points in the middle of the edges of the square carry second and third derivatives as spectral amplitudes. The basis function representation for $fl(x, y)$ is given as

$$fl(x, y) = fl^{lin}(x, y) + fl^{ho,x}(x, y) + fl^{ho,y}(x, y). \tag{11}$$

The basis function representations of the functions in Eq. (11) are

$$
\begin{cases}
fl^{lin}(x, y) = \displaystyle\sum_{i,j=0,1,2,3,\ldots} fl_{i,j}e_{i,j}(x, y), \\
fl^{ho,x}(x, y) = fl_{xx,i+\frac{1}{2},j}b^2_{i+\frac{1}{2},j}(x, y) + fl_{xxx,i+\frac{1}{2},j}b^3_{i+\frac{1}{2},j}(x, y), \\
fl^{ho,y}(x, y) = fl_{yy,i,j+\frac{1}{2}}b^2_{i,j+\frac{1}{2}}(x, y) + fl_{yyy,i,j+\frac{1}{2}}b^3_{i,j+\frac{1}{2}}(x, y).
\end{cases}
\tag{12}
$$

Note that without loss of generality in Eq. (12), we can assume the edge carrying x is in x-direction. The second and third y-derivatives in Eq. (12) are in y-direction.

The 2D basis functions are derived from the 1D functions (see Chapter "Local-Galerkin Schemes in 1D"):

$$
\begin{cases}
e_{i,j}(x, y) = e_i(x)e_j(y), \\
b^2_{i+\frac{1}{2},j}(x, y) = b^2_{i+\frac{1}{2}}(x)e_j(y), \\
b^3_{i+\frac{1}{2},j}(x, y) = b^3_{i+\frac{1}{2}}(x)e_j(y), \\
b^2_{i,j+\frac{1}{2}}(x, y) = e_i(x)b^2_{j+\frac{1}{2}}(y), \\
b^3_{i,j+\frac{1}{2}}(x, y) = e_i(x)b^3_{j+\frac{1}{2}}(y),
\end{cases}
\tag{13}
$$

where $b^2_{i+1/2}, b^3_{i+1/2}, b^2_{j+1/2}, b^3_{j+1/2}$ are defined in Eq. (12) in Chapter 'Local-Galerkin Schemes in 1D'. The transformation between grid point space and spectral space will use the sparse grid. This grid consists of those grid points which are on the edges or corners. The sparse grid is equivalent to the spectral representation. Corner point amplitudes are both spectral and grid point amplitudes and the two edge points can be transformed into the spectral amplitudes $fl_{xx,i+\frac{1}{2},j}$ and $fl_{xxx,i+\frac{1}{2},j}$ using the collocation points on the same edge. Note that along the grid line in x-direction, we can extract a 1D problem and the same transformation equations can be used which in Chapter "Local-Galerkin Schemes in 1D" were derived for one dimension. The same can be repeated in y-direction. As the sparse grid is a subgrid of the full grid, it is possible to transform the full grid to the spectral representation

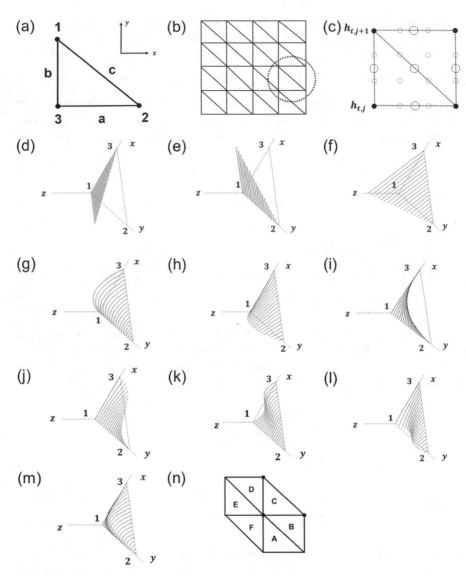

Fig. 2 (a) The rectangular triangular cell with corner points 1,2,3 and edges (sides) a, b, c. (b) and (c) Division of quadrilaterals of the grid shown in Fig. 1f into triangles for the case of a regular grid, solid black: corner points for the definition of corner point amplitudes and corner points derivatives to x and y, small white points: triangular collocation points, large white points: for the definition of second and third derivatives of h serving as spectral amplitudes. The second, third and fourth rows plot all basis functions belonging to (a). 3D perspective plots of the basis functions belonging to the lower of the blown up triangles are shown. (d) to (f) show the three linear basis functions of (a) corresponding to the corner amplitudes $h_{i,j}, h_{i+1,j}, h_{i,j+1}$ (g) to (m) show the second- and third-order basis functions corresponding to amplitudes at mid-edge points carrying the amplitudes h_{ss} and h_{sss}, where s is the coordinate along the edge. (m) shows the center basis function corresponding to the amplitude h_{center}. This basis function is 0 at all edges of the triangle. (n) shows the six triangles in the grid (b) having edge point $\mathbf{r}_{i,j}$ as corner

from the full grid by selecting the sparse grid as a subgrid of the full grid. The back transformation that is to the sparse grid will in this case not give the full grid point amplitudes we have started from. We rather obtain a smoothed version of the original full grid representation. Only on the sparse grid, the original and the back transformed grid representations coincide. Equation (12) can be used to create a full grid representation. This is not identical to the full grid representation and this grid representation is a smoothed representation. For analytic test functions, the smoothed function differs from the original function by terms of fourth order, which is below the accuracy considered here.

As was the case in 1D, the transformation between grid point space and spectral space is local, meaning that the points $fl_{i,j}$, $fl_{i+\frac{1}{3},j}$, $fl_{i+\frac{2}{3},j}$, $fl_{i+\frac{2}{3},j}$ are needed to compute $fl_{xx,i+\frac{1}{2},j}$, $fl_{xxx,i+1,j}$ and no other points are needed for this transformation. The transformation formulas are 1D and given in Chapter "Local-Galerkin Schemes in 1D". The same is true for $fl_{yy,i,j}$, $fl_{yyy,i,j+\frac{1}{3}}$, $fl_{yy,i,j+\frac{2}{3}}$, $fl_{yyy,i,j+1}$. Note that the transformation concerns only the two terms with broken indices. It may be remarked that the constructions given above can be generalized to rather irregular quadrilateral cells, as those shown in Fig. 1f. This will be the subject of the next section.

Euclid's Lemma

The representation formula in Section "Rhomboidal Basis Functions and Sparse Grids for the Regular Grid Case" concerns the case of a regular grid. Now we want to create piecewise polygonal interpolations for arbitrarily shaped quadrilateral cells. The presentations to be derived will be analog to what was done in Section "Rhomboidal Basis Functions and Sparse Grids for the Regular Grid Case" for regular cells. The cells and grids treated will be given by 4 corner grid points for quadrilateral cells. We do not require such grid points to be in a plane and the corner points may be arbitrary in space. These cells may be on a plane and also in 3D space such as occur with polygons inscribed in a sphere (see Chapter "Platonic and Semi-Platonic Solids"). The interpolation on such a cell will be done using linear coordinates x', y' associated with the cells. With a linear coordinate of the dependence, a vector $\mathbf{r}$ on the coordinated x', y' is linear. In this book linear coordinates are preferred. Nonlinear coordinates are often used in order to adapt to the symmetry of the problem, such as polar coordinates for cylindrical geometry and latitude–longitude grids for spherical coordinates (Durran 2010). This is the coordinate transformation method for adapting to spherical geometry and very popular. In this book an alternative method is used. The grid is created using linear coordinates on bilinear surfaces and then is projected to the sphere. Euclid's lemma provides the basis for this proceeding. It states the existence of linear coordinates for arbitrary quadrilaterals and provides a geometric interpretation. Obviously plane coordinates are available for triangular cells and they do not exist for a quadrilateral in 3D. Bilinear surfaces are for quadrilaterals while planes are for triangles. Every quadrilateral can be interpolated on a bilinear surface. In x', y' coordinates, a bilinear surface in 3D space is defined by the following polynomial:

$$z = a_0 + a_{01}x' + a_{01}y' + 2a_{11}x'y'. \tag{14}$$

The linear coordinates to be defined are very powerful in 3D, where they will allow easily to construct grids on the sphere. Let us first consider cells in 2D with $\mathbf{r} = (x, y)$. Let $\mathbf{r}_1, \mathbf{r}_2, \mathbf{r}_3, \mathbf{r}_4$ be the corner points of the quadrilateral. We are looking for linear coordinates $x', y' \in (0, 1)$ with the properties:

$$\begin{cases} \mathbf{r}_1 = \mathbf{r}(0, 0), \\ \mathbf{r}_2 = \mathbf{r}(0, 1), \\ \mathbf{r}_3 = \mathbf{r}(1, 1), \\ \mathbf{r}_4 = \mathbf{r}(1, 0). \end{cases} \tag{15}$$

In the following, x', y' define a 2D coordinate space. They determine a vector $\mathbf{r} = (x, y, z)$ in 3D space. The coordinate x', y' is defined when $\mathbf{r}(x', y')$ is given as a linear function. The edges (such as $\mathbf{r}_2 - \mathbf{r}_1$) of the quadrilateral cell are then defined by $x' = 0$ or $x' = 1$ or $y' = 1$ or $y' = 0$. For quadratic cells on planes, as treated in Section "Rhomboidal Basis Functions and Sparse Grids for the Regular Grid Case", we have $x' = x$ and $y' = y$ and this dependence is given as

$$\mathbf{r}(x, y) = \mathbf{r}_1 + x'\mathbf{n}^x + y'\mathbf{n}^y, \tag{16}$$

where $x' = x - x_i^m$, $y' = y - y_i^m$, $\mathbf{n}^x$ and $\mathbf{n}^y$ are the unit vectors in x- and y-directions and $\mathbf{r}_1$ is the lower left corner of the quadrilateral cell, as shown in Fig. 3a. Note that here x' and y' are coordinates on curved surfaces being oriented arbitrarily in space, while x and y are used for the case that all points are in the same plane for all cells considered. Examples of curved cell systems are given in Fig. 1 in Chapter "Introduction". The square with corners $\mathbf{r}_1, \mathbf{r}_2, \mathbf{r}_3, \mathbf{r}_4$ has coordinate lines defined by x' and y' which is seen from Fig. 3 for the four edges $\mathbf{r}_1 - \mathbf{r}_2, \mathbf{r}_3 - \mathbf{r}_2, \mathbf{r}_4 - \mathbf{r}_3$ and $\mathbf{r}_1 - \mathbf{r}_4$.

While this is entirely trivial for the case of a square, it is seen from Fig. 3b that this generalizes to the case of a general (irregular) quadrilateral cell. Again coordinate lines x'' and y'' cut corresponding coordinate lines with equal dividing relations, and a linear coordinate can be defined, using x'' and y'' as coordinates.

For the general case, given by Eq. (15) and illustrated by Fig. 3c, we have the following alternative but equivalent definitions of cell related normalized coordinates:

$$\mathbf{r}(x', y') = \mathbf{r}_1 + x'(\mathbf{r}_2 - \mathbf{r}_1) + y'\left[\mathbf{r}_4 + x'(\mathbf{r}_3 - \mathbf{r}_4) - \mathbf{r}_1 - x'(\mathbf{r}_2 - \mathbf{r}_1)\right]. \tag{17}$$

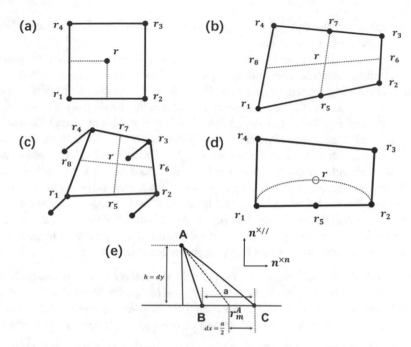

Fig. 3 (**a**) Definition of the normalized coordinates x', y' for the case of a square cells. (**b**) as (**a**), for the case of an irregular quadrilateral: the coordinate lines x' and y' cut the corresponding sides in equal relations of either x' or y'. (**c**) Generalization of (**a**) and (**b**) to 3D quadrilaterals and the creation of a bilinear surface. (**d**) Representation for a curved boundary. (**e**) The two rectangular triangles representing a general triangle

The following definition is equivalent:

$$\mathbf{r}(x', y') = \mathbf{r}_1 + y'(\mathbf{r}_4 - \mathbf{r}_1) + x'\left[\mathbf{r}_2 + y'(\mathbf{r}_3 - r_2) - \mathbf{r}_1 - y'(\mathbf{r}_4 - \mathbf{r}_1)\right]. \tag{18}$$

Equations (17) and (18) are equivalent, as they have the common form:

$$\mathbf{r}(x', y') = \mathbf{r}_1 + x'(\mathbf{r}_2 - \mathbf{r}_1) + y'(\mathbf{r}_4 - \mathbf{r}_1) + x'y'(\mathbf{r}_3 + \mathbf{r}_1 - \mathbf{r}_2 - \mathbf{r}_4). \tag{19}$$

Note that in Eq. (19) the definition of coordinates x' and y' are linear, if the other variable is kept fixed. In x' and y' together, the definition is not linear because of the term proportional to $x'y'$. This is called bilinear dependence and the surface generated by x', y' with Eq. (19) is called a bilinear surface.

Equations (17)–(19) remain valid if the $\mathbf{r}_i$ in Eq. (16) are defined to be 3D vectors $\mathbf{r} = (x, y, z)$. For the 3D case, the set of all coordinate lines defines a bilinear surface in x, y, z space. The situation is illustrated in Fig. 3c, in the 3D case. Coordinate lines defined by either x' or y' divide opposite sides of the quadrilateral with equal dividing relations of x' or y'. Equation (17) means that the sides $\mathbf{r}_3 - \mathbf{r}_4$ and $\mathbf{r}_2 - \mathbf{r}_1$

are divided in the relation $\frac{x'}{|\mathbf{r}_3-\mathbf{r}_4|}$ or $\frac{x'}{|\mathbf{r}_2-\mathbf{r}_1|}$. These points are then connected and divided using y' as dividing relation.

While in 2D all non-parallel lines intersect, this is not necessarily so in 3D. However, the coordinate lines defined by either x' or y' intersect and divide each other with dividing relations of either x' or y'. Note that in classical mathematics, there is some interest in lines cutting other lines and in the dividing relations implied. The situation of Fig. 3b, for the special case that opposite sides of a quadrilateral are parallel, is often treated in elementary mathematics books. The theorem is called the law of parallels and is attributed to Euclid. The normalized cell associated coordinates x' and y' allow to define the basis functions belonging to either linear or high-order representation in analog to Section "Rhomboidal Basis Functions and Sparse Grids for the Regular Grid Case", by using x' and y' instead of x and y.

The geometric relations implied by Eqs. (17)–(19) in the 3D case are the basis of constructing grids and cell structures on bilinear surfaces. Formulated in geometric terms, the above definitions come from the following theorem:

Theorem 4.2 *Euclid's Lemma: Let* $\mathbf{r}_1, \mathbf{r}_2, \mathbf{r}_3, \mathbf{r}_4$ *be the corners of a quadrilateral in 3D space. If opposite sides* $\mathbf{r}_1 - \mathbf{r}_2$ *and* $\mathbf{r}_4 - \mathbf{r}_3$ *are divided by equal relations* x' *and connected by line* l_1 *and sides* $\mathbf{r}_3 - \mathbf{r}_2$ *and* $\mathbf{r}_4 - \mathbf{r}_1$ *are divided by relations* y' *and connected by line* l_2, *then* l_1 *and* l_2 *intersect in 3D space and divide each other with dividing relations* x' *and* y'.

The above theorem provides the geometric interpretation of x' and y' for bilinear surfaces in 3D and it is immediately seen that 3D grids on bilinear surfaces can be created. This will be done in Chapters "Finite Difference Schemes on Sparse and Full Grids and Platonic and Semi-Platonic Solids".

There is one more field representation needing discussion, which is the deformed boundary basis, which is of potential importance when treating outer boundaries of the domain of definition. Figure 3d shows the case, where line $\mathbf{r}_1 - \mathbf{r}_2$ belongs to the outer boundary of the domain of computation. Below this line there may be solid Earth. The simplest case is to use the linear spline resulting from the cell structure to represent the orography. This means a first-order representation of the lower surface. If doing this, the formalism given above is sufficient to define the fields in cells.

If a curved representation is considered, the approximation assumption for fields may still be done for the whole cell. Let us assume that the corner points of the cells bordering the orography are on the orography, and then the orographic curve deviates from the cell boundary only for the inner points of the edge near the lower boundary. This is shown in Fig. 3d. Let in Fig. 3d $\mathbf{r}_5$ be a point at the surface on the boundary edge. As shown in Fig. 3e, a coordinate line for y' cuts through $\mathbf{r}_5$ and a point $\mathbf{r}$ on the curved surface. Let on the boundary curve the field h be described by a basis function $b(s)$ on the boundary line. S is a coordinate along the boundary, and let for a field $h : h(s) = h(\mathbf{r})$ be the value of this basis function. Define then

$$h(\mathbf{r}_5) = h(\mathbf{r}). \tag{20}$$

This defines a function on the lower boundary edge, which generally is not represented by any of the cubic basis function defined above. A function representing the lower boundary and being defined on the whole quadrilateral cell is then:

$$h(\mathbf{r}) = h(\mathbf{r}_5)lin(y'), \tag{21}$$

where lin is a linear function in y', having the values 0 and 1 at the end points. This function can be used to define cells in boundary cells and implement boundary values.

Triangular Basis Functions and Full Grids

When introducing collocation grid points analog to Section "Rhomboidal Basis Functions and Sparse Grids for the Regular Grid Case", this section introduces the triangular basis functions belonging to the full grid. Triangular interpolating functions can be used to construct interpolations for cells of any polygonal forms, in particular for rhomboids, as alternative to the cell interpolation of Section "Euclid's Lemma". The rhomboidal and in particular the quadrilateral cell, as any other polygonal cell can be divided into triangular cells. As shown in Fig. 2c, the triangular cell structure needs the same amplitudes as the rhomboidal cells, and the second and third derivatives $fl_{s's'}$, $fl_{s's's'}$ along the diagonal $\mathbf{r}_{i,j+1} - \mathbf{r}_{i+1,j}$, where s' is the coordinate along one of the edges of the triangles. For example, on the side on the diagonal, we have $\mathbf{r} = s'(\mathbf{r}_{i,j+1} - \mathbf{r}_{i+1,j})$. In this section, we give the triangular basis functions for the case of a general triangle given by its corners $\mathbf{r}_1, \mathbf{r}_2, \mathbf{r}_3$. For understanding and for a number of applications, we will in the next section give the basis functions for the special case of rectangular triangles. Due to the more simple equations, the special case may be easier to understand. The graphic presentation of the triangular basis functions given in Fig. 2 will also use the special case.

Some of the amplitudes (circles shown in Fig. 2c) correspond to more than one triangular cell. For discontinuous representations, such amplitudes may have as many amplitudes as cells corresponding to the collocation point in question. In 2D, the edge amplitudes have two neighboring cells and the corner amplitudes a minimum of 3. For irregular cells, these can be any number. For the regular example given in Eq. (2) and Fig. 1c, there are six associated triangular cells for a corner point. The 6 amplitudes associated with this point for the representation of a discontinuous field can be reduced to any number down to 1.

For the representation of discontinuous functions, we need for the multiple amplitude triangular representation as there are associated cells to this point. This is for the assumption that within each triangle the field definition is continuous, such as occurs when the field inside a triangle is a polynomial. For our example of a regular cell, we have six amplitudes at corner points and two at edge points for each edge amplitude.

The representation of continuous functions is obtained by using the same value for such multiple amplitudes.

Let $\mathbf{r} = (x, y)$ be a point belonging to a triangle and $\mathbf{r}_1, \mathbf{r}_2, \mathbf{r}_3$ be the corners of this triangle. The three sides are called edges in the grid cell system. They are associated with the coordinates $s_1, s_2, s_3 \in (0, 1)$. The edges are $\mathbf{r}_2 + s_1(\mathbf{r}_3 - \mathbf{r}_2)$, $\mathbf{r}_3 + s_2(\mathbf{r}_1 - \mathbf{r}_3)$ and $\mathbf{r}_1 +_3 (\mathbf{r}_2 - \mathbf{r}_1)$. For the interpolation of the field $fl(\mathbf{r})$ between grid points of the special triangle chosen, we have in analogy to Eq. (7) in Chapter "Local-Galerkin Schemes in 1D":

$$fl(\mathbf{r}) = fl^{lin}(\mathbf{r}) + fl^{ho,s_1}(\mathbf{r}) + fl^{ho,s_2}(\mathbf{r}) + fl^{ho,s_3}(\mathbf{r}) + fl^{ho,cent}(\mathbf{r}), \tag{22}$$

where $fl^{ho,s_i}(\mathbf{r})(i = 1, 2, 3)$ are the functions associated with the three edges having local coordinate s_1, s_2, s_3 and $fl^{cent}(\mathbf{r})$ is the central point of the triangle. The functions associated with the triangle sides and coordinates s_1, s_2, s_3 are defined by the property that for example fl^{ho,s_1} is different from 0 at the side s_1 only and 0 on the other two sides. As the function within the triangle is defined as a third-degree polynomial, the function $fl^{ho,s_1}(\mathbf{r})$ restricted to the side s_1 is a 1D polynomial in the variable s_1, similar for s_2, s_3. Using the definitions of 1D piecewise polynomial construction, fl^{ho,s_1} is then defined by second and third derivatives in s_1. This means that for the functions fl^{ho,s_1} we have 2 basis functions with corresponding amplitudes $fl^{ho}_{s_1,s_1}$ and $fl^{ho}_{s_1,s_1,s_1}$. The central point could be defined as the point at the center of gravity of the triangle, but any other point may be used. The linear part is determined by the 3 corner amplitudes of the triangle and each of the edge associated parts in Eq. (22) in this chapter involves 2 basis functions and the center part has one amplitude. This means that for the definition of the third-degree polynomial defining the field $fl(\mathbf{r})$ inside the chosen triangle we have 10 amplitudes: 3 corner points, 2 times 3 edge amplitudes, and the center amplitude. As corners and edges are shared between several cells, the number of dynamic amplitudes in a grid system will be less than 10 times the number of cells.

The basis functions are given for any triangle at corner points $\mathbf{r}_{i,j}, \mathbf{r}_{i+1,j}, \mathbf{r}_{i,j+1}$ with no regularity properties required. We give ten discontinuous functions for each triangle: three corner nodes, six edge nodes, and one center node. The number of principal nodes in the i, j grid is $(i_e + 1)(j_e + 1)$, if $i \in 0, 1, \ldots, i_e$ and $j \in 0, 1, \ldots, j_e$. As each rhomboidal cell is divided into two triangles, the number of triangles in the grid is $2i_e j_e$. The number of grid-amplitude points is according to the above: $2i_e j_e \left(3 \cdot \frac{1}{6} + 6 \cdot \frac{1}{2} + 1\right) = 9i_e j_e$. As the number of rhomboidal cells is $i_e j_e$, the number of points in the full grid is $9i_e j_e$, identical to the number of degrees of freedom in the triangular grid with third-order piecewise polynomial representation. In the following, it will be shown that the full grid description is equivalent to the amplitude description.

Using Eq. (22), cascading interpolation will be used as described in the following: $fl^{lin}(\mathbf{r})$ is determined by the three corner points and the resulting linear interpolation will be subtracted from the seven remaining points. The six edge points are used to compute $f^{ho,s_1}(\mathbf{r}) + f^{ho,s_2}(\mathbf{r}) + f^{ho,s_3}(\mathbf{r})$. The resulting interpolation is subtracted from the center amplitude to obtain $f^{cent}(\mathbf{r})$. Each of the contributions in Eq. (22) has a basis function representation analog to Eqs. (11) and (12):

$$\begin{cases} fl(\mathbf{r}) = \sum_{i=1,2,3} fl_i e_i^{tr} + fl^{ho}(\mathbf{r}), \\ fl^{ho}(\mathbf{r}) = fl^{ho,s_1}(\mathbf{r}) + fl^{ho,s_2}(\mathbf{r}) + fl^{ho,s_3}(\mathbf{r}) + fl^{cent}(\mathbf{r}), \\ fl^{ho,s_1}(\mathbf{r}) = fl_{s_1 s_1, i+\frac{1}{2}} b^{tr,s_1,s_1}(\mathbf{r}) + fl_{s_1 s_1 s_1, i+\frac{1}{2}} b^{tr,s_1,s_1,s_1}(\mathbf{r}), \\ fl^{cent}(\mathbf{r}) = fl_i^{cent} b^{cent}, \end{cases} \tag{23}$$

where $fl^{ho,s_2}(\mathbf{r})$ and $fl^{ho,s_3}(\mathbf{r})$ are computed analog to $fl^{ho,s_1}(\mathbf{r})$. In the case of continuous functions and the example of the regular grid (Fig. 1c), the amplitudes of the first equation of Eq. (23) are used for six triangles, those of the second and third equations in two and of the fourth in one triangle.

The basis functions e_i^{tr} of Eq. (23) are given for the example of $e_1^{tr}(\mathbf{r})$ being the linear function which is 1 at $\mathbf{r}_{i,j}$ and 0 at $\mathbf{r}_{i+1,j}$ and $\mathbf{r}_{i,j+1}$. To obtain the expression of $e_1^{tr}(\mathbf{r})$, we define the normal vector firstly:

$$\mathbf{n}_1 = \left(\frac{\mathbf{r}_{i+1,j} - \mathbf{r}_{i,j+1}}{|\mathbf{r}_{i+1,j} - \mathbf{r}_{i,j+1}|} \right)^{\perp}, \tag{24}$$

with the symbol $\perp$ means the normal orthogonal vector and the direction of $\mathbf{r}$ is from $(\mathbf{r}_{i+1,j} - \mathbf{r}_{i,j+1})$ to $\mathbf{r}_{i,j}$. We define

$$\begin{cases} e_1^{tr}(\mathbf{r}) = \dfrac{(\mathbf{r} - \mathbf{r}_{i+1,j}, \mathbf{n}_{i,j})}{|(\mathbf{r}_{i,j} - \mathbf{r}_{i+1,j}, \mathbf{n}_{i,j})|} & \text{for } \mathbf{r} \in \Omega_{i,j}^{tr1}, \\ e_1^{tr}(\mathbf{r}) = 0, & \text{otherwise}, \end{cases} \tag{25}$$

where $\Omega_{i,j}^{tr1}$ is the triangle with corner points $\mathbf{r}_{i,j}, \mathbf{r}_{i+1,j}, \mathbf{r}_{i,j+1}$. $e_2^{tr}(\mathbf{r})$ and $e_3^{tr}(\mathbf{r})$ are defined in analogy to Eq. (25).

According to Eq. (23), each triangle $\mathbf{r}_1, \mathbf{r}_2, \mathbf{r}_3$ or in the grid $\mathbf{r}_{i,j}, \mathbf{r}_{i+1,j}, \mathbf{r}_{i,j+1}$ has 10 associated basis functions and corresponding amplitudes. The basis functions are shown in Fig. 2. The linear basis functions e_1^{tr}, e_2^{tr}, and e_3^{tr} are given, Eq. (25), and shown in Fig. 2d–f. The corresponding amplitudes are the corner values of $fl(\mathbf{r}_k), k \in (1, 2, 3)$. The corner values define fl^{lin} in Eq. (22). We use the cascading interpolation defined above such that we subtract fl^{lin} from the given third-order representation of fl. After this subtraction, we have a third-degree polynomial being 0 at corners. This means that the third-degree polynomial restricted to the side s_1 is a 1D polynomial in s_1 being 0 for $s_1 = 0$ and $s_1 = 1$. To represent such functions, the 1D basis function described in Chapter "Local-Galerkin Schemes in 1D" can be used. It follows that on the side s_1 the second and third derivatives of fl in the direction s_1 are amplitudes. The same is true for s_2 and s_3 which means that for fl^{ho} we have 6 edge related amplitudes. Again using the cascading interpolation, we subtract the fields corresponding to the edge related amplitudes from fl, and taking this function at the center point $\mathbf{r}^{cent}$, we get the amplitude for the center basis function. The triangular center basis function is 0 on all edges. While the corner related basis functions are defined in Eq. (25), the 6 edge related basis functions and the center basis function will be given in the following. According to Eqs. (22) and (23), there are 2 dynamic amplitudes associated with each of the three triangular edges. The coordinates along these edges are called s_1, s_2, and s_3. The edge line $\mathbf{r}_2 - \mathbf{r}_3$ is associated with the coordinate s_1 and there are two amplitudes associated with this edge, which are the second and third directional derivatives to s_1. The corresponding edge related basis functions are $b^{tr,s_1,2}(\mathbf{r})$ and $b^{tr,s_1,3}(\mathbf{r})$. The basis functions $b^{tr,s_k,k'}(k \in 1, 2, 3, k' \in 2, 3)$ are analog to the second- and third-order b^2 and b^3 basis functions in one dimension being described in Chapter "Local-Galerkin Schemes in 1D". The index s_k indicates the side of the triangle, where here we consider the side s_1 to be opposite to the corner $\mathbf{r}_1$. The second upper index k' is the order of the basis function. The amplitudes belonging to these basis functions in one dimension were second and third derivatives. In the triangular case the amplitudes are second- and third directional derivatives in the direction of the side. To define $b^{tr,s_1,2}(\mathbf{r})$ we assume that the s_1 coordinate is along the line defined by $\mathbf{r}_2 - \mathbf{r}_3$:

$$b^{tr,s_1,2}(\mathbf{r}) = \frac{e^{tr_2}(\mathbf{r})e^{tr_3}(\mathbf{r})}{b^{2norm}}, \tag{26}$$

where b^{2norm} is to be determined such that the second directional derivative along edge $\mathbf{r}_2 - \mathbf{r}_3$ is 1. The definition of $b^{tr,s_2,2}(\mathbf{r})$ and $b^{tr,s_3,2}(\mathbf{r})$ is done in analogy to Eq. (26).

To define $b^{tr,s_1,3}(\mathbf{r})$, we define $\mathbf{r}_m = \frac{1}{2}(\mathbf{r}_2 + \mathbf{r}_3)$ and define $\mathbf{n}_3$ to be the orthogonal normal vector of $\mathbf{r}_1 - \mathbf{r}_m$. Then we define

$$b^{tr,s_1,3}(\mathbf{r}) = b^{tr,s_1,2}(\mathbf{r}) \frac{(\mathbf{r}, \mathbf{r}_m)}{b^{3norm}}, \tag{27}$$

with b^{3norm} being chosen such that $b_{s_1 s_1}^{tr,s_1,3}(\mathbf{r}_m) = 1$ which is the directional third derivative of $b^{tr,s_1,3}$ along the assumed base line of the triangle where s' is the coordinate along side 1, which above was named s_1.

For the center basis functions h^{cent}, we define

$$b^{cent}(\mathbf{r}) = \frac{e_1^{tr}(\mathbf{r})e_2^{tr}(\mathbf{r})e_3^{tr}(\mathbf{r})}{e_1^{tr}(\mathbf{r}^c)e_2^{tr}(\mathbf{r}^c)e_3^{tr}(\mathbf{r}^c)}, \tag{28}$$

where $\mathbf{r}^c$ is the point chosen for center collocation. $\mathbf{r}^c$ can be any point in the interior of the triangle, such as the triangle's center of gravity. For the special case of rectangular triangles, the basis functions are plotted in Fig. 2.

Above the basis functions are given in a coordinate freeway, just using vector operations on the corner points. The triangular basis functions can alternatively be defined by local coordinates. Let the triangular cell, as shown in Fig. 3e, have corners A, B, C and sides a, b, c where side a is opposite to corner A, etc. So side

a is between corners B and C, etc. We already know that from the triangular basis functions three are associated with the corners, two with each of sides a, b, c and one at the center. This leads to $3 + 3 \times 2 + 1 = 10$ basis functions, corresponding to the number of collocation points of a triangle. A local coordinate is associated with each corner and will lead to the definition of the corner related basis function $e^A(\mathbf{r})$ and two edge related basis functions $e^{A_1}(\mathbf{r})$ and $e^{A_2}(\mathbf{r})$. We will define this for the corner A and the other two cases can be done in an analog way. The triangle is shown in Fig. 3e. Apart from the triangle $\triangle_{ABC}$, the height $h = dy$, the mid-point $\mathbf{r}^m = \frac{\mathbf{r}_B + \mathbf{r}_C}{2}$, and the two unit vectors $\mathbf{n}^x = \frac{\mathbf{r}_B - \mathbf{r}_C}{|\mathbf{r}_B - \mathbf{r}_C|}$ and $\mathbf{n}^y$ are needed.

The local coordinates x'' and y'' are then defined as

$$\mathbf{r} = \mathbf{r}_m + \mathbf{n}^{x''} x'' + \mathbf{n}^{y''} y'', \text{ with } x'' \in (-dx', dx'), y'' \in (0, dy), \tag{29}$$

where $dx' = \frac{1}{2} dx, dx = |\mathbf{r}_B - \mathbf{r}_C|$.

As in Section "Euclid's Lemma", we can define normalized coordinates $x' = \frac{x''}{dx}, y' = \frac{y''}{dy}$. For the linear corner function $e^A(\mathbf{r})$ belonging to corner A, we have

$$e^A(\mathbf{r}) = 1 - y', \text{ with } y' \in (0, 1). \tag{30}$$

The linear basis functions corresponding to the other corners are obtained in an analog way. Above we have obtained the higher-order basis functions as products of first-order basis functions. For example, $b^{s_{1,2}}(\mathbf{r}) = e^B(\mathbf{r})e^C(\mathbf{r})$. In the same way, Eq. (28) can be used to obtain the higher-order basis functions. The different first-order basis functions used as factors in the definition of the high-order basis functions may be in different coordinates.

The transformation between grid point space and the standard spectral space is not difficult. For the grid point space, we refer to Section "Rhomboidal Basis Functions and Sparse Grids for the Regular Grid Case". The standard spectral space consists of the corner amplitudes at corner points, which are both spectral and grid point amplitudes and are not transformed. The six remaining edge spectral amplitudes consist of second and third derivatives in the direction of the corresponding edge. As indicated before, this transformation can be done as in 1D. 1D problems can be extracted from the 2D problem. Note that this operation is not dependent on a regular grid. The transformation of the edge amplitudes by 1D extraction can be done for regular grids. The transformation from spectral to grid point space can be done by inserting the collocation points into Eq. (22).

Note that not all collocation points must be dynamic points, meaning that not all such points are transformed. If a fully triangular cell structure is used, all points are used as dynamic points and involved in the spectral/grid point transformation. However, if other polygonal cell shapes, such as quadrilaterals, pentagons, or hexagons, are used, interpolation is done. Some of the collocation points are created diagnostically, such as by interpolation. In this case only a subset of collocation points is involved in transformations and there are fewer spectral amplitudes. The standard third-order spectral representation is only one possibility. There is the possibility to transform the basis functions and in this way describe the same field with different amplitudes. There are many ways to treat transformation to such spaces. The most obvious method is to do the transformation in steps, by first transforming to the standard spectral space and then to other spaces, such as those using one-sided derivatives at corners as amplitudes.

Triangular Basis Functions for the Rectangular Case

Section "Triangular Basis Functions and Full Grids" gave a coordinate free definition of triangular basis functions. For illustration here we give the definition for the special case of rectangular coordinates. Local coordinates x', y' are used and these coordinate lines are assumed to enclose the right angle of the triangle.

The special case of a rectangular is of practical importance, as such cells occur when an external boundary is introduced in a rectangular grid. By introducing a diagnostic diagonal edge, any quadrilateral cell can be divided into two rectangular triangles.

The triangle is shown in Fig. 2a. The 3 corners are called 1, 2, 3 and the opposite sides, which are called edges, are named a, b, c. We call a and b the cell lengths, $a = dx$ and $b = dy$. We introduce the coordinates $x'' = \frac{x'}{dx}$ and $y'' = \frac{y'}{dy}$, and after this transformation we have triangles of side lengths $a = b = 1$. In the following we will call the transformed coordinates again x and y.

The sides a, b, c are defined by the vector $\mathbf{r} = (x, y)$ as

$$\begin{cases} a = (x'', 0) = (s_1, 0) \\ b = (y'', 0) = (s_2, 0), \\ c = (x'', 1 - x'') = (s_3, 1 - s_3), \end{cases} \tag{31}$$

where the sides a, b, c are opposite to the points $\mathbf{r}_1, \mathbf{r}_2, \mathbf{r}_3$. We are free to name the parameter for the sides as we want. So consistent with terminology introduced before, we could in the first equation of Eq. (31) replace x'' by s_1, in the second y'' by s_2, and in the third x'' by s_3. The notation a, b, c is consistent with many elementary books on triangular geometry. It is also used in Stöcker (2008), and so the a, b, c notation is useful when comparing with formula from books. The points called here $\mathbf{r}_1, \mathbf{r}_2, \mathbf{r}_3$ in such books are often called A, B, C. The notation introduced in this section is more suitable for triangles in a Cartesian grid.

Linear basis functions are called $e_1(x, y), e_2(x, y), e_3(x, y)$. They are linear in the coordinates x, y and e_1 is 0 on Edge a and 1 at Point 1. The other 2 e-functions have analog properties.

For any point (x, y) $(x + y < 1)$ in the triangle, the linear basis functions are

$$\begin{cases} e_A(x, y) = y, \\ e_B(x, y) = x, \\ e_C(x, y) = 1 - x - y. \end{cases} \tag{32}$$

The basis functions of Eq. (32) are called corner basis functions, as the natural amplitudes are positioned at the corners 1,2,3 of the triangle (see Fig. 2a). The basis function with index $i \in 1, 2, 3$ is equal to 1 at corner point i and 0 at all other corner points.

Each triangular edge a, b, c has two basis functions $b^{s_m,2}$ and $b^{s_m,3}$ $(s_m = a, b, c)$, characterized by second and third derivatives at the edge center for differentiation along the edge. For notation we use the two corners belonging to an edge, for example, we define $a = s_1(\mathbf{r}_3 - \mathbf{r}_2)$ for second-order basis functions. Now we define

$$\begin{cases} b^{s_1,2} = N^a e_B e_C = -\frac{1}{2}(x - x^2 - xy), \\ b^{s_2,2} = N^b e_A e_C = -\frac{1}{2}(y - y^2 - xy), \\ b^{s_3,2} = N^c e_A e_B = -\frac{1}{2}xy, \end{cases} \tag{33}$$

where the arguments (x, y) of e were dropped. N^m $(m = a, b, c)$ is the norm amplitude to achieve that the basis functions associated with side a have the second derivative 1 for differentiation along this side.

Let the corners of the triangle be $\mathbf{r}_A = (0, 1), \mathbf{r}_B = (1, 0), \mathbf{r}_C = (0, 0)$. For the side $a = s_1(\mathbf{r}_B - \mathbf{r}_C)$, the associated parameter is s_1 and the points of the side a are represented by the curve

$$\mathbf{r} = \mathbf{r}_C + \frac{\mathbf{r}_B - \mathbf{r}_C}{|\mathbf{r}_B - \mathbf{r}_C|} s_1, \tag{34}$$

where $s_1 \in (0, 1)$. For the other sides s_2 and s_3, the directional definition is done in an analog way.

The representation Eq. (34) can be used to define derivatives along the edge of the functions $b^{s_m,2}$ in Eq. (33) or, in analogy, of any other function of the variables x, y. We have $b^{s_1,2}(x, y) = 0$ for (x, y) being on the edges b or c. For side a, defined by $y = 0$, the representation of $b^{s_1,2}$ is

$$b^{s_1,2}(x, 0) = N^a(x - x^2). \tag{35}$$

For the second derivative along edge a, we obtain

$$b_{xx}^{s_1,2} = 1 \tag{36}$$

for the second differentiation along side a. As we request that the mid-edge second derivative is 1, we have $N^a = -\frac{1}{2}$ and analog $N^b = -\frac{1}{2}$, $N^c = -\frac{1}{2}$. There is an alternative form for Eq. (33):

$$
\begin{cases}
b^{s_1,2} = \frac{1}{2}\left[\left(x - \frac{1-y}{2}\right)^2 - \left(\frac{1-y}{2}\right)^2\right], \\
b^{s_2,2} = \frac{1}{2}\left[\left(y - \frac{1-x}{2}\right)^2 - \left(\frac{1-x}{2}\right)^2\right], \\
b^{s_3,2} = \frac{1}{2}xy.
\end{cases}
\tag{37}
$$

Note that in Eq. (37) for a given y, the equation is valid for $x \in (1 - y, 1)$. For other x, $b^{s_1,2}$ is defined to be 0. So Eq. (37) for every y is formed in analogy to the one-dimensional case Eq. (12), where $\frac{1}{2}(1 - y)$ is the dx' from Eq. (12).

Following Eq. (37), we obtain for the third-order edge basis functions:

$$
\begin{cases}
b^{s_1,3} = \frac{1}{3}\left(x - \frac{1-y}{2}\right)b^{s_1,2}, \\
b^{s_2,3} = \frac{1}{3}\left(y - \frac{1-x}{2}\right)b^{s_2,2}, \\
b^{s_3,3} = C^{norm}(x - y)b^{s_3,2}.
\end{cases}
\tag{38}
$$

Amplitudes for the basis functions Eqs. (37) and (38) are the directional second and third derivatives along the sides a, b, c. There is a third basis function belonging to an amplitude in the middle of the triangle. The center of gravity can be chosen:

$$
b^{cent,3} = e_1 e_2 e_3 = xy(1 - x - y) = xy - x^2 y - xy^2.
\tag{39}
$$

For the performance of the L-Galerkin procedure, the integrals of the basis functions are necessary, which in the program MOW_GC are obtained by computer generated polynomial algebra. The integral is named the mass of the basis function and indicated by the prefix m. We have

$$
\begin{cases}
\int\int_{\Omega_{i,j}^\Delta} e_m dxdy = \frac{1}{6}, & \text{for } m = A, B, C, \\
\int_{\Omega_{i,j}^\Delta} b^{s_m,2} dxdy = -\frac{1}{8}, & \text{for } m = 1, 2, 3, \\
\int_{\Omega_{i,j}^\Delta} b^{s_m,3} dxdy = 0, & \text{for } m = 1, 2, 3.
\end{cases}
\tag{40}
$$

A few examples for mass values are given here

$$\begin{cases} me_1 = me_2 = me_3 = \dfrac{1}{6}, \\ mb^{s_1,3} = mb^{s_2,3} = mb^{s_3,3} = 0, \\ mb^{s_1,2} = mb^{s_2,2} = mb^{s_3,2} = -\dfrac{1}{8}. \end{cases} \tag{41}$$

The linear basis functions e according to Eq. (32) are different from 0 on two triangular sides. The second- and third-order edge basis functions Eqs. (37) and (38) are 0 at two edges and different from 0 only on the defining edge. For example, $b^{s_1,3}$ is 0 on edges b and c and different from 0 on edge a only. The definitions of basis functions were normed in such a way that on the defining edge the 1D basis function as defined in Chapter "Local-Galerkin Schemes in 1D" is obtained. For regular grids, where we have lines of grid points in x-direction, for example, we can extract straight coordinate lines in x- and y-directions and on these lines we have a 1D representation as discussed in Chapter "Local-Galerkin Schemes in 1D". The performance of time marching with the RK4 time scheme was discussed in Chapter "Simple Finite Difference Procedures". For regular $x - y$ grids we can achieve grids where the contributions from lines in x- and y-directions are added, making the transition to 2D similar simple as with non-Galerkin FD schemes.

In order to solve the dynamic equation and predict amplitudes in time, we must take spatial derivatives of fluxes. This means taking derivatives of the basis functions to x and y. For all basis functions, this will lead to second-order polynomials. This means that a spatial derivative of a basis function can be represented as a linear combination of the 3 linear basis functions $e_i (i = 1, 2, 3)$ and the quadratic basis functions $b^{s_m,2} (m \in \{a, b, c\})$. Using Eqs. (32) and (33) polynomial formula for the derivatives of basis functions can easily be taken and here only some examples are given:

$$\begin{cases} e_{3,x} = -1, \\ e_{3,y} = -1, \\ b_x^{s_3,2} = y. \end{cases} \tag{42}$$

A complete list of basis function derivatives is given in Tables 2 and 3.

For the representation of a field $h(x, y) = h(\mathbf{r})$, the amplitudes for a basis functions representation in a triangle include:

1. h_i: the values at corner points
2. $h_{xx}, h_{yy}, h_{s's'}$: the second directional derivatives in the directions of the edges a, b, c
3. $h_{xxx}, h_{yyy}, h_{s's's'}$: the third directional derivatives in the directions of the edges a, b, c
4. h_{cent}: the value at a center point

Table 2 Linear derivative of basis functions

Basis Functions	$\frac{\partial}{\partial x}$	$\frac{\partial}{\partial y}$
$e_1 = y$	0	1
$e_2 = x$	1	0
$e_3 = 1 - x - y$	-1	-1
$b^{s_1,2} = -\frac{1}{2}x(1 - x - y)$	$-\frac{1}{2}(1 - 2x - y)$	$\frac{1}{2}x$
$b^{s_2,2} = -\frac{1}{2}y(1 - y - x)$	$\frac{1}{2}y$	$-\frac{1}{2}(1 - x - 2y)$
$b^{s_3,2} = xy$	y	x
$b^{s_1,3} = \frac{1}{6}x(1 - x - y)(1 - y)$	$\frac{1}{6}(1 - 2x - 2y + y^2 + 2xy)$	$\frac{1}{6}(x^2 - 2x + 2xy)$
$b^{s_2,3} = \frac{1}{6}y(1 - x - y)(1 - x)$	$\frac{1}{6}(y^2 - 2y + 2xy)$	$\frac{1}{6}(1 - 2x - 2y + x^2 + 2xy)$
$b^{s_3,3} = xy(x - y)$		
$b^{cent,3} = xy(1 - x - y)$	$y - 2xy - y^2$	$x - 2xy - x^2$

Table 3 High-order derivatives of basis functions

Basis functions	$\frac{\partial^2}{\partial x^2}$	$\frac{\partial^2}{\partial x \partial y}$	$\frac{\partial^2}{\partial y^2}$	$\frac{\partial^3}{\partial x^3}$	$\frac{\partial^3}{\partial x^2 \partial y}$	$\frac{\partial^3}{\partial x \partial y^2}$	$\frac{\partial^3}{\partial y^3}$
e_1	0	0	0	0	0	0	0
e_2	0	0	0	0	0	0	0
e_3	0	0	0	0	0	0	0
$b^{s_1,2}$	1	$\frac{1}{2}$	0	0	0	0	0
$b^{s_2,2}$	0	$\frac{1}{2}$	1	0	0	0	0
$b^{s_3,2}$	0	1	0	0	0	0	0
$b^{s_1,3}$	$\frac{1}{3}(y - 1)$	$\frac{1}{3}(x + y - 1)$	$\frac{1}{3}x$	0	$\frac{1}{3}$	$\frac{1}{3}$	0
$b^{s_2,3}$	$\frac{1}{3}y$	$\frac{1}{3}(x + y - 1)$	$\frac{1}{3}(x - 1)$	0	$\frac{1}{3}$	$\frac{1}{3}$	0
$b^{cent,3}$	$-2y$	$1 - 2x - 2y$	$-2x$	0	-2	-2	0

So the representation of h is the sum of a first-order group, involving the basis functions e_i, the second-order group involving b_m^2, the third-order edge group involving b_m^3, and the center point correction:

$$h^c(\mathbf{r}) = \sum_i h_i e_i(\mathbf{r}) + \sum_i \left\{ \sum_{m=1,2,3} \left[h_{s_m s_m, i+\frac{1}{2}} b^{s_m,2}(\mathbf{r}) + h_{s_m s_m s_m, i+\frac{1}{2}} b^{s_m,3}(\mathbf{r}) \right] \right\}.$$
(43)

The index i is going over all triangles. In Eq. (43), $h^c(\mathbf{r})$ is the third-order representation of h without the contribution of the center basis function b_{cent}^3. $h_{s's',i+\frac{1}{2}}$ and $h_{s's's',i+\frac{1}{2}}$ are second- and third-order directional derivatives along the side m. The directional derivative will be introduced in general terms in Section "Nonconserving Schemes for Full Grids" in Chapter "Finite Difference Schemes on Sparse and Full Grids". For purposes of triangular representation, we assume the points on edge m are derived by the edge coordinate $t' \in (0, 1)$, with $\mathbf{r}(t') = \mathbf{r} + t'\mathbf{n}$

with **n** being the unit vector of the direction where the derivative is taken, such that all points of m are obtained by some value of t'. Then the directional derivative is $\frac{\partial h_{\mathbf{r}(t')}(t')}{\partial t'}$.

To define the amplitude of the center basis function, choose any point $\mathbf{r}_{cent}$ in the interior of the triangle. This means that $\mathbf{r}_{cent}$ is not a corner and not an edge point. $\mathbf{r}_{cent}$ could be the center of gravity of the triangle. Any other point can be chosen, where the center basis function is not 0. Then define the center excess δh as

$$\delta h = h(\mathbf{r}_{cent}) - h^c(\mathbf{r}_{cent}). \tag{44}$$

For the representation of h, we then obtain

$$h(\mathbf{r}) = h^c(\mathbf{r}) + \frac{\delta h}{b^3_{cent}(\mathbf{r}_{cent})} b^3_{cent}(\mathbf{r}). \tag{45}$$

If we have a triangular cell structure, each triangle has a representation according to Eqs. (43) and (45). The corner or edge points belong to more than one triangle. Edge points and edge amplitudes belong to 2 triangles, the center point to one and corner points, depending on the grid, may belong to many cells. Chapter "Finite Difference Schemes on Sparse and Full Grids" will give simple examples.

The Corner Derivative Representation

In 1D schemes being more regular than just continuous were described in Sections "The L-Galerkin scheme: o3o5C^1C^2 and The L-Galerkin Scheme: o4o5C^1C^2" in Chapter "Local-Galerkin Schemes in 1D". In this section we bring the scheme with increased high regularity to two dimensions.

We have seen in Section "The L-Galerkin scheme: o3o5C^1C^2" in Chapter "Local-Galerkin Schemes in 1D" that even in 1D there may be sparseness. Sparseness is obtained by providing the approximating function system with a property which the final solution is supposed to have. In this case it is smoothness of the spline, meaning differentiability. We must require smoothness of the solution because many approximations are based on Taylor series expansion, which is only obtained for smooth fields in reasonable accuracy. The alternative to obtain accuracy with discontinuous representations is followed with DG methods, an approach not investigated in this book.

So sparseness means that a reduced function space for the dynamic field is constructed such that fewer amplitudes can be represented, but the useful resolution remains the same. For serendipity grids in two or three dimensions, high-order polynomials are eliminated which would not contribute to the performance of the approximation according to the approximation order. Another possibility considered here is to require differentiability when using approximations which would not be valid with non-differentiable fields. The latter would in principle allow for schemes nearly having the whole wavenumber range as useful resolution. For the fast meteorological waves, schemes having a positive group velocity for all wavenumbers exist, but so far such schemes have not been found for the advection process. This subject is not sufficiently investigated.

We investigate function systems being more regular than just continuous and we go for such cases to 2D. In this chapter, we are considering function systems which are continuous and differentiable at some points. In 2D this not necessarily implies that the approximative function system is differentiable at all points. In view of the huge number of possibilities, just one example will be supplied.

Differentiable fields representation by splines means the existence of a derivative at corners. In this work, we limit ourselves to differentiable representations at corner points and do not require the existence of a derivative vertical to edges.

Many of the methods to be applied will evolve in a straightforward way from 1D. This makes the 2D application easier. For the example presented, we require differentiability at corner points, not at edges. Edges are 1D sections within the 2D computational area. Therefore, we can always extract 1D sections and apply the results of the sections obtained before. This includes the standard spectral representation as long as we are considering straight lines of grid points. For the standard spectral third-order representation, we can go to the representations using one-sided derivatives at corner points. As a first step, we introduce one-sided derivatives as amplitudes. This was already explained in one dimension and in two dimensions is applied for each of the edged separately.

In general, these one-sided derivatives may be different for all coordinate lines meeting at a corner point. Arbitrarily many lines may meet at such a point. The examples in Fig. 1 show grids where between three and eight edges meet at a corner. The introduction of one-sided derivatives is done for each triangular cell separately. Here we consider one cell and at this point do not consider cellular systems and their associated grids. A triangle contains 10 amplitudes, and in this section we are not worrying about some of the amplitudes belonging to other triangles as well. According to Section "Triangular Basis Functions for the Rectangular Case", the amplitudes are $h_i (i = 1, 2, 3), h_{ss,m}$, and $h_{sss,m}(m = a, b, c)$ and h_{cent} where s represents the direction x, y or the diagonal. The aim of this subsection is to replace the 6 amplitudes $h_{ss,m}, h_{sss,m}$ by other dynamic variables, which are the one-sided directional derivatives $h_{s,i(p)}$ and $h_{s,i(q)}$, where p and q are the two edges meeting at point $i(i = 1, 2, 3)$. If we have the derivatives in two directions p and q, the directional one-sided derivatives in all other directions at the given point are obtained by the rules of differentiation. Each point allows differentiation in all directions of $360°$. We can require the existence of one-sided derivatives just for some sections of the $360°$.

In order to obtain the transformation between the $h_{s,i(p)}, h_{s,i(q)}(i = 1, 2, 3)$ and $h_{ss,m}$ and $h_{sss,m}(m = a, b, c)$, we can derive this transformation for each edge m separately. Then according to Fig. 2, we see that on edge a there are the corners 2 and 3. Figure 2 treats the case of rectangular triangles, where side a has the coordinate x and side $b y$. The variable s along the edge a is $s = x$. We have the amplitudes

$h_2, h_3, h_{xx,a}, h_{xxx,a}$ to be transformed into $h_2, h_3, h_{x,2(c)}, h_{x,3(b)}, h_{x,2(a)}, h_{x,3(a)}$. h_1, h_2 and h_3 are not transformed. So for the transformation of these variables, we have a 1D problem and the 1D basis functions $e_2^+(x), e_3^-(x), b_{\frac{1}{2}}^2(x)$ and $b_{\frac{1}{2}}^3(x)$. We have thus created a 1D problem for the variable $x \in (0, 1)$. Without loss of generality, we have chosen $dx = 1$. This is done as an example. The other directions in Fig. 2a and the general case that none of the edges meeting at point r_2 coincide with the x-direction can be treated in an analog way. On this line we have two equivalent representations for a function h. The end points of this 1D functions are given by r_2 and r_3 (see Fig. 2a), the coordinate for this special case is x, and the mid-point is $\frac{1}{2}$.

The standard representation of h is by amplitudes h_2 and h_3 and high-order derivatives $h_{xx,\frac{1}{2}}$ and $h_{xxx,\frac{1}{2}}$. The basis functions for this representation are e_2^+ and e_3^-, $b^{2,+}$ and $b^{3,-}$. The one-sided derivative representation for the same interval uses amplitudes h_2, h_3, h_x^+, h_x^- with basis functions $e_2^+, e_3^-, b^{2,+}, b^{3,-}$. b^+ and b^- are defined in Eq. (18) in Chapter "Local-Galerkin Schemes in 1D". The transformation between these variables follows from Section "Functional Representations, Amplitudes, and Basis Functions" in Chapter "Local-Galerkin Schemes in 1D". h_2 and h_3 are not transformed.

The transformation equations between the two representations follow from Section "Functional Representations, Amplitudes, and Basis Functions" in Chapter "Local-Galerkin Schemes in 1D":

$$\begin{cases} h_{xx,\frac{1}{2}} b_x^2(r_2) + h_{xxx,\frac{1}{2}} b_x^3(r_2) = h_{x,2}^{ho}, \\ h_{xx,\frac{1}{2}} b_x^2(r_3) + h_{xxx,\frac{1}{2}} b_x^3(r_3) = h_{x,3}^{ho}, \end{cases} \tag{46}$$

where h_x^{ho} means the derivative of the high-order part of h,

$$h_x^{ho}(r_3) = h_x(r_3) - (h_3 - h_2), \tag{47}$$

with $dx = 1$. Equation (46) is the transformation from the amplitudes $h_{xx,\frac{1}{2}}$ and $h_{xxx,\frac{1}{2}}$ to the one-sided derivatives $h_{x,3}, h_{x,2}$. If the one-sided derivatives $h_{x,3}$ and $h_{x,2}$ are given, to obtain the back transformation, Eq. (46) must be solved for $h_{xx,\frac{1}{2}}, h_{xxx,\frac{1}{2}}$:

$$\begin{cases} h_{xx,\frac{1}{2}} = \dfrac{1}{2b_x^2(dx')}(h_{x,3} - h_{x,2}), \\ h_{xxx,\frac{1}{2}} = \dfrac{1}{2b_x^3(dx')}(h_{x,3} + h_{x,2}), \end{cases} \tag{48}$$

where $dx' = \frac{1}{2}$. The solution of Eq. (46) is simplified by observing $b_x^2(\frac{1}{2}) = -b_x^2(-\frac{1}{2})$ and $b^3(\frac{1}{2}) = b^3(-\frac{1}{2})$. If the grid cells are obtained from straight lines, such as for example shown in Fig. 2b, a 1D set of amplitudes can be extracted and the transformations, Eq. (46), are 1D operations on the extracted set. The above equations show that the transformation is also simple when we do not have straight grid lines, as with unstructured meshes.

Now consider that the triangle $\Delta_{1,2,3}$ is part of a cell structure with many neighboring triangular cells. One advantage of this representation of h is that differentiable functions can easily be represented by adjusting the one-sided derivatives. This will be exploited in Section "An Example of a Regularization Operator" in Chapter "Local-Galerkin Schemes in 1D", but the basic idea is the following:

If the approximative function would be differentiable at this point, all first one-sided derivatives could be computed from two data which are the derivatives in two directions, which must not necessarily be the directions of a grid line. This determination is such that the derivative is continuous, when moving to any direction. With the spline approximations considered here, the derivative may be continuous when moving to some directions and not in others. Note that first derivatives are always used to define the regularity of splines. We consider C^1 and no higher. Second derivatives occur only in the middle of edges, where the fields are polynomials and all derivatives exist. Now let us consider the approximations at corner point $\mathbf{r}$ and let $\mathbf{n}^q, q = 1, 2, \ldots i_e$ be the direction for taking a one-sided derivative at the corner point. To define directions, the $\mathbf{n}^q$ are assumed to be unit vectors:

$$\mathbf{n}^q = (cos\phi_q, sin\phi_q),\qquad(49)$$

where q is the direction of the one-sided derivative. Note that the derivatives for two directions p and q are necessary to determine the derivative in any given direction. So we assume that the directional information is given with respect to two orthogonal vectors $\mathbf{n}^x$ and $\mathbf{n}^y$, which are the coordinate direction in the coordinates x and y. Now assume that the approximative function is differentiable and the two directional derivatives of the field h in x- and y-directions are $dh_{\mathbf{n}^x}$ and $dh_{\mathbf{n}^y}$. Then differentiability means that for the one-sided derivative in direction n^i, we obtain

$$\frac{\partial h}{\partial \mathbf{n}^q} = cos\phi_q dh_{\mathbf{n}^x} + sin\phi_q dh_{\mathbf{n}^y}.\qquad(50)$$

Note that if Eq. (50) is satisfied for every vector $\mathbf{n}^q$, h is differentiable. Though we may assume that Eq. (50) is valid for some edge line, we may not require this for other lines. Let us call the use of Eq. (50) "quasi-differentiability", as the differentiability is given only for some directions. This means that if a number of edges meet at a corner, differentiability according to Eq. (50) must not necessarily be assumed for all such edges.

When assuming full differentiability, it is possible to assume Eq. (50) to be valid for all edge directions. For example, when defining hexagons, it is possible not to assume the existence of all directional derivatives for the center point (see Chapter "Full and Sparse Hexagonal Grids in the Plane"). However, Eq. (50) can be required for all edge lines. If this is done, there may be the need to apply Eq. (50) also for other auxiliary lines. For example, when creating an interpolation for square cells, one may want to use only amplitudes on the corners and edges, in order to exploit sparseness. Then the missing amplitudes due to sparseness must be created by interpolation. Linear interpolation can be achieved by connecting the corner points and linearly interpolating the corner amplitudes. Exploiting quasi-differentiability according to Eq. (50) will allow to compute derivatives along the lines and this will enable to bring the interpolation to order 3. This may be explained in more detail:

Let P_1 and P_2 be two corner points within the discretization area. Let s be the coordinate along $P_1 P_2$ and we have $s \in (0, 1)$. Let the two derivatives $dh_{\mathbf{n}^x, P_1}$ and $dh_{\mathbf{n}^y, P_1}$ and $dh_{\mathbf{n}^x, P_2}$ and $dh_{\mathbf{n}^y, P_2}$ in Eq. (50) be given for both points. We also assume in Eq. (50) $q = arccos \frac{(\mathbf{n}^x, \mathbf{r}_2 - \mathbf{r}_1)}{|\mathbf{r}_1 - \mathbf{r}_2|}$. Then Eq. (50) can be used to define directional derivatives at P_1 and P_2. On $P_1 P_2$, h is a function of s whose derivatives at the two corner points are given. This means that 1D formula from Chapter "Local-Galerkin Schemes in 1D" can be used to compute h_{ss} and h_{sss} which is done for all edges of the system. Using this formula Eq. (46) of Section "Triangular Basis Functions and Full Grids" in Chapter "Local-Galerkin Schemes in 1D" for triangular cells and Section "Euclid's Lemma" in Chapter "Local-Galerkin Schemes in 1D" for quadrilateral cells, we can use the third and second derivatives to compute the fields within the cells. In order to illustrate this theory, we set an example in Section "An Irregular Structured Quadrilateral Grid with Triangular Cells" in Chapter "Local-Galerkin Schemes in 1D" for irregular quadrilateral quasi-structured grids, such as shown in Fig. 1f.

So far we have reformulated the center edge amplitudes by one-sided derivatives. The center edge second and third derivatives can be replaced by one-sided derivatives at interval ends. Any method based on high-order edge amplitudes h_{xx} and h_{xxx} can be reformulated using one-sided derivatives as amplitudes and the two formulations would be equivalent. They are just arithmetic alternatives of the same scheme. This is for the formulation of the field h. However, if the fluxes in different cells are derived using Eq. (50), we obtain fluxes being differentiable in the chosen directions.

If the derivatives of h at corner points are changed to obtain a differentiable basis function representation, the method changes fundamentally. We obtain a 2D version of the o3o3$C^1 C^1$ scheme (see Section "The L-Galerkin Scheme: o4o5$C^1 C^2$" in Chapter "Local-Galerkin Schemes in 1D").

When the aim is to create differentiable versions of fluxes, any method can be used to approximate a derivative at corner points. In two dimensions, this must be done for two directions p and q. If this approximation of a continuous function by a differentiable function is desired for h, we must take care to observe mass conservation (or conservation of other linear quantities). Such mass conserving operations are called regularization operators. An example will be given in Section "An Example of a Regularization Operator".

An Irregular Structured Quadrilateral Grid with Triangular Cells

As seen in Section "Triangular Basis Functions and Full Grids" and Fig. 1, there are many irregular cell structures. In order to solve dynamic equations on such grids for each cell, the corresponding corners and edges and neighboring cells must be known. In multiprocessing computer environments, it must be achieved that data are distributed in such a way that message passing is sufficiently small. In this respect for the truly irregular cell structure, a specialized data management scheme will be used by many modelers.

As the use of unstructured programming systems can be difficult for a single researcher with just a PC, a more simple system may be useful. Irregular cell structures occur also with structured grids. In this section a structured program for data storage is given. This is supposed to be a simple accessible test problem for irregular grids. The purpose is that without learning a complicated data management system, test problems of irregular resolution can be solved. The grid is shown in Fig. 1f and is structured with indices i, j which will be disturbed by a perturbation grid $(dx_{i,j}, dy_{i,j})$. Tests on regular grid can be performed.

Define first the regular grid:

$$\mathbf{r}_{i,j} = (x, y)_{i,j} = (idx, jdy).\tag{51}$$

Let $(\delta x_{i,j}, \delta y_{i,j})$ be a stochastic variable with $|\delta x_{i,j}| < 1$ and $|\delta y_{i,j}| < 1$. Then the irregular grid is defined by

$$\mathbf{r}_{i,j}^{ir} = \mathbf{r}_{i,j} + (\delta x_{i,j}, \delta y_{i,j}).\tag{52}$$

The grid is shown in Fig. 1f. The example can be used with quadrilateral cells and basis functions. Triangles are also possible by dividing the quadrilateral cells. From Fig. 2, it is obvious which amplitude belongs to a basis function. The linear basis functions e belong to corner values as amplitudes and each of them is different from 0 at only one corner. The six basis functions per cell belonging to second and third derivatives as amplitudes have their amplitudes defined at the middle of edges and these basis functions are 0 on all edges except one. Euclid's Lemma as described above can be used to construct the quadrilateral basis functions for the irregular cells. Amplitudes for a basis function can be defined at any point where the basis function is different from 0. Therefore, from Fig. 1f, field values at corner points can only be used as amplitudes with the bilinear basis functions e_1, e_2, e_3, as all other basis functions are 0 at corner points. Amplitudes defined on edges can be used with the second and third basis functions, where each of them is different from 0 at just 2 edges. The center basis function is 0 on all edges and corers and therefore needs an amplitude defined in the inner of the triangle.

If a basis function is obtained by dividing the quadrilateral cell into 2 triangles, there are three sets of points for each triangle which alternatively can define h as a third-order polynomial inside this triangle. As a third-order polynomial has 10 polynomial coefficients, there are 10 amplitudes for each triangle. The first set of amplitudes is the standard spectral representation. It consists of 3 grid point values at corner points (black in Fig. 2c), 3 second and 3 third derivatives along the 3 edges of a triangle (large white points in Fig. 2c, one triangle has 3 large white points but each has 2 amplitudes), and the center point (small white point in Fig. 2c). An alternative representation to this standard spectral representation is by collocation points (3 corner points (black in Fig. 2c) and 6 inner points on edges (small white points) and the two center points). A third-order representation is by the 3 corner points (black in Fig. 2c), 6 one-sided derivatives at corner points and the center point. This is called the one-sided derivative representation. Therefore, we have 3 sets of each 10 amplitudes for each triangle. These are equivalent representations. We do not need to give the transformation equations here. Most of the transformations are 1D and the transformations have been given in Chapter "Local-Galerkin Schemes in 1D" or Section "The Corner Derivative Representation".

The two triangles belonging to the square i, j have 20 amplitudes, but the 4 amplitudes positioned on the diagonal are identical (small black and white points). So the two triangles have 16 independent amplitudes. For the collocation representation, these correspond to the full grid with 16 points in the square of grid length $\frac{1}{3}$, when the edge length of the square is 1. Again some of the amplitudes are identical between squares. We have called a set of points which cover all amplitude points uniquely a compact grid. For the collocation grid, an obvious compact grid is obtained when omitting the amplitudes on the upper and right side of the square and store these with the squares to the left and on bottom. This means that the compact grid in collocation space has $3 \times 3 = 9$ points per square. For the full representation independent from the choice of amplitudes, there is always a set of 9 amplitudes for the full representation in the compact grid. For example, in standard representation, the compact grid has 1 corner point, 2 center triangle points, and 6 second and third derivative amplitudes.

The compact grid can further be reduced. When the center points are created by interpolation and not be considered as dynamic points, the compact grid has 7 amplitudes per square. The derivatives along the diagonal of the square can be computed using the derivatives ending at the same corner point. This would bring the number of points of the compact grid in a square down to 5. Note that this computation of a derivative on the diagonal does not mean that we assume differentiability at the corner point. It just means that the one-sided derivatives within the square are assumed to exist, not all derivatives along lines pointing into other triangles.

The one-sided derivative representation shown above can be used to create diagnostic edge amplitudes for situations where many edges meet at a corner point. We may select 3 of the meeting edges. Each two of these determine a sector and requiring one-sided differentiability within this sector. The other edges in the sector have one-sided derivatives determined by the 2 chosen edges. Using the one-sided derivatives at the other end of the edges, the high-order amplitudes of these edges are determined diagnostically. While this already increases the sparseness of the grid, it is obvious that at each point two directional derivatives are sufficient to determine all other edge amplitudes diagnostically which potentially introduces a high sparseness into a triangular mesh. Note that a regular triangular grid has a sixfold value of one-sided derivatives which can be reduced to one value by achieving a differentiable field by introducing sparseness.

An Example of a Regularization Operator

When the fields are represented by a one times differentiable spline, the fluxes can be represented as differentiable splines. Therefore, the corner point derivatives are already known. The spline of the time derivative is continuous but not differentiable. The purpose of this section is to approximate a continuous spline by a differentiable spline with the same mass.

A reduction of the compact grid can be achieved by assuming that the fields are differentiable at corner points. This means that all 6 directional derivatives belonging to the 6 edges meeting at a corner point are derived from the derivatives in two directions p and q which may be assumed to be x and y. This means that for the 2 triangles belonging to a square, we have the corner point value and for each edge two first derivatives as amplitudes. If functions have no further regularity properties than being continuous, we have as many one-sided derivatives and amplitudes at a corner point as the number of edges meeting at the point. In the case that 6 edges are meeting at a point, functions being just continuous have as amplitudes the corner point amplitude and 6 directional derivatives for the 6 edges meeting at this point. For the representation of a one time differentiable function apart from the corner amplitudes, we need two directional derivatives in two directions. These directions may be chosen as x and y. All other directional derivatives can be computed when assuming that the function is differentiable at this point.

For the construction of L-Galerkin operations, it is useful to have an operator which approximates a continuous function by a differentiable function. As we have seen, in the grid of Fig. 2b, a continuous function, in addition to the corner grid point values, is determined by 6 one-sided derivatives of the edge lines meeting at the corner point. The differentiable function needs only derivatives in x- and y-directions to be defined.

For simplicity, we define this operation just in one dimension, as the principle can be seen in 1D. Let a continuous function be given in one-sided derivative representation by grid point values h_i and two one-sided derivatives $h_{x,i}^+$ and $h_{x,i}^-$,

Define the derivative excess as

$$h_{x,i}'^{+} = h_{x,i}^{+} - \frac{h_{i+1} - h_i}{dx_{i+\frac{1}{2}}}. \tag{53}$$

Define $h_{x,i}'^{-}$ analog.

Then we have the representation using the one-sided high-order basis functions of (*.**)

$$h(x) = \sum_i h_i e_i + h'^{+}_{x,i} b^{+}(x) + h'^{-}_{x,i} b^{-}(x). \tag{54}$$

Using Eq. (54), we associate the following mass m_i with grid point i:

$$m_i = h_i + mb^+ + mb^-, \tag{55}$$

with $mb^+ = \int_{\Omega_i} b_i^+(x)dx$, $mb^- = \int_{\Omega_i} b_i^-(x)dx$.

Define now the continuous derivative $h'_{x,i}$ as $h'_{x,i} = w_1 \dots$ fourth-order differencing. Then we can define

$$h'_i = \frac{1}{1 + h'_{x,i}(mb^+ + mb^-)} m_i. \tag{56}$$

Finite Difference Schemes on Sparse and Full Grids

Abstract The notion of FDs on sparse grids is introduced. It means that from a regular distribution of grid points not all are used, which offers the opportunity to make models more efficient for the same resolution. This section aims at transferring some of the 1D schemes defined in Chapter "Local-Galerkin Schemes in 1D" to two dimensions. There is no way the most general L-Galerkin scheme or a class of such schemes can be presented. The number of possibilities is too large. So just examples are presented to show how the schemes work. In particular, most examples are 2D. While the complexity of computer programs normally increases substantially when going to 3D, it is often obvious how to proceed from 2D to 3D.

Keywords Spline differentiation · Conserving schemes · Three leg stencil · Baumgardner's method · Sparse L-Galerkin schemes · Staggered difference schemes · Cut cells

This chapter deals with 2D sparse FD schemes, such as the o3o3 scheme and non-sparse control systems. It is to be appreciated that o2o3 can be brought in a similar way to two dimensions, and there are many variants of these schemes. If grid points are on a line, classical FD schemes can directly be applied to do time prediction. As regular grids do not exist on the sphere, methods suitable for an irregular resolution are needed. If cells are based on the division of quadrilaterals, as shown in Fig. 1 in Chapter "Introduction", it is possible to choose the grids using two sets of great circles. Full adaptivity requires an arbitrary distribution of cellular node points. Both situations will be treated in the following. The alignment of points on straight lines or great circles leads to the direct applicability of 1D schemes in 2D and 3D and will in general be faster on computers. Sparse grids need deviations from classical differencing. An important point is conservation. If conservation laws are integrated, approximations are requested to conserve the same quantities. Some high-order FD schemes do not conserve, and L-Galerkin schemes were introduced to combine a high approximation order with conservation.

A word of warning in the beginning may be in order. To create a FD scheme on a sparse grid, one method to proceed is obvious. One could interpolate the fields from

J. Steppeler, J. Li, *Mathematics of the Weather*, Springer Atmospheric Sciences,
https://doi.org/10.1007/978-3-031-07238-3_5

the dynamic points to the full grid and then apply normal differencing at the dynamic points. Unfortunately, in general, this method is not stable. This has been tried for advection on a regular square grid using serendipity interpolation. The method was unstable for all time steps. Other methods investigated in the following will not use any interpolations to the unused points.

In Chapters "Local-Galerkin Schemes in 1D" and "2D Basis Functions for Triangular and Rectangular Meshes", tools were provided to define a large number of schemes. Only a small fraction of the potential schemes is treated here. As is the case with the simple schemes treated in Chapter "Simple Finite Difference Procedures", the number of possibilities to formulate polygonal spline schemes based on order consistent polynomials is huge. This is in contrast to the SE scheme. When the polynomial order and the cell structure are given, the SE scheme is uniquely defined (Steppeler et al. 2008). Even the collocation grid is unique with SE. So we are giving only examples of discretizations. The presented schemes are not necessarily optimal choices within SE. We just give examples for the construction of such schemes.

Non-conserving Schemes for Full Grids

For the full grid, differentiation is done along coordinate lines, and the only unusual feature is the non-orthogonality of the grid lines, determined by the angle φ between the two coordinates. We assume the x- and y-directions to be orthogonal and denote the corresponding unit vectors by $\mathbf{n}^x, \mathbf{n}^y$, and they satisfy $\mathbf{n}^y = \mathbf{n}^{x\perp}$, where the symbol $\perp$ means the orthogonal vector. For the regular rhomboidal grid, we assume the unit vectors $\mathbf{n}^{x'}, \mathbf{n}^{y'}$ for the coordinate directions in s^x and s^y. For the special case of orthogonal grids, we have $x = s^x$ and $y = s^y$. Differentiation along coordinate lines will give directional derivatives in x' and y' in Eq. (1) in Chapter "2D Basis Functions for Triangular and Rectangular Meshes".

The directional derivative is defined for any unit vector $\mathbf{n}^s$, where s can be defined by the angle φ of $\mathbf{n}^s$ to $\mathbf{n}^x$ and $\mathbf{n}^y$: $\mathbf{n}^s = cos\varphi' \mathbf{n}^x + sin\varphi' \mathbf{n}^y$. If $\mathbf{n}^s$ is any unit vector in 2D, the directional derivative of a field h in direction $\mathbf{s}$ is defined as

$$\frac{\partial h(\mathbf{r})}{\partial \mathbf{n}^s} = \frac{\partial h(\mathbf{r} + s\mathbf{n}^s)}{\partial s}, \tag{1}$$

where s is called the directional parameter and $\mathbf{n}^s$ is the direction of differentiation (see above, $\mathbf{n}$ is defined by φ). For inner points of a cell, the derivative Eq. (1) always exists, as for inner points h is defined to be a polynomial. In the following, the directional derivative will often be used at corner points, where its existence must be postulated and may be different for different cells belonging to the corner. We will assume that Eq. (1) exists as a one-sided derivative where Eq. (1) may assume multiple values, representing discontinuities of the derivative for the representation of non-differentiable functions. Such different directional derivatives occur when different edges meet at a corner point. Now let x', y' be a non-orthogonal coordinate system with $x' = x$ and angle φ' between the two coordinate directions. We obtain

$$\frac{\partial h(\mathbf{r})}{\partial \mathbf{n}^{y'}} = cos\varphi' \frac{\partial h(\mathbf{r})}{\partial \mathbf{n}^x} + sin\varphi' \frac{\partial h(\mathbf{r})}{\partial \mathbf{n}^y}, \tag{2}$$

and

$$h_x = h_{x'}, h_y = \frac{h_{y'}}{sin\varphi'} - \frac{h_{x'}}{tan\varphi'}. \tag{3}$$

We first consider the rhomboidal cell structure as shown in Fig. 1e in Chapter "2D Basis Functions for Triangular and Rectangular Meshes". If the grid lines are straight, we can extract 1D data sets and obtain the directional derivatives in the directions $\mathbf{n}^{x'}$ and $\mathbf{n}^{y'}$. The dynamic equations are often given for orthogonal coordinates. Equation (2) can be used to obtain the derivatives in the orthogonal directions x and y. The sphere can be mapped to a plane, using the polar stereo-graphic projection (Durran 2010). Any mapping from the sphere to a plane necessarily creates singularities that are not easy to handle (Arakawa and Lamb 1977). Therefore, in this book (see Chapter "Platonic and Semi Platonic Solids" and Steppeler et al. 1990; Yamazaki and Satomura 2010; Steppeler 1993; Sadourny 1972), we map the sphere to more than one patch of nearly regular rhomboidal grid similar to that defined in Eq. (2) in Chapter "2D Basis Functions for Triangular and Rectangular Meshes". We need to combine several patches of grids, as shown in Fig. 1 in Chapter "Introduction". This matching of grids necessarily creates corner points surrounded by irregular grids, such as shown in Fig. 1 in Chapter "Introduction". Compared to the very simple proceeding considered so far, this leads to points not having coordinate lines being straight at these points where several rhomboidal patches meet. One way of proceeding used in this situation is to create a great circle line continuation of the two main coordinate directions and create the missing points by interpolation (Steppeler et al. 2008). Such points outside the area of computation are called *phantom points*, and the set of phantom points is called *the halo of the area in question*. Figure 11 in Chapter "Platonic and Semi-Platonic Solids" gives a graphic illustration of phantom points obtained by interpolation. The differencing with phantom points is given for the second-order case:

$$h_x = h_{i-1,j}^* w_{-1}^i + h_{i,j} w_0^i + h_{i+1,j} w_1^i, \tag{4}$$

where $h_{i-1,j}^*$ is an interpolated point being outside the dynamic area. It is positioned on the connection line of $\mathbf{r}_{i-1,j}$ and $\mathbf{r}_{i,j}$. The interpolation of such phantom points is possible when the continued line is in a cell, and line extrapolation $h_{i-1,j}^* = 2h_{i,j} - h_{i+1,j}$ can be used with the interpolations defined in Chapter "2D Basis Functions for Triangular and Rectangular Meshes" to create the grid point values.

Dealing with rhomboidal or cubic full grids, we have four points joining at a corner point. On the sphere, there may be exceptions at the poles (Satoh et al.

2014). The totally irregular grid shown in Fig. 1f in Chapter "2D Basis Functions for Triangular and Rectangular Meshes" may create irregularities at every point, even though this is a structured grid. For unstructured meshes, it may happen that more than four or only three points meet at a corner. The moderately irregular but structured grids according to Fig. 1f in Chapter "2D Basis Functions for Triangular and Rectangular Meshes" are simple to program irregular systems. The increased irregularities possible with unstructured grids are more difficult but accessible by methods presented here. The aim of this chapter is to provide spatial discretizations for irregular points. Non-conserving schemes use just rhomboidal or triangular cells within the icosahedral patch distribution, and the amplitudes are simply positioned at the corners. One older method is Baumgardner's Cloud method (Baumgardner and Frederickson 1985). This method selects a cloud of points in the surrounding of the target point and will be described in Section "Baumgardner's Cloud Derivative Method" .

For the triangular full grid at each point, there meet six coordinate lines. There are a number of possibilities to compute x- and y-derivatives there. The cases are trivial when there are four half-lines that are on two straight lines. We consider here the case that three lines meet. For a full grid as shown in Fig. 1c in Chapter "2D Basis Functions for Triangular and Rectangular Meshes", these three lines may be selected among the 6 edge lines meeting at each corner. Let the target point be $\mathbf{r}_0$. For an irregular scheme, we consider the case that we have selected three edges meeting at $\mathbf{r}_0$. This is the minimum stencil considered to compute finite differences in two directions. Two edges meeting would be sufficient, but then we would have one-sided derivatives, which according to Chapter "Simple Finite Difference Procedures" and Durran (2010) are not convenient for time discretization with RK4. Let the three points surrounding $\mathbf{r}_0$ be $\mathbf{r}_1$, $\mathbf{r}_2$, $\mathbf{r}_3$. Then we define the three unit vectors

$$\mathbf{n}_i = \frac{\mathbf{r}_i - \mathbf{r}_0}{|\mathbf{r}_i - \mathbf{r}_0|}, i = 1, 2, 3, \tag{5}$$

where the $\mathbf{r}_i (i = 1, 2, 3)$ are the corner points. For each line l_i, two edge points on this line are $\mathbf{r}_i$ and $\mathbf{r}_0$, which for simplicity we assume to be regularly spaced. This assumption excludes SE schemes that in this book are considered in one dimension only. Therefore, the three stencil contains ten points $\mathbf{r}_{i,k}, i = 1, 2, 3, k = 0, 1, 2, 3$, because the points $\mathbf{r}_{i,0}$ are identical to $\mathbf{r}_i$. Figure 1 is an illustration of this situation.

$$\mathbf{r}_{i,k} = \frac{k}{3} (\mathbf{r}_i - \mathbf{r}_0) . \tag{6}$$

The local line parameter is called $s_i (i = 1, 2, 3)$ and the line $\mathbf{L}_i$ includes the set of points:

$$\mathbf{L}_i = s_i (\mathbf{r}_i - \mathbf{r}_0) , s_i \in (0, 1). \tag{7}$$

On this line $\mathbf{L}_i$, the field function $fl(\mathbf{r})$ is defined as

$$fl(s_i) = fl\left[\mathbf{r}_0 + (\mathbf{r}_i - \mathbf{r}_0)s_i\right], \tag{8}$$

where the basis function representation can be used to create a sufficient number of points to be used for a third- or fourth-order difference equation at the corner points. Note that Eq. (6) defines the difference vector between two stencil points, and Eq. (7) extends this line outside the interval $(\mathbf{r}_0, \mathbf{r}_i)$.

The edge amplitudes of second order, such as $fl_{s_i,s_i}(\mathbf{r}_m)$, are obtained as done in one dimension from the request of mass conservation. The third-order derivative amplitudes, such as $fl_{s_i,s_i,s_i}(\mathbf{r}_m)$, have no mass associated with them, as was the case in one dimension. These amplitudes are obtained by FD method using the second derivative amplitudes. This procedure has so far been applied only for semi-regular grids with irregularities arising only when different patches of the grid need to be fitted together (see Section "Test of the o3o3 Scheme on the Cubed Sphere Grid Using the Shallow Water Version of the HOMME Model" in Chapter "Numerical Tests" and Steppeler et al. 2008).

Alternative Methods to Compute Derivatives

Let us assume that we want to compute the derivative at a corner point $\mathbf{r}_1$ where these points are part of a grid. We assume a minimum of three neighboring corner points $\mathbf{r}_2, \mathbf{r}_3, \mathbf{r}_4$. The methods proposed in this section are possible in any grid, where to obtain interpolations a minimum of 3 points are necessary. We assume that on the connection lines being edges $\overline{\mathbf{r}_1\mathbf{r}_2}$, $\overline{\mathbf{r}_1\mathbf{r}_3}$, $\overline{\mathbf{r}_1\mathbf{r}_4}$, more amplitudes will be given. In the example of Fig. 1, we assume that each of the edges has 2 higher-order edge points in addition to the corner points. So we have a total of 10 points to compute the derivative of h at point $\mathbf{r}_1$. Without loss of generality, we may assume that the edge $\overline{\mathbf{r}_1\mathbf{r}_2}$ is in the x-direction. This set of 10 points is called the stencil for the derivative calculation. As this stencil consists of three edges, we call it a 3-stencil or a three legged stencil. This example is enough to show the principle. The considerations to be given will be applicable to stencils with more legs. Quadrilateral grids will have stencils with 4 legs. Triangular grids will typically have a larger number of legs for the corner stencils, for example, 6 for a regular triangular grid. Other polygonal cells are obtained by treating some triangular amplitudes as diagnostic. The example of Fig. 1a shows a case where points $\mathbf{r}_1, \mathbf{r}_2, \mathbf{r}_3$ are part of a hexagonal cell.

In our example, the points on the single legs of the stencil are sufficient to compute one-sided derivatives at $\mathbf{r}_1$ in the directions of the different legs. Even if we want to compute the derivative in the x-direction, the one-sided derivative along leg $\overline{\mathbf{r}_1\mathbf{r}_2}$ may not be what we want. Many time marching schemes, including the RK4 used in this book, are unstable when combined with one-sided derivatives. Figure 1a considered ways to combine one-sided derivatives of the different legs of the stencil.

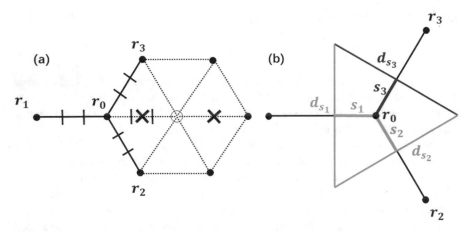

Fig. 1 The 3 legged stencil for the computation of derivatives at point $\mathbf{r}_0$. (**a**) The two points marked as × are obtained by interpolation within a cell, which for this example is assumed to be hexagonal. (**b**) The infinitesimal triangle method to average derivatives

Another way to obtain an x−derivative of h is to create phantom points. So assume 2 phantom collocation points points $\mathbf{r}_1^c$ and $\mathbf{r}_2^c$ on the prolongation of edge $\overline{\mathbf{r}_1\mathbf{r}_2}$, which are indicated by x in Fig. 1a. The interpolation should be third order. In the example of Fig. 1a, it is interpolation in a hexagonal cell, which will be treated in Chapter "Full and Sparse Hexagonal Grids in the Plane". Any other cell structure, such as triangular or quadrilateral, can be used to obtain the interpolated point. Let the edge points on leg $\overline{\mathbf{r}_1\mathbf{r}_2}$ be $\mathbf{r}_{-1}^c$ and $\mathbf{r}_{-2}^c$. Positive indices are indicated by × in Fig. 1a. We form the 1D coordinate s with grid point values s_i and obtain a 1D stencil in the coordinate s with grid point values $s_{-2}, s_{-1}, s_0, s_1, s_2$:

$$\begin{cases} s_{-2} = |\mathbf{r}_1 - \mathbf{r}_{-2}^c|, \\ s_{-1} = |\mathbf{r}_1 - \mathbf{r}_{-1}^c|, \\ s_0 = 0, \\ s_1 = |\mathbf{r}_1 - \mathbf{r}_1^c|, \\ s_2 = |\mathbf{r}_1 - \mathbf{r}_2^c|. \end{cases} \tag{9}$$

The s_i may be points where amplitudes in the grid are given or where such amplitudes are interpolated. As in this book the emphasis is on basis function methods, there is always a standard interpolation method using the basis functions. However, any other interpolation method may be used. As s is a 1D coordinate, weights w_i can be obtained, and the derivative of h is taken in complete analogy to equation Eq. (15) from Chapter "Simple Finite Difference Procedures":

$$h_{x,0} = w_{-2}h_{-2} + w_{-1}h_{-1} + w_0h_0 + w_1h_1 + w_2h_2, \tag{10}$$

where the weights depend on the choice of the interpolated collocation points. In Eq. (10), h_i are grid point values of the field h, being either given as amplitudes or created by at least third-order interpolation.

The specific choices, in particular the choice of the phantom collocation points, will have a profound impact on the allowed time step of a model. To explore the impact of such choices on CFL, the von Neumann method (Section "Von Neumann Analysis of Some o*nom* Schemes" in Chapter "Simple Finite Difference Procedures") will often not be suitable, as the grids are irregular. So we need to use simpler methods, such as described in Section "1D Homogeneous Advection Test for o*nom* Methods, SEM2 and SEM3" in Chapter "Numerical Tests". Equation (9) can be used efficiently if the points of a stencil are on a straight line and only a few points require phantom collocation points. Such an example is given in Steppeler et al. (2008) for the grid shown in Fig. 1 in Chapter "Introduction".

With a dense grid, such as a regular triangular or rectangular grid, Eq. (9) can be used at every point, not distinguishing between edge and corner points. An example of such grid is shown in Fig. 1c in Chapter "2D Basis Functions for Triangular and Rectangular Meshes". Such schemes are of third order, and often super-convergence to fourth order is observed. These schemes are not formally conserving mass. These are generalizations of the simple homogeneous schemes of Chapter "Simple Finite Difference Procedures" to 2D and 3D. For the L-Galerkin schemes, Eq. (10) is used at corner points only, and edge and inner amplitudes are obtained from considerations of conservation. This will be best understood in Section "The Full Triangular o3o3 Method".

Baumgardner's Cloud Derivative Method

This method (Baumgardner and Frederickson 1985) allows taking derivatives for 2D fields on a totally irregular grid. It was developed for second-order approximations and 2D fields only without formal mass conservation. The L-Galerkin methods described in this book allow third-order approximations and a 3D irregular grid, if desired. In fact, it is possible by obvious generalization to have L-Galerkin of any desired order, and for the case of SEs, such higher than third-order schemes have been explored. However, such options, beyond the 2D second-order case, are computationally very inefficient with full grids. It was found (Herrington et al. 2019) that currently for practical modeling considerations, there is no point to go beyond third order. Second- or third-order approximations are suitable from the practical point of view. For first order, it was shown by practical considerations that the accuracy is not suitable for atmospheric modeling (Thuburn et al. 2001).

The importance of Baumgardner's method was that it solved the problems arising from the irregular grids with icosahedral discretization for the first time. The icosahedral grid is shown in Fig. 1 in Chapter "Introduction", and a more complete treatment will be given in Chapter "Platonic and Semi-Platonic Solids". In this way, the Baumgardner's method made a spherical discretization on the basis of the icosahedral grid possible. The icosahedral grid was first suggested in the 1970 (Williamson 1968; Sadourny and Morel 1969). Until around 2005, this discretization was not used due to problems with internal boundaries of the icosahedral grid. The Baumgardner's cloud method was the first FD method resulting in a uniform second-order approximation meaning that there are no exceptional points, such as poles where the order of approximation drops below 2. A uniform order > 1 has turned out essential to avoid problems with internal boundaries.

Internal boundaries for icosahedrons are lines in the grid where the structure of the grid changes discontinuously. In Fig. 1c in Chapter "Simple Finite Difference Procedures", an internal boundary is illustrated. In Fig. 1 in Chapter "Introduction", internal boundaries can be seen as the boundaries between the 10 rhomboidal patches of the icosahedron. It was shown that error generation at internal boundaries can be strong (Thuburn et al. 2001). In all computational examples known so far, noise generation at internal boundaries is avoided by a uniform order of approximation n, where n must be greater than 1. The Baumgardner's method is limited to 2D. Most weather forecast models have a smoothly varying vertical grid structure. By implementing this method for weather forecast applications, Baumgardner demonstrated that problems and inaccuracies at the particular internal boundaries of the icosahedron disappeared when this uniformly second-order scheme was used (Baumgardner and Frederickson 1985).

The computation of a derivative in an irregular grid such as shown in Fig. 1c in Chapter "Simple Finite Difference Procedures" by the cloud method is not difficult. Let us consider the icosahedral grid in Fig. 1c in Chapter "Introduction". It can be seen that most points are surrounded by 6 nearest neighbors. Some points, such as the poles, have 5 neighbors. Including the center point, we have 7 or 6 point stencils.

In the surrounding of the target point, we represent the field by a second-degree polynomial $p(x, y)$:

$$p(x, y) = c_{0,0} + c_{1,0}x + c_{0,1}y + c_{2,0}x^2 + 2c_{1,1}xy + c_{0,2}y^2. \tag{11}$$

For the case of 5 neighbors, the 5 neighbors' amplitudes plus the center point of a field generate 6 equations that allow the computation of the $c_{i,j}(i, j \in 0, 1, 2)$. For the case of 6 neighbors to the target point, a least square method will give the polynomial coefficients.

With any time marching procedure (such as RK4), the approximated derivatives of fluxes will result into a time-stepping procedure for the dynamic field. The method can in principle be generalized to third order. However, a third-order polynomial has 10 polynomial coefficients. In order to have enough collocation points to compute them, we need to go to the next neighbors and the number of points involved would be too big for the method to remain practical. A lack of practicability concerns the generalization to three dimensions. In atmospheric models, this is avoided by using a structured grid in the vertical. In the following, the problem of too many neighbors is avoided by sparse grids, which brings the number of points in a stencil down to a reasonable amount. SEs are not using sparse grids.

This method achieves reasonable stencil sizes by having the grids aligned on great circles or straight lines. This allows to obtain stencils similar as for differentiation on a plane.

Third-Order Differencing for Corner Points with a Second-Degree Polynomial Representation

When the field $h(x, y)$ is represented in third order, as with the method o3o3, the generalization to an irregular 2D grid is straightforward. As in the 2D regular grid, some of the differences are taken on straight lines. For irregular grids, some points of such linear stencils must be created by interpolation. For the grid of Fig. 1 in Chapter "Introduction", this is described in Section "A Simple Non-conserving Homogeneous Order Discretization on the Sphere" in Chapter "Platonic and Semi-Platonic Solids". For the stencils in this grid that require the creation of point amplitudes by interpolation, this is illustrated in Fig. 11 in Chapter "Platonic and Semi-Platonic Solids". In the example to be given in Chapter "Platonic and Semi-Platonic Solids", the grid is not totally irregular but rather has regular features. This reduces the need to interpolate points and thus leads to a reduced computational effort. However, the method of Section "A Simple Non-conserving Homogeneous Order Discretization on the Sphere" in Chapter "Platonic and Semi-Platonic Solids" is suitable for a totally irregular grid. In Section "A Simple Non-conserving Homogeneous Order Discretization on the Sphere" in Chapter "Platonic and Semi-Platonic Solids", a non-conserving scheme without sparseness will be described. However, as for schemes on the plane, the o3o3 scheme is suitable to create conserving sparse grid schemes for irregular grids. A less trivial example is the o2o3 scheme because the assumed interpolation for $h(x, y)$ is not third order, and therefore, the creation of phantom points in stencils is less obvious. As there exists little research concerning L-Galerkin schemes on irregular grids, in the following, a rough sketch will be given of how to generalize schemes to irregular grids.

The o2o3 method represents fields by second-order splines and creates fluxes represented by third-order splines. Essential for the construction of such third-order splines is a computation of flux derivatives in third or higher order at corner points (see Section "The L-Galerkin Scheme: o2o3" in Chapter "Simple Finite Difference Procedures"). In 1D, this is done by collecting node points that are on a straight line. When the grid is regular, in 2D or 3D, the same can be done, as straight coordinate lines can be extracted. As indicated in Fig. 1 in Chapter "Full and Sparse Hexagonal Grids in the Plane", even a regular hexagonal grid allows the extraction of such linear stencils. For a totally irregular grid, it is necessary to form derivatives at corner points in the situation when there is no straight coordinate line with a sufficient number of node points. This will be explained for the case of the irregular quadrilateral grid shown in Fig. 1f in Chapter "2D Basis Functions for Triangular and Rectangular Meshes".

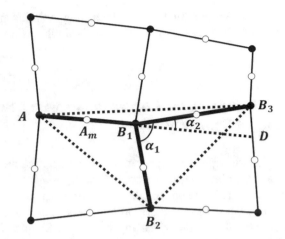

Fig. 2 Blowup of Fig. 2f in Chapter "2D Basis Functions for Triangular and Rectangular Meshes" B_1 is the target point for differentiation. Grid cells are shown together with auxiliary lines. The directional derivative in the direction $\overline{B_1 A}$ is to be computed. The control triangle $\Delta_{AB_2B_3}$ used from third-order interpolations is shown as a dotted line. The black points are corner points of the quadrilateral grid, where field values are given. The white points are edge points, where for the second-order representation the second directional derivatives along the grid line are given as dynamic variables. Alternatively, grid point values may be given at these points. The dashed line is the continuation of line $\overline{A B_1}$, the line where the third-order derivative is to be taken. Using the vector point B_1, the triangle can be divided into 3 sub-triangles, which will be used for forming derivatives. Second-order amplitudes and these triangles can be obtained diagnostically, but for the determination of derivatives at the central point B_1 only the amplitudes on the stencil are needed

To compute the derivative of a flux in x- or y-directions, Fig. 2 shows the four surrounding quadrilateral cells of the target point for the differentiation. The four surrounding cells have amplitudes on the corners (black points) and edges (white points), and the total number of these neighbors is 21. As in 2D, there are 10 polynomial coefficients, these are over-determined by the 21 surrounding points. So Baumgardner's cloud method would be applicable to obtain the derivatives at the target point, but rather inefficient. We still show that the points on the stencil (thick solid line in Fig. 2) are sufficient to determine the derivatives. As a preparatory step, the three second-order directional derivatives along the legs of the stencil are computed. The third-order accurate derivatives can be determined as a linear function of the three second-order accurate first derivatives. As was the case in one dimension, it is important to use such points for interpolation, where the third-order contribution is 0.

In the 4 quadrilaterals surrounding the target point B_1, we have a second-order accurate representation of all fields, such as flux components. This immediately leads to second-order accurate derivatives at all corners and edges, where at edges the derivatives may be one-sided. The second-order representation is given by field grid point values at the 9 corners shown in Fig. 2 and second directional derivatives along the 12 edges. This means that at the target point, the second-order field is one-sided differentiable for all four directions going through B_1, and the derivative is obtained as a linear function forming the one-sided derivative along the two neighboring edges. As we are not content with second-order schemes, we need to compute the one-sided derivatives at corner points in third-order accuracy, and this will be the objective of the following pages.

Figure 2 shows 4 cells of the irregular grid structure. This involves 21 corner and edge points. Note that with the third-order representation introduced in Chapter "2D Basis Functions for Triangular and Rectangular Meshes", the 21 corner and edge points are also corner and edge points of a third-order representation. There are a number of possibilities that all have not been implemented and their properties explored. Consider in Fig. 2 the computation of the derivative of h in the direction $\overrightarrow{AB_1}$. The third-order stencil is on line $\overline{AD}$, and the stencil Eq. (15) in Chapter "Simple Finite Difference Procedures" uses points A, B_1, D and the mid-points A_m of $\overline{AB_1}$ and D_m of $\overline{B_1D}$. These A, A_m and B_1 in Fig. 2 are given. Note that these collocation points are also collocation points in a third-order representation. D_m and D need to be obtained by third-order interpolation. Note that all points indicated in white or black in Fig. 2 are also collocation points of a third-order representation. A third-degree polynomial in two dimensions has 10 coefficients, and so we normally need at least 10 points to compute them by interpolation. It is normally considered convenient to use points surrounding the target point in all directions. D_m is surrounded by 12 points, from which 10 may be selected by interpolation. In a similar way, D is surrounded by 11 points. Such interpolation schemes lead to differencing stencils at point B_1, and the numerical properties of such stencils have not yet been explored for the irregular grid case.

An alternative method is based on triangular cell representations for the three triangles in Fig. 2 consisting of dotted and solid lines, which combine to the larger triangle of dotted lines. We construct a new second-order representation based on triangles, which are not necessarily identical to the quadrilateral representation, but can be computed from it, and both representations are second-order approximations of an analytic function. The triangles are the 3 sub-triangles $(\Delta_{B_1B_2B_3}, \Delta_{B_1B_2A}, \Delta_{B_1B_3A})$ of the triangle $\Delta_{AB_3B_2}$ indicated in Fig. 2 as the thick dotted line. The three small triangles have a second-order representation that can be obtained from the quadrilateral representation. Each of the small triangles has 6 second-order amplitudes. This amounts to 10 amplitudes for the three small triangles, as some of the amplitudes are identical. The third-order representation of the large triangle representing the combination of the three small ones needs three amplitudes. Only four amplitudes of this third-order amplitude must be determined, as 6 are identical to amplitudes of the small triangles. The three amplitudes are the third directional amplitudes along the dotted line in Fig. 2 and the center point amplitude. The determination of these amplitudes can be done using the least square method. Once the third-order representation is determined in the large triangle surrounding the point B_1, the third-order derivative at point B_1 can be obtained by differentiating this third-order representation.

A further example is to determine point D by third-order interpolation and in the middle of the line $\overline{B_1 D}$ and to determine the second directional derivative from the quadrilateral second-order representation. This means that we have second-order representations for the lines $\overline{A B_1}$ and $\overline{B_1 D}$. We can then by 1D least square representation determine a third- or fourth-order representation along the line $\overline{A D}$, which can be used to compute the first directional derivative at B_1. Once the derivatives at corner points are determined, the second-order amplitudes at the edges can be obtained using the principle of mass conservation, as was done in the regular grid case.

These examples constitute suggestions to create third-order accurate difference schemes on irregular cells. It is obvious that this is more expensive than the case of a regular grid, as the computations for doing interpolations come in addition. There are many applications, such as the flow in a complicated terrain, which requires an irregular grid. Very often such applications require the high resolution only for a small area, which means that the irregular resolution can be very efficient. Another application is global medium range forecasts, which is often done using a regular grid, as all parts of the globe are important for the medium range forecast at any place. For example, it was found that in 2 days a measurement at the east coast of the USA has an impact on the weather forecast in Europe. Therefore, for global forecasting in the medium range, models of uniform resolution on the globe can be used. Using grids being approximately regular can save the numerical effort associated with an irregular grid. Examples of test integrations using such semi-regular grids will be shown in Chapter "Numerical Tests".

Enhanced Stencil Order

Many numerical schemes being based on a third-order spline representation come out as fourth-order convergent when analyzing the results by systematically changing the resolution. This method is similar to those used with Fig. 2 in Chapter "Simple Finite Difference Procedures". In Section "Test of the o3o3 Scheme on the Cubed Sphere Grid Using the Shallow Water Version of the HOMME Model" in Chapter "Numerical Tests", an example will be given. One of the methods for computing the derivative at corner points is to use the fourth-order collocation derivative Eq. (10).

In this section, we try to obtain a fourth-order term on an edge line with coordinate s meaning the amplitude h_{ssss} of the basis function $b^4(s) = s^4 - \left(\frac{ds}{2}\right)^4$. ds is the length of the edge line or another line on which a derivative is taken. Similarly, when the representation is a second-order spline, we may explore an amplitude for the basis function of $b^3(s)$, with b^3 being defined in Tables 2 and 3 in Chapter "2D Basis Functions for Triangular and Rectangular Meshes".

If our 2D grid is such that the edge line with coordinate s meets another grid point, this point can be used to determine the amplitude h_{ssss} of b^4 that is the fourth-order correction on this line. For this purpose, we extend the third-order representation:

$$h'(s) = h(s_0) + h_s e(s) + h_{ss} b^2(s) + h_{sss} b^3(s), \qquad (12)$$

outside the edge interval $(-\frac{1}{2}ds, \frac{1}{2}ds)$.

Let s_4 be the s-coordinate of the point $\mathbf{r}$ outside the edge interval and $h(\mathbf{r}) = h(s_4)$. We define the function $h''(s)$ by

$$h''(s) = h(s) - h'(s). \qquad (13)$$

The fourth-order correction term is then obtained by

$$h_{ssss} = \frac{h''(s_4)}{b^4(s_4)}. \qquad (14)$$

When the surrounding points are too irregular, this point $\mathbf{r}$ must be obtained by interpolation. This fourth-order correction $h_{ssss} b^4(s)$ can be used to compute fourth-order corrections $h_{ssss} b_s^4(s)$ to the directional derivative along the edge to which the coordinate s belongs. In one dimension, this correction can be applied when computing the spectral form of the derivative computation at corner points. For computing corner point derivatives in two dimensions using the three legged stencil, the fourth-order correction can be used to increase the accuracy at 3 legged corner points (see Fig. 1). Currently, it is not well investigated to what degree such fourth-order corrections are necessary and useful.

The Full Triangular o3o3 Method

The o3o3 scheme described in Chapter "Local-Galerkin Schemes in 1D" is suitable for irregular resolution. An arbitrary division of the computational area into triangles is possible.

In Sections "Non-conserving Schemes for Full Grids" and "Alternative Methods to Compute Derivatives", the computation of derivatives within a 2D mesh is described. This means that the o3o3 scheme in two dimensions can be done in analogy to the 1D case. At corner points, any difference scheme can be used to compute the time derivative of a dynamic field. At this point, we do not have to be concerned with conservation. In one dimension, the second derivatives of the fileds are high-order amplitudes and are computed using the request of mass conservation. In two dimensions, these high-order amplitudes are the second directional derivatives along edges, which again have to be obtained by considerations of mass conservation. The third directional derivatives are again computed by finite differences using the already computed second directional derivatives. For the special case of a regular triangular mesh, it is possible to extract 1D grids and do the differentiation there. As the case was in one dimension, changing the value of the third derivative amplitude does not change the mass of the system.

It is important to know all neighborhood relations, such as how many edges a particular triangle has, how many triangles meet at a corner point, and what are the positions of collocation points or spectral amplitudes for each triangle. For a given corner point, any number of triangles can meet. For amplitudes at an edge, there are always two triangles that share this amplitude. In analogy to the 1D case, the edge amplitudes are second and third directional derivatives along the edge. The representation of fields in third order will require 10 amplitudes for the description in each triangle. These amplitudes, taken independently at each triangle, will describe a discontinuous field, where the discontinuities may occur at corners or edges. When a point or edge is shared between more than one triangle, we have many partial amplitude values at the definition point of the amplitudes. For example, when 6 triangles meet at a corner point, the 6 amplitudes describe a discontinuous field. The field is continuous when all such amplitudes are the same for each corner. Within a time step, such discontinuous functions may arise. It is the task of L-Galerkin operations to approximate this by a continuous function.

Neighborhood relations can be rather complicated, and in general they require a storage management system to administrate them. Therefore, for ease of understanding, we will here describe the regular case. The grid shown in Fig. 2b in Chapter "2D Basis Functions for Triangular and Rectangular Meshes" has 6 triangles belonging to each corner point. A generalization to irregular cells is shown in Fig. 1f in Chapter "2D Basis Functions for Triangular and Rectangular Meshes", which is a quadrilateral grid, but triangular cells can be obtained by dividing each quadrilateral into two triangles.

The triangular cells for the regular case are shown in Fig. 1c in Chapter "2D Basis Functions for Triangular and Rectangular Meshes". The corners of the triangles are indicated as black points. Other collocation points are indicated in white. There are ten collocation points per cell, and within each cell, the field $fl(x, y)$ is represented as a third-degree polynomial. A third-degree polynomial in x, y is defined by the ten polynomial coefficients to $1, x, y, x^2, xy, y^2, x^3, x^2y, xy^2, y^3$. Section "Triangular Basis Functions and Full Grids" in Chapter "Local-Galerkin Schemes in 1D" treats the case of field representations for a single triangular cell. It was shown that the polynomial and the collocation point representations are equivalent. Section "Triangular Basis Functions and Full Grids" in Chapter "2D Basis Functions for Triangular and Rectangular Meshes" gives the basis function representations Eqs. (22)–(23). The corner points determine the linear part of the basis function representation. For each of the triangular sides, there exist two basis functions being 0 at the two other sides. The corresponding amplitudes are determined by the two collocation points on this side (see Fig. 1c in Chapter "2D Basis Functions for Triangular and Rectangular Meshes"). According to Section "Triangular Basis Functions and Full Grids" in Chapter "2D Basis Functions for Triangular and Rectangular Meshes", the two collocation points belonging to a side are equivalent to the second and third directional derivatives along this side to use as amplitudes. In the grid of Fig. 1c in Chapter "2D Basis Functions for Triangular and Rectangular Meshes", each cell has one center point to be used as the center amplitude in Eq. (23) in Chapter "2D Basis Functions for Triangular and Rectangular Meshes".

The 10 amplitudes for each cell are not independent dynamic amplitudes. The corner points are shared by 6 cells, and the edge amplitudes by two cells. The center amplitude is specific for each cell. This means that amplitudes for the description of a field $h(x, y)$ are the center amplitude, the corner amplitudes of h, and the second and third directional derivatives along all edges for one cell. These are amplitudes for the spectral representation. Their definition is shown in Fig. 2c in Chapter "2D Basis Functions for Triangular and Rectangular Meshes" as black (corner) points and large white (second derivatives) points. The black points are both collocation points and spectral amplitudes. Corresponding to 10 independent third-order polynomial coefficients, there are 10 amplitudes per triangle. The large white points on edges in Fig. 2c in Chapter "2D Basis Functions for Triangular and Rectangular Meshes" carry 2 amplitudes. The collocation grid of the full grid is shown in Fig. 1c in Chapter "2D Basis Functions for Triangular and Rectangular Meshes" and also contains 10 grid point value amplitudes. So the number of degrees of freedom for the spectral and collocation grid representations is the same. Note that all such amplitudes are chosen independently. A discontinuous function is represented, which will occur when forming the RHS of the equation of motion Eq. (1) in Chapter "Introduction". We have seen in Chapter "2D Basis Functions for Triangular and Rectangular Meshes" that there is a one-to-one transformation between the two representations. Therefore, the grid of triangular grid representation corresponds to the full collocation grid, and this is called the full triangular representation. At each corner point, 6 edges meet for our particular choice of the grid. For an alternative option, see Fig. 1a in Chapter "2D Basis Functions for Triangular and Rectangular Meshes".

When representing a continuous field, each amplitude point shared between two or more points must be assigned a unique value. We have already pointed out that for efficient computer programming, it is important that we assign only one storage to such points, rather than having the identical value stored with different triangles. This means that we have to use a compact storage system.

For a better understanding of the computation of the time derivative and the order-consistent field representation, the third-order basis functions are shown in Fig. 2d–m in Chapter "2D Basis Functions for Triangular and Rectangular Meshes". When using Eq. (1) in Chapter "Introduction" to obtain the time derivative, we obtain the time derivative as a second-order triangular spline representation. This representation is mass conserving. The method can easily be done for the irregular grid case which adds complexity, but no essential new problems arise. Here we limit ourselves to the regular grid case. For the rectangular case, the formula for the triangular basis functions is simple. For this special case, Tables 2 and 3 in Chapter "2D Basis Functions for Triangular and Rectangular Meshes" give the derivatives of the triangular basis functions and can be used when practically computing the second-order discontinuous representation.

We have a FD scheme, if we approximate the discontinuous second-order representation by a continuous field. Using the already obtained discontinuous representation, we can get amplitudes of the time derivative for corner amplitudes by averaging the discontinuous amplitudes in a mass conserving way; in analogy we get the second directional derivatives at edges. This procedure is a 2D version of the 1D o2o3 with spectral corner point computations. The third directional derivatives used as amplitudes do not contribute to the mass. Therefore, we can compute them by extracting the 1D scheme along a coordinate line. For the time derivative at central point, we obtain 0. This means that the center point is in the 0-space of the approximation. The disadvantage of this situation was discussed in Chapter "Simple Finite Difference Procedures" for one dimension and also a possible remedy using the scheme o2o3. In complete analogy to one dimension, when the time derivatives at node points are given, we can perform the time step using RK4. In one dimension, the standard o3o3 scheme uses an ordinary FD scheme at corner points, which is not necessarily conserving. On a regular grid, this procedure is available in the same way. As we can extract 1D problems along coordinate lines, we can do the same with triangular cells. To compute the mass content of a corner or edge amplitude, the mass of the basis functions shown in Fig. 2 in Chapter "2D Basis Functions for Triangular and Rectangular Meshes" must be used.

Homogeneous advection in two dimensions is a rather trivial test. A better test is advection in a curved velocity field. We assume that we solve an advection problem, and the generalization of the 1D homogeneous advection to 2D is

$$\frac{\partial h(\mathbf{r})}{\partial t} = u(\mathbf{r})\frac{\partial h(\mathbf{r})}{\partial x} + v(\mathbf{r})\frac{\partial h(\mathbf{r})}{\partial y} = \nabla \cdot (\mathbf{u}h(\mathbf{r})), \tag{15}$$

in which it is assumed that $\mathbf{r} = (x, y)$ and $\mathbf{u} = (u, v)$ is a non-divergent velocity

$$\nabla \cdot \mathbf{u} = \nabla \cdot (u, v) = 0. \tag{16}$$

When using a curved velocity field, we assume that u and v are given in approximations that satisfy Eq. (16) or approximate it. We also need to take the product of two functions in polynomial representations. These more complicated for o3o3 have not yet been used in tests.

Sparse Grids

The triangular o3o3 scheme described in Section "The Full Triangular o3o3 Method" uses the full grid. Limited test results for sparse grids are available only for quadrilaterals and hexagons. We describe the quadrilateral serendipity sparse grids. Even though this method works for irregular grids, the description will be for regular grids.

Sparseness occurs naturally with the serendipity interpolation at quadrilateral grids. The natural set of collocation points for a triangular mesh is the full grid. However, by treating some points as diagnostic, sparse triangular meshes are possible. Figure 1 in Chapter "Introduction" suggests that triangular grids can be transformed into a hexagonal cell scheme by treating some points as diagnostic. With FD in a sparse grid, standard FD schemes can be applied at corner points. At edge points (see Fig. 1 in Chapter "2D Basis Functions for Triangular and Rectangular Meshes"), there are not enough points to do high-order differences in the standard way for some directions of differentiation or rather coarse grids need to be used. For simplicity in the following, we will concentrate on the simple case of a square grid shown in Fig. 2b in Chapter "2D Basis Functions for Triangular and Rectangular Meshes", where the cell division into triangles is not used.

Here we consider the example of a sparse grid based on the square shown in Fig. 2 in Chapter "2D Basis Functions for Triangular and Rectangular Meshes". From the points belonging to a square, we do not use the two points on the diagonal and the two inner points of the triangles forming the square in Fig. 2b in Chapter "2D Basis Functions for Triangular and Rectangular Meshes". The 4 points omitted to achieve the sparse grid can be interpolated using the quadrilateral interpolation from Chapter "2D Basis Functions for Triangular and Rectangular Meshes" based on Euclid's Lemma. This method of achieving the quadrilateral sparse grid is also called the serendipity grid, and the interpolation is called serendipity interpolation. As the square has 9 points in the compact grid, the sparseness factor is 5/9. Note that in 3D the sparseness factor in 3D is 7/27, which would translate into a considerable computer economy. This point will be further explored in Section "CPU Time Used with a 3D Version of o3o3 Scheme" in Chapter "Numerical Tests". More complicated situations are encountered when combining more triangles for polygonal cells, such as 5 triangles for a pentagon. While sparseness is an option for all such polygonal grids, here we treat only the most simple example of a square cell.

In 2D, we use as a test problem the 2D form of the homogeneous advection equation:

$$fl_t = u_0 fl_x + v_0 fl_y. \tag{17}$$

The more general velocities of Eq. (15) have so far not been used for tests. We need to obtain time derivatives at each point of the sparse grid to do the time marching. One way of doing this is to compute some of the full grid points diagnostically and in this interpolated full grid do normal differencing, but use only the points of the sparse grid. This is a rather obvious way of proceeding, but it turns out to be unstable with RK4 time-stepping for the square grid.

There is another simple alternative that works using a non-conserving FD scheme at corner points. For $fl_{i,j}(i, j = 1, 2, 3, \ldots)$, straightforward fourth-order differencing at corner point is possible, and we can use the third-level points $fl_{i+\frac{1}{3},j}$, $fl_{i+\frac{2}{3},j}$, $fl_{i,j+\frac{1}{3}}$, $fl_{i,j+\frac{2}{3}}$.

We get the x-derivative fl_x^x of $fl(\mathbf{r})$ at (x_i, y_j) with standard fourth-order FD scheme and obtain one-sided derivatives along the diagonal by adapting Eq. (15) in Chapter "Simple Finite Difference Procedures" to the line of grid points with constant j:

$$fl_{x,i,j}^x = w_{m2} fl_{i-\frac{2}{3},j}^x + w_{m1} fl_{i-\frac{1}{3},j}^x + w_{p1} fl_{i+\frac{1}{3},j}^x + w_{p2} fl_{i+\frac{2}{3},j}^x, \tag{18}$$

which is obtained by extracting the 1D scheme for each j. This also delivers $fl_{xx,i+\frac{1}{2},j}^{x,1}$. In an analog way, we obtain $fl_{y,i,j}^{y,1}$ and $fl_{yy,i,j+\frac{1}{2}}$. The upper index 1 indicates that this is the result of the extracted 1D problem. For the divergence of (fl^x, fl^y), we obtain as examples for some of the terms to be computed:

$$\begin{cases} \nabla(fl_{i,j}^x, fl_{i,j}^y) = fl_{x,i,j}^x + fl_{y,i,j}^y, \\[2mm] \nabla(fl_{i+\frac{1}{2},j}^x, fl_{i+\frac{1}{2},j}^y) = fl_{x,i+\frac{1}{2},j}^{x,1} + \dfrac{1}{2dy}(fl_{i+\frac{1}{2},j+1}^y - fl_{i+\frac{1}{2},j-1}^y), \\[2mm] \nabla(fl_{i,j+\frac{1}{2}}^x, fl_{i,j+\frac{1}{2}}^y) = \dfrac{1}{2dx}(fl_{i+1,j+\frac{1}{2}}^{x,1} - fl_{i-1,j+\frac{1}{2}}^{x,1}) + fl_{y,i,j+\frac{1}{2}}^y. \end{cases} \tag{19}$$

where the index x, xx indicates the second-order part of the 1D approximation for the derivative of fl. The high-order basis functions used are $b_{i+\frac{1}{2}}^2(x)e_j(y)$ and $e_i(x)b_{i+\frac{1}{2}}^2(y)$ (analog for the third-order functions). Note that the linear basis function for point i, j can be written as $e_{i,j}(x, y) = e_i(x)e_j(y)$.

For the third derivative of fl_x^x used as amplitude, we obtain analog to Eq. (55) in Chapter "Local-Galerkin Schemes in 1D":

$$\begin{cases} h_{t,xxx,i+\frac{1}{2},j} = \dfrac{1}{2dx}(h_{t,xx,i+\frac{3}{2},j} - h_{t,xx,i-\frac{1}{2},j}), \\[2mm] h_{t,xxx,i,j+\frac{1}{2}} = \dfrac{1}{2dx}(h_{t,xx,i,j+\frac{3}{2}} - h_{t,xx,i,j-\frac{1}{2}}). \end{cases} \tag{20}$$

There are many further possibilities beyond quadrilateral serendipity grids to define sparse grid schemes. In Williamson (1968), an example of a sparse hexagonal grid is given. In Section "L-Galerkin Schemes for Sparse Triangular Meshes", there are few computational tests done for sparse grids.

L-Galerkin Schemes for Sparse Triangular Meshes

Beyond the quadrilateral serendipity method described in Section "Sparse Grids", there is a large number of further possibilities some of which will be mentioned here. The o2o3 sparse method on hexagons is described in Steppeler et al. (2019), including a simple computational example. While this subject is largely unexplored, it was apparent that the potential of saving computational effort with sparse hexagons is larger than that with sparse quadrilaterals.

Sparseness with polygonal spline methods is possible, and when starting from the full grid, the functional representation is made lower dimensional in phase space by requesting a property from the function space whose violation would imply functional modes of low accuracy. For the serendipity grid discussed in Section "Sparse Grids", these are basis functions representing in terms of lower accuracy than the target order of approximation of the scheme. For fourth-order SEs (SE4), for example, polynomials $x^n y^m$ with $n, m < 5$ are used. The term $x^4 y^4$ is used with SE4. This is an eighth-order term that may be expected to contribute little to a scheme designed to be of fourth order. Therefore, the o*nom* scheme drops some of such terms without losing accuracy.

Sparseness can be achieved in one space dimension. So far, we considered continuous field representations. We may also require fields to be one time differentiable at corner points. As many of the time translation methods depend on the spatial Taylor development and the existence of a derivative at corner points, it may be expected that the additional resolution is achieved when allowing the fields to be non-differentiable at corner points will not make the solution more accurate. The numerical technique can be handled when using the representation of the high-order part of fields by one-sided derivatives. Sparseness by requesting one time differentiability is in one dimension achieved by requesting the two one-sided derivatives to be equal. In two space dimensions, any number of triangles can meet at a corner. For the regular triangular grid shown in Fig. 1c in Chapter "2D Basis Functions for Triangular and Rectangular Meshes", 6 triangles meet at each corner. This means that the limiting value of one-sided derivatives at a corner can have 6 values, representing a rather irregular function. Note that two directional derivatives are sufficient to represent a differentiable function. Creating sparseness by requiring functional representations to be differentiable is called the QUASAR system (see Appendix "The Quasi-arithmetic Rendition QUASAR to Obtain a Sparse Field Representation"). We discuss a few options and consider the potential of saving by the sparseness factor. This is the relation of the dynamical amplitudes of a cell for the sparse and the full grid:

1. One dimension grid: The third-order representation uses 3 amplitudes per grid interval. The QUASAR system uses just one corner point amplitude and one field derivative. Sparseness factor = $\frac{2}{3}$.
2. Triangular grid: The full grid uses on average $4\frac{1}{2}$ points per triangular cell (9 for two triangles). The QUASAR uses the node point amplitude and derivatives in two directions for 6 triangles, meaning on average $1\frac{1}{2}$ amplitudes per triangle. Sparseness factor $1\frac{1}{2}/4\frac{1}{2} = \frac{1}{3}$.
3. Hexagonal grid: Amplitudes are defined only on corners and edges. The full hexagon contains 27 points. Sparseness factor = $\frac{6}{27}$.

Totally Irregular Triangular and Quadrilateral Mesh:
Hexagons and Other Polygons

We consider a third-order field representation. If several cells meet at a corner point and the piecewise cubic representation is considered continuous, the directional field derivative can be different in every triangle meeting at the point. Already for the regular grid considered in Section "L-Galerkin Schemes for Sparse Triangular Meshes", a way to construct one time differentiable field representation of third order was considered. In this section, we consider ways to define differentiable 2D splines in arbitrary triangular grids. We also define ways to define center cell amplitudes using the corner derivatives as amplitudes. Let us first consider a heuristic argument for our assumption that by requesting the function system to be differentiable, we obtain no loss of resolution, even though our grid system becomes sparse by this request. For this purpose, we consider the 1D case.

Continuous third-order representations at lowest resolution may appear rather noisy in 1D space. We may argue heuristically that using a basis function representation that does not include such noises will not reduce accuracy, even though with such noisy basis functions formally some resolution is sacrificed. It appears as if resolution (from 2 to 3 amplitudes) is rendered to the system by using the algebraic request of differentiability. Such systems are called *"QUASi-Arithmetically Rendered (QUASAR)"* (see Appendix "The Quasi-arithmetic Rendition QUASAR to Obtain a Sparse Field Representation"). The QUASAR system allows to reduce redundancy by making fields differentiable at corner points. The basic tools for discretization in the QUASAR system are given in this book. However, the dynamics of QUASAR scheme has not been tested, and currently, it is unknown if these schemes are stable.

We consider the triangular grid such that at each corner point, 6 edge lines meet, and they may have different one-sided derivatives. As the time prediction of fields relies on spatial differentiation, the request of a differentiable field at corner points will result into many of the edge amplitudes to be diagnostic. For example, a hexagonal discretization is achieved by making all inner amplitudes of the cell diagnostic.

The discretization of regular quadrilateral or triangular cells on regular grids can be achieved by extracting the 1D discretizations along coordinate lines. An irregular conforming grid, as shown in Fig. 1f in Chapter "2D Basis Functions for Triangular and Rectangular Meshes", can be seen as two triangles dividing each quadrilateral cell i, j. When we have amplitudes consisting of fields values and derivatives in two directions at all points $\mathbf{r}_{i,j}$, the corner point field values and derivatives are called corner amplitudes. The three corner point amplitudes have all index i, j. In addition, for each triangle, we have the center point amplitudes that lead to the sparseness factor of $\frac{5}{9}$. The sparseness factor $1\frac{1}{2}/4\frac{1}{2} = \frac{1}{3}$, which was reported in Section "L-Galerkin Schemes for Sparse Triangular Meshes", is obtained by not treating the center point amplitudes as dynamic points but rather as diagnostic. This means that the center amplitude is computed diagnostically from the corner amplitudes. This diagnostic determination can be achieved in the structured irregular grid of Fig. 1f in Chapter "2D Basis Functions for Triangular and Rectangular Meshes". The corner points $r_{i,j}$ and $r_{i+1,j+1}$ are used to determine a third-order interpolation along the line $r_{i,j} - r_{i+1,j+1}$; the two directional derivatives along $r_{i,j} - r_{i+1,j+1}$ and the point values can be obtained from the corner amplitudes at this point and these 4 points; this is sufficient to determine the center points of the two involved triangles.

For the o3o3 schemes, third-order representations according to Chapter "2D Basis Functions for Triangular and Rectangular Meshes" are given on the edges of cells. For a triangular mesh, an amplitude in the inner of the triangle must be given. The fully dynamic treatment of o3o3, as described in Section "An Example of a Regularization Operator" in Chapter "Local-Galerkin Schemes in 1D", for the regular grid case transfers to the irregular case. Grid sparseness factors for the irregular conforming grid in Fig. 1f are the same as for the regular grid case given in Section "L-Galerkin Schemes for Sparse Triangular Meshes". For the irregular case, similar relations hold. The irregular triangular cells can be treated in analogy to the regular case. We consider the sparseness obtained by combining 2 triangular cells to a quadrilateral and 6 triangles to a hexagon. When we define a new cell structure, such as irregular quadrilateral or irregular hexagons, some of the amplitudes of the triangular grid become diagnostic, and this defines the sparseness. For the hexagon, the case has been treated that the center of the hexagon is a dynamic amplitude (Steppeler et al. 2019). For simplicity here, we treat the case that dynamic amplitudes are on the cell boundaries of the hexagon only. With the terminology introduced in Chapter "2D Basis Functions for Triangular and Rectangular Meshes", we call such boundaries edges of the hexagon. For computing the sparseness factors of a mesh, we count the dynamic amplitudes on the cell edges in relation to the sum of all inner and dynamic amplitudes. A point on the edge is divided by 2, as in two dimensions it is shared by two cells, as long as hanging nodes are avoided. For the corner points, we count this after dividing by the number of cells meeting at this point. When for quadrilaterals we assume that 4 cells meet at each corner and for hexagons that 3 cells are meeting at a corner, we obtain the same sparseness factors as given for the regular case in Table 1 in Chapter "Full and Sparse Hexagonal Grids in the Plane".

Note that for some irregular grids, the similarity to the regular case is such that they can be performed in complete analogy. The methods obtained for a plane can be transferred to the irregular case. These are the grids on the sphere, for which examples were discussed in Chapter "Introduction" and more examples will be given in Chapter "Platonic and Semi-Platonic Solids". The grids on the sphere are obtained by projecting grids on planes or bilinear surfaces to the sphere and thus consist of two families of great circles. These great circle grids are not regular. On the sphere, no regular grids exist except for the 5 Platonic solids (see Chapter "Platonic and Semi-Platonic Solids"). However, on such great circle grids, o3o3 can be performed in analogy to the plane grid case. Corner derivatives on o3o3 are obtained by doing finite differences along coordinate lines. This is on great circles the same as on the plane. The fourth-order FD formula Eq. (15) in Chapter "Simple Finite Difference Procedures" for irregular grid must be used on the sphere. As on the sphere the great circles forming the coordinate lines will not be perpendicular, this must be accounted for by applying weights. The second derivative edge amplitudes for the time derivative follow in the same way as in the plane case. The third derivative edge amplitudes are obtained again by differentiating along great circles. If amplitudes for other basis functions are desired, they can be obtained by transformation. Note that this example is of considerable practical importance. Weather prediction models for 5–10 days have often a homogeneous resolution with a high-resolution part for the target area only. This means that plane o3o3 or other L-Galerkin schemes can be applied to this problem using slightly modified plane models. Methods of irregular and unstructured grids can, but not necessarily, be applied to this problem. The problem of combining several patches on the sphere and patch boundaries will treated in Chapter "Platonic and Semi-Platonic Solids".

o2o3 (see Chapter "Local-Galerkin Schemes in 1D") needs numerical differentiation of second-order fluxes in order to create the third-order version of the fluxes. In this way, third derivatives are obtained when only a second-order field representation is given. A possibility is to take field values at points where the third-order correction is 0. From the infinity of possible schemes, these three (o3o3, o2o3, and QUASAR) are taken as examples in this book, and a short account of the implementation of the time-stepping procedure of these three schemes for dimension > 1 is given. Sparse systems may use other cell structures, such as quadrilaterals, hexagons, pentagons, etc.

The description is given for 2D. The transfer to 3D is straightforward, and in Table 1 in Chapter "Full and Sparse Hexagonal Grids in the Plane", some sparseness factors for 3D are given. For some applications in environmental flow, a totally irregular grid in 3D is useful. For global weather prediction, keeping a structured vertical grid has advantages, and the sparseness factor in Table 1 in Chapter "Full and Sparse Hexagonal Grids in the Plane" for the hexagonal cell assumes a vertically structured grid. In this case, the values given in Table 1 in Chapter "Full and Sparse Hexagonal Grids in the Plane" for a regular grid apply also to irregular hexagons, in particular hexagons on great circle grids.

We describe this without using cell related coordinate systems:

(1) Full Triangular Grid
The time step is performed in the following way:

(a) *Form the functional representation of fluxes fl^x and fl^y.* Often this involves multiplication. For example, when velocities and density field are (u, v) and $\rho(\mathbf{r})$, the flux fl^x for ρ is $u\rho$. This product can be formed in grid point space and transformed to the spectral space (see Chapter "Local-Galerkin Schemes in 1D"). For the spectral element, this is by far the dominant method. The product can also be formed in spectral space, which offers advantages when performing the physics part of the model on corner points only. Doing this makes an o*nom* scheme totally in spectral space possible. In such schemes, transformations to a grid point space are not performed.

(b) *Form derivative of at least third-order approximation in x- and y- directions at corner points using methods indicated above.* The Baumgardner's cloud method to form derivatives has an abundance of neighboring points when used with third-order schemes. This would potentially make the method less efficient. Another option is to form a stencil at corner points by using not all legs meeting at a corner to form the stencil. For the great variety of schemes along this line, the possibility of 0-spaces should be investigated (0-spaces are sub-spaces of the spectral space, on which the time derivative is 0). The derivative of fl^x is necessary only in x-direction. This gives some guidance for the choice of the stencil.

(c) *Form low-order derivative orthogonal to the edges to correct the high-order part of edge amplitudes.*

(2) Sparse Triangular Grid

The steps (a), (b), (c) are performed similar as with (1). Sparse grids often occur using larger polygons as cells, such as quadrilaterals or hexagons, as shown in Fig. 3. The advantage over (1) is that there is no need with step (b) to select a small stencil. For the hexagonal case (Fig. 3), the number of points is just right for a third-order stencil. It can be shown that the differentiation at corner points implies super-convergence to fourth order.

(3) o2o3

(a) *Form fluxes similar as with (1)(a).* Note that in this case we start from a second-order representation of fields, which implies a second-order representation of fluxes, denoted as $fl^{2,x}$ and $fl^{2,y}$. These second-order fluxes are continuous but not differentiable.

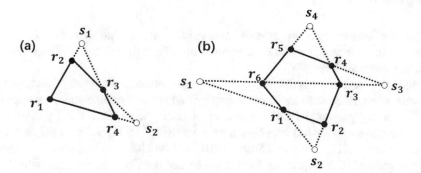

Fig. 3 The geometric meaning of the local coordinates $(x'y')$ for a quadrilateral (**a**) and a hexagon (**b**). The irregular hexagon is defined by two irregular quadrilaterals. To define the coordinate $x' \in (0, 1)$ and $y' \in (0, 1)$, additional points s_1 and s_2 are defined for the quadrilateral (**a**) and s_1, s_2, s_3, s_4 for the hexagon (**b**). The corner points for the quadrilateral (**a**) are r_1, r_2, r_3, r_4, and for the hexagon $r_1, r_2, r_3, r_4, r_5, r_6$. The x'-coordinate for the quadrilateral (**a**) is the cut-off of a line through s_1 with the line $\overline{r_2 r_3}$. For y', it is the cut-off of lines through $\overline{r_3 r_4}$. For the hexagon, the definitions are analog. To define the point $\mathbf{r}(x', y')$, consider the lines $\mu(\mathbf{r}_2 + x'(\mathbf{r}_3 - \mathbf{r}_2) - s_1)$ and $\nu(\mathbf{r}_3 + y'(\mathbf{r}_4 - \mathbf{r}_3) - s_2)$. $\mathbf{r}(x', y')$ is the point where these two lines intersect

(b) *Approximate fl^2 by third-order differentiable versions.* As seen above, this is achieved by forming third-order accurate first derivatives at corner points. Using Eq. (18) in Section "Sparse Grids", derivatives of fluxes at corner points together with corner field amplitudes define a third-order accurate representation of the field, which is differentiable.

(c) *Using the derivatives computed in (3) (b), differentiation will lead to an approximation to the time derivatives of the dynamic fields.* By construction, these are continuous at corners. At edge amplitudes, some mass adjustment vertical to the edges is necessary to obtain continuity.

(4) QUASAR

(a) *The fields are assumed to be differentiable at corners.* The fluxes computed from these are differentiable as well. So we can compute the time derivative immediately, which will be a continuous spline.

(b) *The continuous spline resulting after step (a) is mass adjusted at edge amplitudes.* The continuous representation for the time derivative is approximated by a differentiable field.

Note that this book gives the theory of computing the numerical derivatives in the desired order, but the practical performance may be complicated. The weights used in the examples given are obtained by the Galerkin Compiler MOW_GC.

Staggered Grid Systems and Their Basis Function Representation

Staggered systems concern models involving velocities and densities. Here we consider the example of shallow water equations (Eq. (58) in Chapter "Simple Finite Difference Procedures").

In the context of basis function of L-Galerkin methods, we speak of a staggered system when different fields (such as density and velocity components) are approximated using polynomial representations of different degrees. Such polynomial spline-based staggered schemes have been investigated for high-degree polynomials (Melvin et al. 2019; Cotter and Shipton 2012). With classic FD schemes, we speak of staggering if the amplitudes for velocities and density functions are defined at different positions of a grid cell. In the L-Galerkin world, this corresponds to basis functions of the lowest order meaning piecewise constant and piecewise linear basis functions (Steppeler 1990).

Let $\mathbf{r}_{i,j} = (x_i, y_j)$ be the cell centers of a grid, shown in Fig. 4. Low-order systems often use a staggered field representation. This means that when the field h is stored at cell centers $\mathbf{r}_{i,j}$, the fluxes fl with components fl^x and fl^y in x- and y-directions are stored with grid points at cell boundaries, as shown in Fig. 4. In this book, we use the Arakawa C-grid staggering (Steppeler et al. 2019; Mesinger

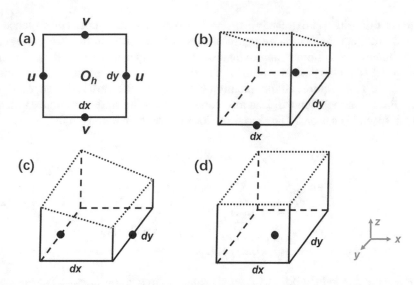

Fig. 4 Low-order basis functions used to emulate the Arakawa C-grid. (**a**) The grid square with the positions of amplitudes for the 2 velocity components u, v and the density field h. (**b**), (**c**), (**d**) The definitions of u, v, h. All basis functions (dotted) for this low order have discontinuities. The black points are the points of definition for the corresponding amplitudes. Background are dashed. In the horizontal plane, the cell shown in (**a**) is plotted, and the vertical coordinate shows the functional value, which is v for (**b**), u for (**c**), and h for (**d**)

1981), where the flux fl^x is defined at point $(x_{i-\frac{1}{2}}, y_j)$, the amplitude is called $fl^x_{i-\frac{1}{2},j}$, and analogously the amplitude for fl^y is $fl^y_{i,j-\frac{1}{2}}$. As a test problem, we take the shallow water equations (Eq. (58) in Chapter "Simple Finite Difference Procedures") with variables u, v, h, where u and v are the velocity components and h may be interpreted as density.

Arakawa C-grid FD schemes (Steppeler 1990) can be described by a conformal piecewise spline representation, where the polynomials are low order in comparison with the conformal schemes described in Steppeler et al. (2008). The basis functions are piecewise constant and piecewise linear. In order to create the low-order staggered representation, at first for regular grids, we assume the cell $\Omega_{i,j}$ possesses of the point $\mathbf{r}_{i,j}$ at the lower left corner. The center of cell is $\mathbf{r}_{i+\frac{1}{2},j+\frac{1}{2}}$. Then, we introduce the characteristic function $\chi_{i,j}(\mathbf{r})$ of the cell $\Omega_{i,j}$:

$$\chi_{i,j}(\mathbf{r}) = \begin{cases} 1, & \text{for } \mathbf{r} \in \Omega_{i,j}, \\ 0, & \text{otherwise}. \end{cases} \tag{21}$$

In Eq. (21), the indices $i + \frac{1}{2}$, $j + \frac{1}{2}$ were suppressed.

This representation, same as the FD C-grid applications, is mainly used on regular grids. With unavoidable irregularities, modelers have gone to great lengths to make the grid approximately regular. On regular grids, super-convergence to second-order approximation normally occurs. The application to a truly irregular mesh allows second-order approximation only in special cases (Steppeler and Klemp 2017). At the positions of the u and v, fluxes $fl^x = uh$ and $fl^y = vh$ must be defined. This requires an interpolation of χ. Again, the most naive approach of using linear interpolation will lead to second order for the whole difference scheme only in special occasions. The representations for the field h and the fluxes are

$$
\begin{cases}
h(\mathbf{r}) = \displaystyle\sum_{i,j=1,2,3,\ldots} h_{i,j}\chi_i(x)\chi_j(y), \\[2ex]
fl^x = \displaystyle\sum_{i=\frac{1}{2},\frac{3}{2},\frac{5}{2},\ldots,j=1,2,3,\ldots} fl^x_{i,j}e_i(x)\chi_j(y), \\[2ex]
fl^y = \displaystyle\sum_{i=1,2,3,\ldots,j=\frac{1}{2},\frac{3}{2},\frac{5}{2},\ldots} fl^y_{i,j}\chi_i(x)e_j(y).
\end{cases}
\tag{22}
$$

Even for a regular grid and a rectangular discretization area, it is necessary to observe boundary conditions for advection. In x-direction, we use periodic boundary conditions:

$$
fl_{i,j} = fl_{i+i_e,j},
\tag{23}
$$

where $i_e dx$ is the length of the interval: i element of $(0, 1, \ldots, i_e)$. In Eq. (23), fl can stand for h, fl^x, or fl^y. (i, j) has its appropriate range: for example, for $fl = fl^x$, we have $i = \frac{1}{2}, \frac{3}{2}, \frac{5}{2}, \ldots$.

For the top and bottom boundaries in our tests, we use the condition $h(\mathbf{r}) = 0$ for h being on the boundary. This is a possible boundary condition when the vector (u, v) is pointing into x-direction, meaning $v = 0$. For the tests with constant velocity vectors $(u, v) = (u_0, v_0)$, the condition is wrong, and in practice, the boundary may become a source of errors or numerical noise. For simple tests, one gets away with such wrong boundary condition for advection, when $h = 0$ in the vicinity of the boundary. Of course, it is possible to use periodic boundary conditions also for the top and bottom boundaries.

The cell system used for discretization is shown in Fig. 4. One cell is the square cell of size $dx \cdot dy$, where $\mathbf{r}_{i+\frac{1}{2},j+\frac{1}{2}}$ is in the middle of the cell $\Omega^0_{i,j} = (x_{i-\frac{1}{2}dx}, x_{i+\frac{1}{2}dx}) \times (y_{j-\frac{1}{2}dy}, y_{j+\frac{1}{2}dy})$. We call the support of an amplitude, the set of points where the corresponding basis function, such as $e_i(x)$, $e_j(z)$, is different from zero. The support of $\chi_{i,j}$ is a square of length dx and dy. It is formed for each of the cells $\Omega_{i,j}$ by the four corner cells, which have $\mathbf{r}_{i,j}, \mathbf{r}_{i+1,j}, \mathbf{r}_{i,j+1}, \mathbf{r}_{i+1,j+1}$ as four corners. A similar cell system is obtained for grid points $\mathbf{r}_{i-\frac{1}{2},j}$ for fl^x and $\mathbf{r}_{i,j-\frac{1}{2}}$ for fl^y. This grid system is shown in Fig. 4.

We assume the lower boundary as a straight line:

$$y = y_0. \tag{24}$$

From Eq. (22), we obtain the following discretization formula for the mass M of the system:

$$M = \sum_{i,j=1,2,3,\dots} \int_{y_0}^y \int_{x_0}^{x_{ie}} \chi_{i+\frac{1}{2}}(x) \chi_{j+\frac{1}{2}}(y) h(\mathbf{r}_{i+\frac{1}{2},j+\frac{1}{2}}) dx \, dy \tag{25}$$

$$= h(\mathbf{r}_{i+\frac{1}{2},j+\frac{1}{2}}).$$

The discretization given above is valid for the inner points of the computational area. Boundaries must be considered at all sides of the computational domain. Of particular interest is the lower boundary as this represents the orography, which has a direct impact on the weather forecast.

The dynamic equations for a dry atmospheric model will be given in the next section. The RHS of the dynamic equations consists of two parts: one part is the advection in a velocity field that is the test problem used a lot in the preceding sections. Another part describes the fast waves of the atmosphere, such as sound waves. A full overview about waves of the atmosphere is given in any book on theoretical meteorology, such as Pedlosky (1987). The problem of posing boundaries for fast waves is quite different than that for advection. The main difference is that fast waves in x-direction have two speeds of propagation, which differ in sign. Both reflecting and open boundaries are therefore possible with fast waves. For a given velocity, advection has only one speed, which is in the direction of the velocity vector. Open boundaries are possible with advection, but no reflection. Attempts to create boundary conditions for the advection process often create very noisy or even unstable solutions. Conservation of mass is only possible when the velocity vector is in the direction of the boundary. A velocity with a component vertical to the boundary will lead to a loss of mass through the boundary that is called the open boundary formulation. In Section "A Conserving Version of the Cut-Cell Scheme", more information will be given. The open boundary formulation will automatically lead to mass conservation and an advection following the boundary when the velocity is parallel to the boundary. A suitable scheme will be given in Section "A Conserving Version of the Cut-Cell Scheme". A lack of care in the formulation of boundary conditions for advection can lead to very noisy solutions even when the mountain representing the lower boundary condition is smooth (Steppeler and Klemp 2017).

Triangular cells of irregular shape can be used to describe a curved boundary when the orography is represented by a linear spline. The boundary condition of advection is that the velocity vector is parallel to the piecewise linear spline representing the orography. A pointwise representation of the boundary condition is possible. This means that the boundary condition is satisfied at every point of the boundary, rather than only by integrals along the boundary. Note that a pointwise representation of the boundary condition necessarily implies the representation of one of the velocity components by a discontinuous spline.

Another option is illustrated in Fig. 5 for the case of a staggered grid. With this method, the grid is assumed to be regular or approximately regular. The square cells containing a piece of the orography are called cut cells. Within the model area, the fields are considered to be polynomials within a cell. For the case of a low-order staggered system treated above, the polynomials are up to first degree. This means that when they are known for a part of a cell, they can be extended to the whole cell. They can be represented by amplitudes in the same way as for uncut cells. As we need to satisfy boundary conditions, some of these amplitudes must be created by the process of posing boundary values. This will be the subject of the following sections. So far, only very simple cases of posing boundary values have been explored.

A Simple Cut-Cell System Based on the Staggered Low-Order Basis Functions

This section explores the representation of a curved lower boundary in models for the representation of mountains. The overwhelming majority of models uses numerical quadrilateral grids. To represent mountains, a transformation of the vertical coordinate is done. This method is described in Durran (2010) and called the terrain-following coordinate method. Here a special example is given for a large number of similar approaches (Adcroft et al. 1997). Another scheme will be described in Section "A Conserving Version of the Cut-Cell Scheme". The model presented is very simple and not mass conserving. However, it is suitable to avoid the problems of noise generation at the lower surface (Shaw and Weller 2016). An even more simple scheme would be obtained by assuming the mountains to be represented by steps, meaning by piecewise constant basis functions. This latter approach, however, has been shown to be non-convergent (Gallus and Klemp 2000) and is not treated in this book.

Here we want to describe the mountains represented by irregularly shaped cells. A simple method is to use regular cells and represent the mountain by a linear spline, which cuts polygons out of the regular quadratic cells, as shown in Fig. 5. Cells that are not cut by the orographic line are squares. As seen from Fig. 5, the orographic line may cut out triangular, quadrilateral, and pentagonal cells. The 2D example is sufficient to describe the principle. The transition to 3D adds complexity, but no essential additional difficulty. When in 2D we have 3 cell shapes that are triangle, quadrilateral, and pentagon, for the case of 3D the number of shapes of the cut cubes is about 50. Here we use the very simple approach of bringing the linear function system in Section "Staggered Grid Systems and Their Basis Function Representation" to cut cells (Steppeler et al. 2006, 2002).

Cut cells are here described within the framework of L-Galerkin schemes. In this section, we describe a simple cut-cell method (Steppeler et al. 2006) within the simple FD schemes described in Chapter "Simple Finite Difference Procedures".

For the mountain height $o(x)$, we assume

$$o(x) = \frac{o_0}{1 + \left(\frac{x}{a}\right)^2}, \tag{26}$$

where the parameter a determines the half-width of the mountain and o_0 is the highest height of the mountain. This problem is solved by the nonhydrostatic equations (Dudhia 1993). The name nonhydrostatic means that the hydrostatic approximation (Braun 1958) is not applied, and thus the equations are the same as the Euler equations used in fluid dynamics when there is no diffusion. Our example does not contain physical diffusion (see Chapter "Local-Galerkin Schemes in 1D"), as this is only a simplified test. In computational fluid dynamics as well as in meteorology, most of the time a physical diffusion is applied.

The nonhydrostatic dynamic equations are

$$\begin{cases} u_t = -\mathbf{u} \cdot \nabla u - \frac{1}{\rho} \frac{\partial p}{\partial x}, \\ w_t = -\mathbf{u} \cdot \nabla w - \frac{1}{\rho} \frac{\partial p}{\partial z} - g, \\ p_t = -\mathbf{u} \cdot \nabla p' + g\rho_0 w - p\frac{c_p}{c_v} \nabla \cdot \mathbf{u}, \\ T_t = -\mathbf{u} \cdot \nabla T - \frac{p}{\rho c_v} \nabla \cdot \mathbf{u}, \end{cases} \tag{27}$$

in which $\mathbf{u} = (u, w)$ are the velocities in x- and z-directions, p is the pressure, T is the temperature, and ρ is the density. The thermodynamic variables are connected by $p = \rho RT$, with R being the gas constant and c_p and c_v are the specific heats for constant pressure and volume. This form is called the non-conservation form of dynamic equations, as on the RHSs there are not divergences. The name comes from the fact that conserving forms of the equation of motion are often used to derive conserving approximations. The $'$ on the variable p means that this is the deviation of pressure to a reference pressure p_0, which is a horizontally layered reference field p_0 that depends on z only. For our test solution, all fields u, w, p, T are small deviations from a reference field that is constant in x-direction. So we can define the linearization background, which is the constant field that approximated the solution best. For realistic fields meaning the observed weather, the deviations of the field to the linearization background are huge. This means that these fields cannot accurately be linearly approximated. The linearization background field is reasonably defined only for the idealized solution such as considered here.

Note that by having a larger and realistic difference between the background state and the atmospheric field, this test can become more severe. For example, a perfectly noise-free test solution with a background field being as near as possible to the atmospheric solution may become noisy and wrong after adding a constant of

20° Celsius to the background temperate profile. For the analytic solution, there is no difference between these two solutions.

For the cut-cell solution discussed here, the reference state does not matter as the solution does not depend on it. For terrain-following models (Durran 2010), the choice of this reference field matters. In fact, it was introduced to make terrain-following solutions more accurate. For the gravitational test case considered here, the background state is often chosen as the linearized background state. This means that the p' occurring in Eq. (27) is a small field. For the real atmosphere, this choice is not available. The mountain wave test used here with the cut-cell approach can therefore be made more demanding by choosing the background field more general and not as the linearization field. An example will be given in Chapter "Numerical Tests".

It will be shown in the end of this chapter that taking the linearization profile as background field makes the mountain wave test less severe. This concerns the terrain-following models, as the cut-cell solution is independent of the choice of the background state.

Equation (27) is called the dry adiabatic equation, as no moisture variable is used and therefore no heating by condensation can be observed. Our test problem uses initial values of constant $u = u_0$ and $w = 0$, which for the analytic solution is not possible, as the boundary conditions at the mountain are not fulfilled. The surface boundary conditions are created numerically, and the solution is integrated until a stationary state is achieved.

The solutions of Eq. (27) are well-known and depend on the width of the mountain determined by the parameter a in Eq. (26). If the mountain width is larger than 10 km, the so-called hydrostatic mountain wave is produced, consisting of areas of positive and negative vertical velocities above the mountain. The analytic solution is known, and this consists of a flow following the mountain. This is because we have a solution without friction (Euler equations). If the Navier–Stokes equations (Stull 2018) were used, the shedding of fluid vortices, as known in fluid mechanics, may occur. For mountains of moderate height up to 700 m, these solutions are regularly seen in numerical approximations (Saito and Doms 1998).

To define an approximation for Eqs. (26)–(27), the different terms of these equations are treated differently. First, we define the fast wave part of u, w, p, T as

$$\begin{cases} u_t^{fa} = -\frac{1}{\rho}\frac{\partial p}{\partial x}, \\ w_t^{fa} = -\frac{1}{\rho}\frac{\partial p}{\partial z} - g, \\ p_t^{fa} = g\rho_0 w - p\frac{c_p}{c_v}\nabla \cdot \mathbf{u}, \\ T_t^{fa} = -\frac{p}{\rho c_v}\nabla \cdot \mathbf{u}. \end{cases} \tag{28}$$

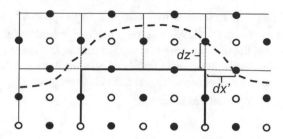

Fig. 5 The cut-cell grid structure. The flux limiters dx' and dz' are used for the approximation of the fast waves only. The cells are assumed to be hollow for the approximation of the fast waves. This will change the frequency of the fast waves, which does not have much impact on the meteorological solution. For the advection terms, the flux limiters are not used. Using the flux limiters for advection would lead to very noisy solutions. For advection, a change of oscillation frequency is not acceptable. The advection uses the velocity adjustment from the fast waves but blocks mountains, which in this example means that the mountain is represented by the middle cell in the lower row only. The cells are indicated by thin black lines, the thin walls by thick dotted lines, and the block orography by thick black lines. The definition points of u and w are black, and the white points are the definition points of T and ρ

Using these definitions, we can write Eq. (27) as

$$
\begin{cases}
u_t = -\mathbf{u} \cdot \nabla u + u_t^{fa}, \\
w_t = -\mathbf{u} \cdot \nabla w + w_t^{fa}, \\
p_t = -\mathbf{u} \cdot \nabla p' + p_t^{fa}, \\
T_t = -\mathbf{u} \cdot \nabla T + T_t^{fa}.
\end{cases}
\tag{29}
$$

The grid for the solution is shown in Fig. 5. It shows a quadratic cell structure. The corners of the cells have indices i and k. For the cell centers and edges, half-indices are used. Note that with the cut-cell approach, different cell structures and mountain representations are used. For the fast wave part, the thin-wall approximation is used, which means that the mountains are considered to be hollow, and thus all cut cells have the same volume $dxdz$ (Steppeler et al. 2002). The advection terms are treated as if the thin walls were not present. So for the advection terms, the mountains are represented as block mountains. A refinement of this advection treatment would pose boundary conditions for u and w such that at the surface the flow would follow the mountain. In this simple treatment, we do not use the conservation form of equations and the approximation is non-conserving. Another reason for non-conservation is that boundary conditions for the time derivatives from uncut cells are posed. This latter difficulty can be solved by posing boundary conditions for fluxes in x- and z-directions instead. This will be described in Section "A Conserving Version of the Cut-Cell Scheme".

p and T are defined at cell centers and have indices $i + \frac{1}{2}, k + \frac{1}{2}$ that are coordinates for the cell center. We use the Arakawa C-grid structure (Arakawa and Lamb 1977; Gresho et al. 1977), where u and w are defined at the edges of the cells, $u_{i,k+\frac{1}{2}}$, $w_{i+\frac{1}{2},k}$, as defined in Section "Staggered Grid Systems and Their Basis Function Representation". We assume a piecewise constant representation of p and T, and we use the thin-wall approximation for the fast wave tendencies. This means that we consider the cut cells as totally filled with fluid and the amplitudes of p and T defined at the cell centers. To get the cell average of $\overline{D}_{i+\frac{1}{2},k+\frac{1}{2}}$, integrate over a cell $\Omega_{i,k}$ and apply Stokes theorem:

$$
\begin{aligned}
\overline{D}_{i+\frac{1}{2},k+\frac{1}{2}} &= \frac{1}{V_{\Omega_{i+\frac{1}{2},k+\frac{1}{2}}}} \int_{\Omega_{i,k}} (\nabla \cdot \mathbf{u}) \, dx dz \\
&= \frac{1}{V_{\Omega_{i+\frac{1}{2},k+\frac{1}{2}}}} \left(dx'_{i+\frac{1}{2},k+1} w_{i+\frac{1}{2},k+1} - dx'_{i+\frac{1}{2},k} w_{i+\frac{1}{2},k} \right) \\
&+ \frac{1}{V_{\Omega_{i+\frac{1}{2},k+\frac{1}{2}}}} \left(dz'_{i+1,k+\frac{1}{2}} u_{i+1,k+\frac{1}{2}} - dz'_{i,k+\frac{1}{2}} u_{i,k+\frac{1}{2}} \right),
\end{aligned}
\tag{30}
$$

in which dx' and dz' are the lengths of the parts of the corresponding edges that are not inside the lower boundary. For uncut cells, we have $dx' = dz' = 1$, and Eq. (30) becomes the ordinary FV approximation for square grids (Durran 2010). We assume that the volume is not that of the part of the cell outside the orography. We imagine that the cell is hollow, and we can imagine that the orography is formed of thin cardboard walls. Then we have

$$
V_{\Omega_{i+\frac{1}{2},k+\frac{1}{2}}} = dx' dz'. \tag{31}
$$

And Eq. (30) is an approximation of the divergence. This allows to compute the fast time derivatives from Eq. (28). For the advection term, for example, it is possible to make the same thin-wall approximation, but this is very inaccurate and leads to bad approximations.

For the advection term, we consider that for each cell center of a cut cell, there is a neighbor being uncut. The next neighbor in the same direction i or k is also an uncut cell. In fact, we can go away from the boundary until approximations of the advection terms are found which use cut cells. From these terms, we obtain the time derivatives of advection in Eq. (29). This approximation, called the thin-wall cut-cell approximation, is easy to implement. In the way used here, it is non-conserving, and it does not enforce the boundary condition for u and w that the velocity is in the direction of the surface for points x, y on the surface. A lack of conservation happens because nothing requires the velocities inside a cut cell to be parallel to the orography. Conservation is achieved when the velocities in cut cells are parallel to the orography. In realistic integrations, this is approximately the case. A conserving scheme would be obtained when boundary conditions to the velocities in the cut cells are imposed such that they are parallel to the orographic line. No investigations of such conserving schemes have been done so far.

The corresponding cut-cell approximations based on L-Galerkin methods are supposed to result in mass conserving approximations. Boundaries for velocities such as those at the surface are requested. This follows from the fact that the fields are represented by functions within cells. Therefore, many conversation properties of the undiscretized problem transfer to the L-Galerkin method. When realizing this example on the computer, some dirty tricks are necessary. We prescribe $u = u_0$, $w = 0$ as lateral boundary conditions. This is not fully correct. In particular, the advection process does not have these boundary conditions. As the fields are defined at different places of the cell, linear interpolation is used whenever a multiplication is necessary. This is usual for Arakawa schemes. As here we use the staggered grid for advection only in uncut cells, these interpolations are always with equal weights. For the divergence terms, multiplication is necessary only for cell centers where divergence and p and T have their point of definition. It is not known if a correct open boundary formulation would help to avoid some of such tricks.

With a small mountain height, all numerical approximations are accurate. So for this case the terrain-following approximation produces good results that coincide with the analytic solution well. The computed vertical velocities for the cut-cell solution and the terrain-following solution (Durran 2010) are shown in Fig. 6. The solutions are rather similar and accurate (Saito and Doms 1998).

Very often mathematicians of the weather are also enthusiastic about the weather itself. Therefore, a remark about the physical interpretation of Fig. 6 may be in order. When the horizontal scale of mountains comes into the order of $10\,km$ and less, marked wave systems involving vertical velocities are created which reach high into the atmosphere. This explains why small mountains often produce strong rain patterns, which cannot be explained by the air just going up and down along the mountain. Due to an increase in available computer power, around 2005 nonhydrostatic models became available for daily forecasting (Steppeler et al. 2006; Giraldo et al. 2003; Wilson and Ballard 1999), which were able to compute flows on

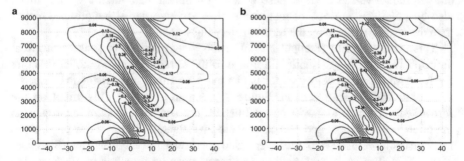

Fig. 6 The gravitational wave caused by a mountain of $400\,m$ height. (**a**) Computer by the cut-cell method (**b**) computed using the terrain-following coordinate. For this shallow well-resolved mountain using the reference temperature identical with the horizontally averaged solution, both methods are simulating the analytic solution very well (Reproduced from Steppeler et al. 2002, © American Meteorological Society, used with permission)

this scale within the framework of operational weather models. The flows computed in this scale had a direct bearing on local precipitation and what the non-weather specialist experiences as weather, and this change of modeling was noticed by the general public.

The transition to the 3D case adds a lot of complexity but no real difficulty. The cut-cell model presented here has been implemented in 3D and provided with interfaces to real data and a full physics package (Doms and Schättler 2002). The results shown in Figs. 3, 4 in Chapter "Introduction", are from this model. The cut cells perform much better than the terrain-following coordinate approximation. This is not indicated by Fig. 6, where both methods are accurate because of the smooth and shallow mountain used as test. Figure 6 shows the highly idealized case of a shallow smooth and well-resolved mountain. For this case, all methods have good results.

In real-life models of the atmosphere, mountains are less smooth than in our test example. In fact, the results of models tend to be more accurate when very small smoothing of the orography is done (Cannon et al. 2017). Also, when the mountain becomes high, terrain-following models produce vertical velocities that are unrealistic. This may lead to strong thunderstorms even if the atmosphere is stably stratified. A stable atmosphere is horizontally homogeneous and layered in such a way that the upward movement of an air parcel means a loss of energy of this parcel. Figure 7 shows the result of a 3D version of the model presented here. The test was made less lenient, as the linearization background was not chosen as the numerical background field. The difference of linearization background and numerical background was $20\,°C$. The atmosphere is supposed to be at rest, and velocities of $0.25\,m/s$ were produced, which are unrealistic. The cut-cell version produced the atmosphere at rest.

As this 3D model had a full physics implementation, we could switch on radiating package. The cooling in the night produces a valley wind, whereas the heating during the day produces a mountain wind. This is shown in Fig. 7 for the cut-cell model version. While these winds with the cut-cell version were realistic also quantitatively in order of magnitude, they are an order of magnitude smaller than the errors produced by the terrain-following version.

This cut-cell model (Steppeler et al. 2006) was extensively tested using real atmospheric data. A systematic difference to the terrain-following model was that the forecasted vertical velocities for land areas were much more realistic compared to the terrain-following version. Figure 7 gives an example. This effect on the forecasted vertical velocities was established in a statistically significant way, as the effect is rather large and 100 cases were investigated. Within a few days, these more realistic vertical velocities were also seen over the large oceans (Steppeler et al. 2013). The impact on medium forecasts, meaning forecasts of 10 days, was also established, which is a consequence of the systematically improved forecasted vertical velocities.

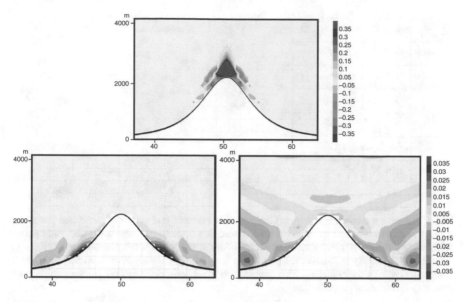

Fig. 7 The atmosphere at rest. Top: the vertical velocity when the initial values are a horizontally stratified stable atmosphere (Reproduced from Steppeler et al. (2006), © American Meteorological Society, used with permission)

A Conserving Version of the Cut-Cell Scheme

The cut-cell system described in Section "A Simple Cut-Cell System Based on the Staggered Low-Order Basis Functions" treats the terms for the fast waves by the "thin-wall approximation," and this approximation is conserving. For the advection part of the dynamic equations (Eq. (27)), a different approximation is used. The discretization is done for the uncut cells only, and the advection time derivatives are obtained by posing boundary conditions. The extrapolation is not done by linear extrapolation but rather by using the constant functional extension. Because of this lack of conservation in the jargon of numerical modelers, this method is "quick and dirty" and was used to create the results shown in Chapter "Introduction". When designing discretizations for cut cells, some numerical schemes produce noisy results (Steppeler and Klemp 2017). In this section, a noise-free discretization of the advection term is given. It will be given for the low-order field representation given in Section "A Simple Cut-Cell System Based on the Staggered Low-Order Basis Functions". The example is given for pure advection and has to be combined with the discretization of the fast waves on Section "A Simple Cut-Cell System Based on the Staggered Low-Order Basis Functions" to get the discretization for the atmosphere. As we are treating the pure advection problem, we consider the velocity field as given.

We assume a C-grid representation for velocity components u, w and density h. h is the dynamic field. u and w must be defined to create a non-divergent field.

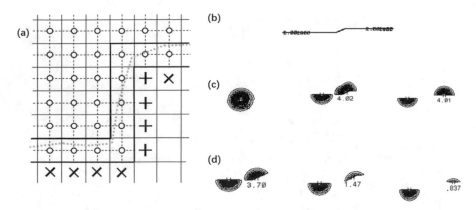

Fig. 8 Advection along orography. (**a**) the grid. Thin dotted: grid cells, "o" definition points for h, thick solid line: the area of cut cells, thick dotted line (green): the orographic line, "×": points for boundary values of fl^z, "+": points for boundary values of fl^x. (**b**) The orographic line, (**c**) a forecast of a density field along the orographic line, (**d**) as (**c**), with $w = 0$. At the ramp, mass disappears through the boundary. This case has no mass conservation. This amounts to an open boundary formulation for advection

The test example uses a constant u field and w being constant in z-direction. The orography $o_{i+\frac{1}{2}}$ ($i = 0, 1, 2, 3, \ldots$) is defined at u-points, and $o(x)$ is defined as a piecewise linear spline. The grid and orography are shown in Fig. 8.

The advection equation is

$$h_t = \nabla \cdot (\mathbf{u}h), \tag{32}$$

where

$$\begin{cases} u_{i+\frac{1}{2},k} = u_0, \\ w_{i,k+\frac{1}{2}} = u_0 \frac{o_{i+1}-o_i}{dx}. \end{cases} \tag{33}$$

For all cut cells, the field representation belonging to Eq. (33) is

$$\begin{cases} u(x, z) = \sum_{i,k} u_{i+\frac{1}{2},k+\frac{1}{2}} e_{i+\frac{1}{2}}(x) \chi_{k+\frac{1}{2}}(z), \\ w(x, z) = \sum_{i,k} w_{i+\frac{1}{2},k+\frac{1}{2}} \chi_{i+\frac{1}{2}}(x) e_{k+\frac{1}{2}}(z), \\ h(x, z) = \sum_{i,k} h_{i+\frac{1}{2},k+\frac{1}{2}} \chi_{i+\frac{1}{2}}(x) \chi_{k+\frac{1}{2}}(z). \end{cases} \tag{34}$$

Assume a spatially constant density field of value 1. Then the fluxes in x- and z-directions are $-uh$ and $-wh$. For points $(x, z) = (x, o(x))$ on the orographic line, the flux vector must be parallel to it. At every point of the boundary, there is no flux component vertical to the orographic line. This approximative boundary condition is called the pointwise boundary condition, as it is valid at every point of the boundary.

For simplicity, the example given here uses the pointwise boundary condition and low-order field representations. An alternative is the interval-averaged boundary condition. This means that the boundary condition is not valid at every point of the boundary, but it is rather required that for certain intervals of the orographic line there is no mass flux through it. This method is called the interval-based boundary condition, and it is not used here. In our approximation, the flux is not differentiable at points $x_{i+\frac{1}{2}}$. It follows that for a general density h the fluxes are defined as discontinuous functions.

From the discussion above, it is seen that there are many ways to formulate boundary approximation. Here just the most simple option is discussed.

We define the flux $\mathbf{fl}(x, z) = (fl^x(x, z), fl^z(x, z))$. In order to follow the most simple method, we obtain flux approximations using amplitudes of u and w at cell centers:

$$
\begin{cases}
u_{i+\frac{1}{2},k+\frac{1}{2}} = \frac{1}{2}(u_{i,k+\frac{1}{2}} + u_{i+1,k+\frac{1}{2}}), \\
w_{i+\frac{1}{2},k+\frac{1}{2}} = \frac{1}{2}(w_{i+\frac{1}{2},k} + w_{i+\frac{1}{2},k+1}).
\end{cases}
\tag{35}
$$

Using Eq. (35), we can define an approximation for the flux:

$$
\begin{cases}
fl^x(x, z) = \sum_{i,k} u_{i+\frac{1}{2},k+\frac{1}{2}} h_{i+\frac{1}{2},k+\frac{1}{2}} \chi_{i+\frac{1}{2}}(x) \chi_{k+\frac{1}{2}}(z), \\
fl^z(x, z) = \sum_{i,k} w_{i+\frac{1}{2},k+\frac{1}{2}} h_{i+\frac{1}{2},k+\frac{1}{2}} \chi_{i+\frac{1}{2}}(x) \chi_{k+\frac{1}{2}}(z).
\end{cases}
\tag{36}
$$

With the cut-cell approach, we assume that the definitions Eqs. (34) and (36) of the fields are considered being above the orographic line. Within each cell, the fields or fluxes are polynomials. For our simple example, the fluxes are constant within each cell. As seen from Fig. 8a, the area of definition may be quadrilaterals, pentagons, or very small triangles. A polynomial is known for all x and z when it is known for a small area. This concerns the representations of density, fluxes, and time derivative of density. So if we have a FD approximation for fields above the orographic line, the field is known for the whole cell. This means in Fig. 8 the cells above and to the left of the thick line are known and determined by boundaries. It also means that we can use the amplitudes to describe fields and fluxes as with uncut cells, even if their position is outside of the dynamic area (above the orographic line).

Using Eqs. (32) and (36), we obtain the following representation of the time derivative of h:

$$
\frac{\partial h}{\partial t} = \sum_{i,k} (fl^x_{i+\frac{1}{2},k+\frac{1}{2}} - fl^x_{i-\frac{1}{2},k+\frac{1}{2}}) \delta(x - x_i) \chi_{k+\frac{1}{2}}(z)
$$
$$
+ (fl^z_{i+\frac{1}{2},k+\frac{1}{2}} - fl^z_{i+\frac{1}{2},k-\frac{1}{2}}) \delta(z - z_k),
\tag{37}
$$

where δ is the Dirac delta function (see Section "Polygonal Spline Solutions Using Distributions and Discontinuities" in Chapter "Simple Finite Difference Procedures"). In order to obtain a FD scheme, it is necessary to approximate the δ functions such as to obtain a representation of the form Eq. (34). The delta function contains mass.

Let us call the area above the orographic line in Fig. 8 the dynamic area, the area above the lower thick line the extended area. The extended area consists of all cut and uncut cells. Even for points below the orography, the field is defined by extension of the polynomials. In all points of the extended area, the density field and the center cell flux amplitudes are defined. For the dynamic approximation Eq. (37), cells outside the extended area are needed. Such outside points are indicated in Fig. 8 by "$\times$" or "+." These cells outside the extended area are called the outside region. These cells of the outside area are used to pose boundary values. From Eq. (37), a representation of h_t with center cell amplitudes $h_{t,i+\frac{1}{2},k+\frac{1}{2}}$ is obtained by integrating the Dirac δ function in the model area and distributes the mass obtained to the corresponding cells. This remapping corresponds to the global spectral method to the forward and backward Fourier transformation. With the global spectral method, we are limited to regular geometries, such as the sphere of square. Other geometries have to be done by coordinate transformation. The L-Galerkin method used here in lowest approximation order is more flexible. The choice of remapping operations is as large as the choice of alternative FD operations. In order to obtain results for the δ functions on the lower thick line in Fig. 8, we need to pose boundary values for the fluxes of fl^x at the points indicated as "+" and for fl^z at points indicated as "$\times$". As the thick line is completely outside the dynamic area, one could think of extending the fluxes at points indicated as "+" or "$\times$" as constants. Doing this normally leads to noise generation at the orographic line (Steppeler and Klemp 2017). Rather popular cut-cell methods put the fluxes to 0 at the boundary of the extended area. This is highly arbitrary and can lead to strong noise generation. For the example shown in Fig. 8, we use linear extrapolation to points "$\times$" and "+". The results given in Fig. 8b show that for a smooth mountain there is no noise generation at the surface.

Mass conservation depends on the velocity vector $\mathbf{u} = (u, w)$ being parallel to the orographic line. If this condition is not fulfilled, the model still works and has no noise generation at the boundary. As there is a flux component vertical to the orographic line, mass will disappear through the boundary. Figure 8b and d shows the result with a velocity $u = u_0$ and $w = 0$. This means that at the orography, there is flux through the boundary and mass is disappearing. In this case, this is an open boundary formulation for advection and could be applied in situations where a limited area model with external boundaries is needed (Davies 1976).

Let the point $\mathbf{r}_{i+\frac{1}{2},k+\frac{1}{2}}$ be a point indicated in Fig. 8a as "+". Then the center point flux boundary value for $fl^x_{i+\frac{1}{2},k+\frac{1}{2}}$ is

$$fl^x_{i+\frac{1}{2},k+\frac{1}{2}} = fl^x_{i-\frac{1}{2},k+\frac{1}{2}} + fl^x_{x,i,k+\frac{1}{2}}dx. \tag{38}$$

The points indicated in Fig. 8a as "×" are treated in an analog way. Figure 8b–d shows the result of an advection experiment using the cut-cell method described above. The orography is a orography where its height changes form one constant value to another in a few grid points. The orography is shown in Fig. 8b. Figure 8c shows the advection along the orography using a velocity $u = u_0 = const$. The w component of the velocity is constant for each index $i + \frac{1}{2}$ and defined in such a way that the velocity vector is parallel to the orographic line. As the orography is assumed to be a piecewise linear spline, the functional representation of w is necessarily discontinuous. The structure is to be advected along the orographic line. The noise generation found with some methods to pose boundary values is absent in Fig. 8c. The method described above can be considered as a generation of Steppeler and Klemp (2017) to the case of curved boundaries. Steppeler and Klemp (2017) used a mountain consisting of a straight line and found noisy advection for some methods. Figure 8d is the same as Fig. 8c, but the w velocity is put to 0. As then for some part of the mountain the velocity is not parallel to it, there is a flux through the boundary and mass disappears as shown in Fig. 8d.

Full and Sparse Hexagonal Grids in the Plane

Abstract Hexagonal grids are constructed by combining triangular patches. On the sphere, a few pentagonal cells are used.

Keywords Hexagonal grids · Hexagon indices · Hexagonal basis functions · Hexagon differentiation · Grid isotropy

When looking at the plotted triangular grid (Fig. 1 in Chapter "Introduction"), it is hard to avoid seeing hexagonal cells. In this chapter, we assume that the Earth's radius in Fig. 1 in Chapter "Introduction" is large, so that we have patches of grid that are practically on a plane. For $\varphi = \frac{\pi}{3}$, the hexagons as well as the triangles are regular, and the grid consists of regular triangles if patch of triangles is regular. Note that in Fig. 1 in Chapter "Introduction" some cells are pentagonal, which leads to small irregularities at neighboring hexagonal cells. Even for $\varphi = \frac{\pi}{2}$ (irregular), hexagonal cells can be constructed. Together with the full grid, the situation is rather similar to the case of triangular grids. In the model MPAS (Skamarock et al. 2012), the density amplitudes are defined at the centers of the hexagonal cells. This is equivalent to a triangular grid where the amplitudes are at the edges. A low-order grid on square cells was treated in Section "A Simple Cut-Cell System Based on the Staggered Low-Order Basis Functions" in Chapter "Finite Difference Schemes on Sparse and Full Grids". In this chapter, we treat the high-order grid, where all fields are defined at the same point. As opposed to this, the MPAS model (Skamarock et al. 2012) is staggered.

In this chapter, we treat the case of a sparse hexagonal grid as an example where all dynamic points are on cell surfaces, being either corner or edge points. Successful programming depends on a good way of indexing the points inside a hexagon. The grid of hexagonal cells is shown in Fig. 1. Figure 1 in addition to the hexagonal cell boundaries shows auxiliary lines that are the coordinate lines used to index the hexagonal cells. The division of a cell into triangles is not shown. This is obtained by connecting all opposing corners of the hexagonal cells. The reader should compare Fig. 1 with Fig. 1c in Chapter "2D Basis Functions for Triangular and Rectangular Meshes", which shows the triangular cells without highlighting hexagons. Note

J. Steppeler, J. Li, *Mathematics of the Weather*, Springer Atmospheric Sciences,
https://doi.org/10.1007/978-3-031-07238-3_6

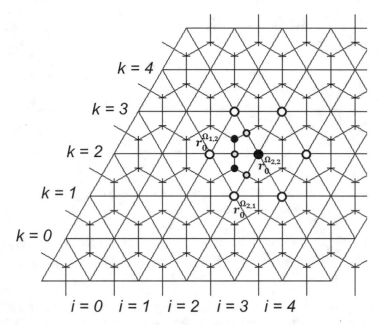

Fig. 1 A hexagonal cell grid for rhomboidal geometry. Corners are indicated as "+." The lines orthogonal to the edges are indicated, and the cell centers are where the orthogonal lines cross. The large black point is $\mathbf{r}_0^{\Omega_{2,2}}$, which is chosen as the example to compute the divergence. The small black corner points are the compact corner points belonging to $\mathbf{r}_0^{\Omega_{2,2}}$. The small white points are the compact edge points for $\mathbf{r}_0^{\Omega_{2,2}}$. Note that for each small white point, there exist 2 amplitudes, as for the sparse o3 representation we have two amplitudes in the sparse grid. These are either the second and third directional derivatives or two grid points. For the QUASAR system, amplitudes are defined at only the small black points, and these are filed with amplitudes and derivatives in two directions, meaning 6 amplitudes per cell. The large white points are neighbors of $\mathbf{r}_0^{\Omega_{2,2}}$ needed to compute the stencil to compute time derivatives for the dynamic points belonging to $\mathbf{r}_0^{\Omega_{2,2}}$

that Figs. 1 and 1c in Chapter "2D Basis Functions for Triangular and Rectangular Meshes" show the cells with center and corner points. We here use a collocation grid for the full collocation grid version with grid length $\frac{ds}{3}$ for the third-degree polynomial representation. This full collocation grid corresponding to third order is shown in Fig. 1c in Chapter "2D Basis Functions for Triangular and Rectangular Meshes".

Indices and Basis Functions of Hexagonal Grids in a Plane

In this chapter, we use the center point indexing and we treat the regular cell case, even though the method is applicable for irregular cells. An example of homogeneous advection with the o3o3 method on regular hexagons is given in Steppeler et al. (2019). The grid is shown in Fig. 1. The indexing of cells is done for rhomboidal coordinate lines, and this fits well for applications on the sphere according to Fig. 1 in Chapter "Introduction". Here, we use a square shape for the position of the center points in the area covered by cells. The indices i, k run in $i \in 0, \ldots i_e, k \in 0, \ldots, k_e$. Note that from Fig. 1, it can be seen that at the boundary of the area covered by hexagonal cells some half cells occur. When imposing periodic boundaries in x-direction, two half cells come together to form a complete hexagon. Call a hexagon $\Omega_{i,k}$, and let the length of a hexagonal side be s and call the height of a triangle $h = \frac{\sqrt{3}}{2}s$. For the position of the hexagon's center point $\mathbf{r}_0^{\Omega_{i,k}} = (x_0^{\Omega_{i,k}}, y_0^{\Omega_{i,k}})$, we have for the rhomboidal numbering:

$$
\begin{cases}
x_0^{\Omega_{i,k}} = 2ih + 2h, & \text{for } i, k = 0, 1, 2, 3, \ldots, \\
y_0^{\Omega_{i,k}} = \frac{3}{2}ks + \frac{3}{2}s, & \text{for } i, k = 0, 1, 2, 3, \ldots.
\end{cases}
\tag{1}
$$

In Steppeler et al. (2019), the o3o3 method was applied on the hexagon. Here we use the o3 field representation. The situation is shown in Fig. 1. For o3o3, compared with Fig. 1 in Chapter "2D Basis Functions for Triangular and Rectangular Meshes", we have 4 collocation points on each edge of the hexagon (boundary of the hexagon) and on the edges of the triangles from which the hexagon is formed. Some points are shared with other edges. For the full grid, the hexagonal center is a collocation point. For each of the 6 triangles of a hexagonal cell, an inner triangular point and two collocation points on 12 triangular edges and 6 corner points are used. This results in 37 collocation points (6 corner + 2 × 12 edge + 6 inner + 1 center points). From these, 27 are independent and form the compact grid for the full grid. From the 27 points, some are diagnostic with a sparse grid. For the sparse grid, only points on the surface of a hexagonal cell are chosen as dynamic points, which are 6 corners and 12 edges meaning 18 dynamic points per cell. Some of these are shared between cells, so that there are less than 18 compact points per hexagonal cell. Corner points are shared between three hexagons and edge points between two. This results in $2 + 6 = 8$ independent dynamic amplitudes per cell, which in the QUASAR system are reduced to 6 amplitudes. For the example of the o3o3 method, this means 8 points of the compact grid will form the sparse compact grid and will be used as dynamic points. These are corner and edge points on the outer boundary. In the non-compact and redundant grid, 8 of the 37 points form the dynamic amplitudes of a hexagon with o3o3. Relative to the center point, we define the 6 corner points of the hexagon and all other collocation points as spectral amplitudes. The sparseness factor for hexagons is larger than that of rhomboids. The full grid has in each hexagon 27 points. With 8 dynamic compact points per cell for the sparse grid, the sparseness factor is $\frac{27}{8}$. This already promises a computing saving, but in 3D the sparseness factor is $27\frac{3}{12} = 6.7$, which becomes $3\frac{27}{8} = 10$ if a differentiable field representation (QUASAR) is used (see Section "1D Example for the QUASAR System" in Chapter "Numerical Tests").

For some operations, such as plotting, the full grid may be used, and for easy programming, it may be useful to create the full grid of a hexagon. For efficiency on multiprocessing computers, it is necessary to obtain a compact storage. This means that a data parcel can be defined such that there are no unused points in it. Unused points are those that are not dynamic points, but rather diagnostic. For plotting and other operations, it may be necessary to transform between full and compact representations.

The cell center storage system to be defined in the following allows to have compact storage, full dynamic point storage, and full grid storage. The full dynamic system is collected from different hexagons, and the points on the full grid are interpolated from dynamic points. The position of each point in a hexagon is defined by the vector relative to the cell center point $r_0^{\Omega_{i,k}}$. From now, the superscript $\Omega_{i,k}$ will be dropped. A point r_p in the grid is written as

$$r_p = r_0 + d_p(cos\varphi_p, sin\varphi_p), \tag{2}$$

where φ_p is the intersection angle of the vector r_p, and the direction of a coordinate line defined by the user, d_p, is the distance from the point to the center of the hexagon. For the corner points, $d_p = s$; for the edge points, $d_p = \frac{\sqrt{3}}{2}s$. As seen above, for the full grid, we have to define 37 points per hexagonal cell, some of them being redundant and most of them being diagnostic with the sparse grid. So a point of a hexagonal cell will be defined by the cell index i, k and a pointer index p, where p runs through the range of collocation points used. This will be 37 or $p \in [0, 36]$ for the full redundant grid. As discussed above, for the compact sparse grid, the range of p will be $[0, 7]$. For an illustration, see Fig. 1c and h in Chapter "2D Basis Functions for Triangular and Rectangular Meshes".

In the full grid, we denote r_0 as the center point, $r_1, r_2, \ldots, r_6$ as the corner points, $r_7, r_8, \ldots, r_{18}$ as the edge points, and $r_{19}, r_{20}, \ldots, r_{36}$ as the interior points of the hexagon. As all inner points are unused or non-dynamic with the sparse grid, the points $r_1, r_2, \ldots, r_{18}$ are a representation of the dynamic points, from which some points are identical to a point in another cell.

The definition of corner and edge amplitudes is illustrated in Fig. 1, in particular the cell $\Omega_{2,2}$, where the compact amplitudes are shown. The corner points are defined as follows:

$$\varphi_p = -\frac{1}{3}\pi p - \frac{5}{6}\pi, \text{ for } p = 0, 1, 2, 3, 4, 5, \tag{3}$$

with $d_p = s$ in Eq. (2). Equation (3) gives all corner amplitudes, including the redundant. If using just the corner amplitudes for $p = 0, 1$, we obtain the set of compact corner amplitudes.

The positions of the edge points and the inner points can be computed in analogy to Eq. (3). This is done for the compact representation that means that in Eq. (3) p has the range $p \in [0, 1]$. For $p = 2, 3, 4$, we define the mid-edge points. At these points, second and third field directional derivatives are defined as amplitudes:

$$\varphi_p = \frac{2}{3}\pi + \frac{2}{3}\pi(p - 2), \text{ for } p = 2, 3, 4. \tag{4}$$

Note that in Eq. (4) similar as in Eq. (1), we can use values of p between 5 and 7 to obtain positions of redundant amplitudes. These are amplitudes that in a compact representation belong to other hexagons. Using $p = 2, 3, 4$ in Eq. (4) and $p = 0, 1$ in Eq. (3) will lead to a compact representation. Note that the positions Eq. (4) are used to store 2 amplitudes, and we have 8 amplitudes per cell for the compact representation.

On an edge, the alternative representations introduced in Chapter "Local-Galerkin Schemes in 1D" for one dimension can be introduced. These are the representation by two collocation points on each edge and first directional derivatives at corner points. These alternative representations using amplitudes defined at different locations are equivalent and can be transformed into each other.

A grid representation of the function $fl_p^{\Omega_{i,k}}$ is obtained if all points in Eqs. (3)–(4) have an associated amplitude. The full grid representation is obtained with $p \in [0, 1, 2, \ldots, 36]$. The dynamic point representation for the sparse grid of o3o3 represents only dynamic points, and we have $p \in [1, 2, \ldots, 8]$. The dynamic points are sufficient to obtain the full grid by interpolation. The full grid representation is non-redundant. Every corner amplitude occurs in three hexagons, and every edge point occurs in two hexagons. The missing points for a sparse representation are to be obtained from other hexagons. The compact representation $fl_p^{c,\Omega_{i,k}}$, $p \in [1, 2, \ldots, 6]$ is for functions being differentiable, and the sparseness factor is reduced. In complete analogy to the 1D case, it is possible to assign amplitudes to all 37 amplitudes of the hexagon. This leads to the representation of discontinuous functions, and such functions may occur by differentiation in an intermediate step when solving the dynamic equations. The property of non-redundancy of the representation becomes clear from Fig. 1c in Chapter "2D Basis Functions for Triangular and Rectangular Meshes" where for the purpose of Chapter "Full and Sparse Hexagonal Grids in the Plane" triangles can be grouped into hexagons.

For efficient programming, it may be convenient to use the redundant representation. This means that we have amplitudes at all collocation points of a cell, even though some of the points on the cell boundary belong also to other cells. For continuous fields, this representation may be obtained by transferring data from other cells to the target cell. If all amplitudes of a redundant representation are chosen independently, discontinuous functions are represented in analogy to the situation in one dimension (see Chapter "Local-Galerkin Schemes in 1D"). Such discontinuous functions may arise by differentiating a continuous function. The compact representation is obtained from the full grid representation by taking only part of the full data set. There are other forms of field representations. The one-sided derivative method can be used in analogy to the quadrilateral case. This consists of storing three directional derivatives for every corner point. These derivatives are dynamic variables, as known from the quadrilateral case. The derivatives are in the direction of the three stencil lines of a corner point (see Fig. 1c in Chapter "2D Basis Functions for Triangular and Rectangular Meshes"). Such alternative amplitude representations are obtained by using the alternative representations described in Chapter "Local-Galerkin Schemes in 1D" for one dimension to each of the edges.

Before we continue with the description of the 2D o3o3 scheme, we give the
sparse 3D representation for the differentiable case, which has the sparseness
factor of 10. The one time differentiable representation is differentiable at corner
points. For the edge boundaries between two cells, the representation will be just
continuous. To obtain the one time differentiable representation, let us start from
the representation of the field within the cell by one-sided derivatives. For each
edge belonging to a corner, we have the field amplitude at this corner as amplitude
plus the directional derivatives for all directions of edges meeting at this corner.
From Fig. 1, we see that at each corner point 3 edges meet in 2D. So the one-sided
derivative representation means that at the corner we have 4 amplitudes: the field
value at the corner and the derivatives in the three directions. For a differentiable
function, one-sided derivatives in two directions are sufficient to compute all
directional derivatives at the corner. Let $\varphi_1, \varphi_2, \varphi_3$ be the angles belonging to the
three edges meeting according to Fig. 1 at the corner. Let s_1, s_2, s_3 be the directional
parameters belonging to these directions such that the directional derivative can be
written as $\frac{\partial fl}{\partial s_i}(i = 1, 2, 3)$. Then we have for a differentiable function for a field
$fl(\mathbf{r})$:

$$\frac{\partial fl}{\partial s_3} = \alpha_1 \frac{\partial fl}{\partial s_1} + \alpha_2 \frac{\partial fl}{\partial s_2}, \tag{5}$$

where α_i are defined in Eq. (50) in Chapter "2D Basis Functions for Triangular
and Rectangular Meshes". According to Eq. (5), the directional derivative belonging
to φ_3 is not an independent derivative, as it depends on the directional derivatives
belonging to φ_1 and φ_2. This means that for a one time differentiable function in
two dimensions, the amplitudes of the compact representation according to Fig. 1
are the two field values at points $\mathbf{r}_1$ and $\mathbf{r}_2$ and the two directional derivatives for
two edges, meaning $2 + 2 \times 2 = 6$ amplitudes for the compact representation.
Compared to 8 amplitudes, we found a further reduction of the sparseness factor for
the differentiable representation. In three dimensions for the simplest case, we have
two further directions up and down at each corner. This means $2 + 3 \times 2 = 8$
compact amplitudes per cell, as compared to $8 + 2 = 10$ amplitudes for the
continuous case. Observing that we have $3 \times 27 = 81$ full grid points per cell,
this explains the sparseness factor 10 for the differentiable case in 3D. To perform
an L-Galerkin scheme or create the full grid for plotting, we need an interpolation
of a representation of the filed $fl(\mathbf{r})$ inside the cells using the dynamic amplitudes.
The amplitudes are given on corners and edges, and for the one-sided derivative
representation, they are corner values of $fl(\mathbf{r})$ and directional derivatives of $fl(\mathbf{r})$.
The interpolation process for a cell needs the full grid representation, meaning for a
field all 6 corner field amplitudes and the 12 edge amplitudes. This means that we
need the full grid representation. The target cell and positions of field amplitudes are
shown in Fig. 2. A field amplitude is given by the cell index $\Omega_{i,k}$ (for the special case
of a structured cell grid) and the pointer parameter p (p points to all 37 points of a
hexagon). For a sparse grid, we have a range of p of 18 points, as with the sparse grid
all amplitudes are on the surface of the hexagon. For the compact representation, we

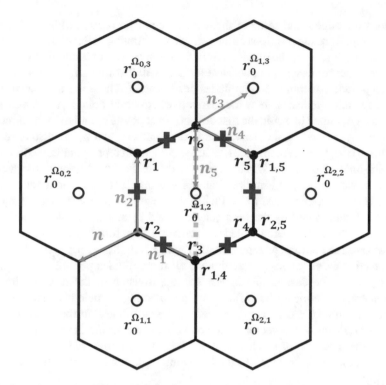

Fig. 2 One of the cells of Fig. 1 with neighboring cells and the positions of some corner point amplitudes (black points) and edge amplitudes (red crosses). The cell is defined by the center point $r_{0,i}$ (white points) of a cell. For the target cell, we have omitted the cell index i. For the compact field representation, we define for each cell a subset of amplitudes, which for corner amplitudes are positioned at $\mathbf{r}_{1,i}$ and $\mathbf{r}_{2,i}$. When using the one-sided derivative representation at these points, the field values and three directional derivatives are stored. In the hexagon at the center, the amplitudes for the sparse grid representation are shown for the case that directional derivatives at edge centers are used as dynamic amplitudes. For the representation using the second and third derivatives along the edges, the positions are in the middle of the edges, the positions being shown as "+." The field values at corners are also amplitudes. For the representation by one-sided derivatives, the positions of all amplitudes are the corner points, indicated as small black points. In this representation, there are 4 amplitudes per point (the field value at the point and one-sided derivatives in three directions). For the representation of differentiable fields, as used with the QUASAR system, only two one-sided derivatives are stored at each black point. The interpolation of a field fl inside a hexagon can be done using triangles, but here we use a division into quadrilaterals. For the cell in the middle, this division is shown by an orange dotted line

store the amplitudes in such a way that points belonging to more than one cell are represented only once. The definition of the compact points of cell $\Omega_{i,k}$ in relation of the center point $\mathbf{r}_{i,k,0}$ of a cell is given using the positions of amplitudes indicated in Fig. 2. The pointer index p is used as lower index. From Fig. 2, it is seen that using the definitions of neighboring cells, we can obtain all corner amplitudes. For example, we have $\mathbf{r}_3 = \mathbf{r}_{1,4}$ or $\mathbf{r}_4 = \mathbf{r}_{2,5}$. This is similar for the edge amplitudes. Edge amplitudes can be second and third directional field derivatives. The directions are along the edges, and some of them are named $\mathbf{n}_1, \ldots, \mathbf{n}_3$ in Fig. 2. The white points in Fig. 2 are the positions of the cell centers. Analog to the corner amplitudes, the edge amplitudes can be obtained from the compact representation of other cells. In the compact representation, we have for each cell 2 corner amplitudes and 3×2 edge amplitudes, meaning 8 amplitudes per hexagon. There are alternative forms of the edge amplitudes, such as first derivatives at corner points. A compact set of variables is obtained by using the first derivatives at points $\mathbf{r}_1$ and $\mathbf{r}_2$ in the directions $\mathbf{n}_3$ and $\mathbf{n}_2$. At $\mathbf{r}_3$ and $\mathbf{r}_6$, the derivatives in directions $\mathbf{n}_3$ and $\mathbf{n}_1$ are also used. This amounts to 6 high-order amplitudes, as in the equivalent representation with second- and third-order derivatives. When we require the fields to be differentiable at corner points, the derivative at $\mathbf{r}_2$ in direction $\mathbf{n}$, for example, is a linear function of the derivative at $\mathbf{r}_2$ in directions $\mathbf{n}_1$ and $\mathbf{n}_2$. This means 6 amplitudes per hexagon, as the field is determined by the field values $\mathbf{r}_1$ and $\mathbf{r}_2$ and 2 directional derivatives at these points. The differentiable representation therefore has greater sparseness compared to the continuous representation with 8 amplitudes per cell. When using first directional derivatives as amplitudes, all 6 amplitudes per cell are defined at corner points. The dashed line in Fig. 2 is the division of the hexagon into two quadrilaterals to define the basis functions interpolating the field inside the hexagon. The center point of a cell, $\mathbf{r}_{i,k,0}$, is used as the cell index. In the following for a function $fl(\mathbf{r})$ and a unit vector $\mathbf{s}$, the notation $fl_{\mathbf{s}}$ means the directional derivative of fl in the direction of $\mathbf{s}$. Therefore, we have for a non-redundant and compact representation $fl^{c,d,\Omega_{i,k}}$ of $fl(\mathbf{r})$ the following representation, which is equivalent to Eq. (5):

$$fl_p^{c,d,\Omega_{i,k}} = fl_{\mathbf{r}_{i,k,p}}, \text{ for } p \in \{1, 2\}, \tag{6}$$

where d indicates the amplitude is directional and

$$
\begin{cases}
fl_3^{c,d,\Omega_{i,k}} = fl_{\mathbf{n}_2,\mathbf{r}_{i,k,1}}, \\
fl_4^{c,d,\Omega_{i,k}} = fl_{\mathbf{n}_3,\mathbf{r}_{i,k,1}}, \\
fl_5^{c,d,\Omega_{i,k}} = fl_{\mathbf{n}_6,\mathbf{r}_{i,k,1}}, \\
fl_6^{c,d,\Omega_{i,k}} = fl_{\mathbf{n}_1,\mathbf{r}_{i,k,2}}, \\
fl_7^{c,d,\Omega_{i,k}} = fl_{\mathbf{n}_2,\mathbf{r}_{i,k,2}}, \\
fl_8^{c,d,\Omega_{i,k}} = fl_{\mathbf{n}_1,\mathbf{r}_{i,k,2}}.
\end{cases}
\tag{7}
$$

We call the regularized representation the representation requiring a differentiable field representation. The regularized representation can be used in the same way as with quadrilaterals and leads to a further increase of the sparseness factor.

Assume that $fl(\mathbf{r})$ at a corner point is differentiable and that φ_1, φ_2 are the angles corresponding to the derivatives $\mathbf{n}_1$, $\mathbf{n}_2$. Let the derivatives to be used as dynamic variables be fl_x and fl_y, with x and y being orthogonal directions. Let $\varphi_1, \varphi_2, \varphi_3$ be the angles of $\mathbf{s}_1, \mathbf{s}_2, \mathbf{s}_3$ with the x-direction. Then for regular hexagons, we have

$$\begin{cases} fl_{\mathbf{n}_1} = cos\varphi_1 fl_x(\mathbf{r}) + sin\varphi_1 fl_y(\mathbf{r}), \\ fl_{\mathbf{n}_2} = cos\varphi_2 fl_x(\mathbf{r}) + sin\varphi_2 fl_y(\mathbf{r}), \\ fl_{\mathbf{n}_3} = cos\varphi_3 fl_x(\mathbf{r}) + sin\varphi_3 fl_y(\mathbf{r}). \end{cases} \tag{8}$$

Directional derivatives at corner points are used as dynamic amplitudes. It is sometimes convenient to compute the high-order part using such derivatives. Let the line $(\mathbf{r}_1, \mathbf{r}_2)$ have the directional derivative $fl_{\mathbf{s}_1}$ defined at $\mathbf{r}_1$ in the direction of $\mathbf{r}_2 - \mathbf{r}_1$, and let the length be $|\mathbf{r}_1 - \mathbf{r}_2| = s$. Then the high-order part of the directional derivative is defined as

$$fl_{\mathbf{s}_1}^{ho}(\mathbf{r}_1 + s(\mathbf{r}_2 - \mathbf{r}_1)) = fl_{\mathbf{s}_1}(\mathbf{r}_1) + s * \frac{fl(\mathbf{r}) - fl(\mathbf{r}_1)}{fl(\mathbf{r}_2) - fl(\mathbf{r}_1)}, \tag{9}$$

in which we have dropped the cell index i, k. $\mathbf{r}_1$ and $\mathbf{r}_2$ are named consistent with the pointer index p according to Fig. 2.

If the high-order part of the one-sided derivative is given at both ends of a line $(\mathbf{r}_1, \mathbf{r}_2)$, then the functional form of $fl(\mathbf{r})$ on this line $(\mathbf{r}_1, \mathbf{r}_2)$ is a third-degree polynomial. As the hexagon is formed by triangles, we could define triangular amplitudes and use the triangular interpolation. This would require the definition of more triangular amplitudes by interpolation. Here we suggest to use the quadrilateral interpolation as a simple method that does not introduce extra triangular corner points, such as the middle of the hexagon. A very simple method is to divide the hexagon by one diagonal and treat the arising quadrilaterals as irregular rhomboids. This division is shown in Fig. 2 as a dashed line.

For each degree of freedom, we can define the natural basis function. We have different representations as described above, such as full grid, sparse redundant grid, and these representations differ only by the range of the variable p used above. For the full redundant grid, it has the range 37, and for the compact grid, the range is 8 with o3o3. For the QUASAR method with functions being differentiable at corners, there are 6 amplitudes per cell. For efficient use of computers and easy programming, the compact representation is important. For compact sparse grids, the example of o3o3 was used for representation. Other schemes, such as o2o3, can be treated in analogy.

Each hexagon has 18 dynamic amplitudes for the redundant grid, which are partly redundant. For each of the 18 amplitudes, there exist basis functions that are obtained by putting one amplitude to 1 and all the others to 0. The natural basis functions are defined in each cell to be discontinuous and 0 outside the target hexagon. When the redundancy of the amplitudes is considered, the same amplitude defines basis functions in more than one hexagon. These combined basis functions are continuous. In particular, the natural basis functions allow to compute the mass associated with an amplitude.

For an amplitude, we can define the boundary values associated with this point. If the interpolated function is a flux, these boundary values define a partial flux into the cells associated with this amplitude. The amplitude associated fluxes of selected points can be combined to compute the time derivative of one of the associated amplitudes. Hexagonal basis functions are not defined in such detail as was done for the triangular basis function. Also the o3o3 scheme is not defined and investigated in detail. Only the o2o3 method was tested for hexagons. This allows to collect the data for a stencil from neighboring hexagons.

Numerical Methods of Hexagonal Grids on the Plane

Many methods can be implemented on a hexagonal grid on the plane, and the programming of such methods may be complicated. It is important to find a method being sufficiently easy to implement. Since the scientific literature has only one example of a sparse hexagonal grid implementation, no detailed recipe of hexagonal sparse implementation can be given. The one existing implementation (Steppeler et al. 2019) came to the conclusion that the choice of indices was a problem for an implementation being sufficiently easy to use. The authors believe that the index system introduced in Figs. 1 and 2 is a step forward toward practical implementation.

Here a rather special method is described, o3o3, which is known in 1D from Chapter "Local-Galerkin Schemes in 1D". The tools provided will allow to implement variants of the described method. It is assumed that in a hexagonal cell there is more than one grid point to form the collocation grid.

The full grid will be described based on the definitions of Section "Indices and Basis Functions of Hexagonal Grids in a Plane". A special sparse grid will be derived from it. Only the principle of defining numerical schemes on sparse hexagonal grids will be described for the example of the o3o3 method. As the implementation of such schemes is in its infancy, the sparse grid used here will have 8 independent points per cell. It should be noted that a hexagon can be composed of triangles where some points are treated diagnostically. So if the full grid is used, the hexagonal grid can also be considered as a triangular discretization. When high-order methods are planned and a set of nearest neighbors are used for the differencing stencil, it comes out that for second-order full grid there is a convenient number of neighboring points where the cloud of points is too big to be practical for third order (Baumgardner and Frederickson 1985). In the version of sparse grids considered here, all collocation points are on the hexagonal surface, and the cloud of points is not too big in the case of orders 3 or 4. This was shown in Section "Indices and Basis Functions of Hexagonal Grids in a Plane".

The sparse hexagonal grid has only a small number of dynamic points neighboring to the target point and thus makes the construction of third-order schemes easier. The computational stencil is small and therefore an increase of efficiency with high order can be expected. Also, when subdividing the hexagonal cell into triangles due to sparseness, each triangular cell has few grid points that are not symmetrically around the triangle. Figure 2 shows the positions for amplitudes in the sparse grid system. So with the sparse grid defined here, a triangular grid is obtained by including diagnostic points. When considering triangular cells with equal sides, it is intuitive to combine them to hexagons. Most people do this intuitively when seeing such a grid (given in Fig. 1 in Chapter "Introduction"). It appears rather natural to construct the hexagons from this and by introducing sparseness to achieve a reasonably small number of nearest neighbor points for each target point to make FD stencils smaller and more efficient.

A hexagonal grid is shown in Fig. 1. We describe a structured or semi-structured approach even though some new models (Park et al. 2014) use an unstructured approach. The transition from a structured to an unstructured approach is a matter of computer software technology rather than a question of mathematics. A structured grid in 2D used indices $\mathbf{r}_{0,i,k}$ to describe the location of the hexagon as shown in Figs. 1 and 2. The location of the hexagon with index i, k is indicated by its center $\mathbf{r}_{0,i,k}$. In the notation used here, the cell center point $\mathbf{r}_{0,i,k}$ defines the cell $\Omega_{i,k}$, where Ω defines the cell as a geometric feature. For the grid points and other locations of amplitudes describing a field $fl(\mathbf{r})$, we use the pointer variable p introduced in Section "Indices and Basis Functions of Hexagonal Grids in a Plane". For unstructured gridding, data are stored using i'. This means that lists of the addresses of neighboring points must be carried. As this concerns mainly problems of informatics, the advantages and disadvantages of unstructured gridding are not discussed in this book, and tests have been done only for the simpler cases. Even though we use the structured notation, the methods to be described will be suitable for unstructured programming. A structured way of programming, as described here, is still a good option for research models created by a single person. Also the structured program suggests a simple way of multi-tasking if more than one processor is to be used. This is because the message passing must take account of neighborhood of grid points.

Figure 1 shows the cells. Let s be the side length of the edges and $dy = \frac{\sqrt{3}}{2}s$ be the distance of the hexagonal center from the edge. The mid-point of a cell is called $\mathbf{r}_{0,i,k}$, $(i, k = 0, 1, 2, 3, \ldots)$, and some of the indices are indicated in Fig. 1. $\mathbf{r}_{0,i,k}$ is assumed to be the center of a cell $\Omega_{i,k}$. The grid points are defined in Eq. (1) where the use of structured indices i, k is shown.

From Fig. 1 in Chapter "2D Basis Functions for Triangular and Rectangular Meshes" and Section "Indices and Basis Functions of Hexagonal Grids in a Plane", it can be seen that the full grid in each cell has 37 points in which 19 inner points belong to one hexagon only, 6 corner points belong to 3 hexagons, and 12 edge points, each belonging to 2 hexagons. Without counting any points double, there are 6 + 12 points on the hexagonal boundary of which 2 + 6 = 8 are independent. The full grid has 19 + 8 = 27 independent points per hexagonal cell. The sparse grid consists of the 18 points on the boundary only, and 8 points per cell are independent. For the sparse grid method, we use only the selected points on the cell boundary as dynamic amplitudes, which are predicted in time. This means that we have 8 dynamic amplitudes per cell, while the full grid has 19 + 8 = 27 amplitudes. The relation of the number n^{full} of dynamic amplitudes in full to the number of sparse systems is called the sparseness factor R^{SP}:

$$R^{SP} = \frac{n^{full}}{n^{sparse}} = \frac{27}{8}. \tag{10}$$

The sparseness factor is a measure of the potential saving of computer resources under the assumption that we can use the same times and that the numerical effort per point is the same. It can easily be applied for a number of schemes, though only for a number of schemes there is a limited indication of practical viability. For 3D, no sparse scheme has yet been implemented. Table 1 gives the R^{SP} factors for a number of schemes with the indication if the stability has already been investigated.

The general principle of sparseness can be described in the way that for a collection of points, which may belong to one or a set of neighboring cells, the description of an analytic function can be done by a set of variables associated with reduced points with the same accuracy as that of the full grid. The same accuracy in this connection means that any analytic function is approximated to the target order (here 3) using the reduced number of amplitudes. In the examples described so far, the reduced number of amplitudes associated with a corner point consists just of the dynamic grid points. This limits sparseness to the 2D and 3D cases. In a sparse grid system, the corner points with the associated amplitudes are called *QUASAR points* (see Chapter "Finite Difference Schemes on Sparse and Full Grids"). In Fig. 2 in one of the hexagons, the amplitude positions for a sparse representation are indicated.

The concept of QUASAR points and increased sparseness are described in Appendix "The Quasi-arithmetic Rendition QUASAR to Obtain a Sparse Field Representation". Note that with the advanced QUASAR concept, a reduced grid called QS-1-1 scheme (see Appendix "The Quasi-arithmetic Rendition QUASAR to Obtain a Sparse Field Representation") is also possible in 1D that allows an approximation of analytic functions by 2 amplitudes per cell, while o3o3 needs 3 points per 1D cell. The simple scheme described here can be considered as an example of the QUASAR concept, but it may also be simply described as a sparse grid.

The o3o3 scheme on hexagons will be described in the following, and some indication of the practical usefulness will be given. With this example, the reader is provided with the tools to investigate other variants of the *onom* family of L-Galerkin schemes. Other sparse schemes are given in Section "Totally Irregular Triangular and Quadrilateral Mesh: Hexagons and Other Polygons" in Chapter "Finite Difference Schemes on Sparse and Full Grids", and the more advanced QS-1-1 is described in Appendix "The Quasi-arithmetic Rendition QUASAR to Obtain a Sparse Field Representation". Figure 1 in Chapter "2D Basis Functions for Triangular and Rectangular Meshes" shows examples of sparse grids in 2D, some of them going beyond the scope described in this chapter. Only some of the schemes listed are described in this book. To describe 3D schemes in detail for advanced applications would go beyond a mathematical book like the present but rather fit into a technical note of a model developing institute. The present

Table 1 Sparseness factors R^{SP} for different grid options. Note that the improved interface to physics may lead to an increase of the sparseness factor. This depends on the relative computational expense of physics to dynamics. If this is 0.5, the increase of the sparseness factor is in the order of a factor 2

Scheme investigated	R^{SP} for factor 0.5	Information on stability of scheme available
o2o3 2D quad[a]	$4/3 = 1.3$	Yes
o2o3 3D quad	$8/4 = 2$	No
o3o3 2D quad	$9/5 = 1.8$	Yes
o3o3 3D quad	$27/7 = 3.8$	No
o2o3 2D hexa[b]	$(5+7)/5 = 2.4$	Yes
o2o3 3D hexa	$24/9 = 3.7$	No
o3o3 1D (QS-1-1)	$3/2 = 1.5$	No
o3o3 2D hexa	$27/8 = 3.4$	No
o3o3 3D hexa	$81/10 = 8.1$	No
o3o3 2D quad (QS-1-1)	$9/3 = 3$	No
o3o3 3D quad (QS-1-1)	$27/4 = 6.7$	No
o3o3 2D hexa (QS-1-1)	$27/6 = 4.5$	No
o3o3 3D hexa (QS-1-1)	$81/8 = 10$	No

[a] "quad" represents quadrilateral
[b] "hexa" represents hexagon

book wants to give an introduction to the technical tools for the numerical part of a model. In Table 1, the sparseness factor R^{SP} goes up to 10, indicating that the potential saving of computer time is worth considering. It should be noted that the saving in a realistic model to be achieved is dependent on the program implementation and an effective message passing. So the saving could be larger or smaller than that indicated in Table 1. In Section "CPU Time Used with a 3D Version of o3o3 Scheme" in Chapter "Numerical Tests", a 3D computer realization on a PC is given with timings. The scheme used was the o3o3 3D quad scheme of Table 1, and the timings were compared to those of the full grid classical o4 scheme of Chapter "Simple Finite Difference Procedures". In Table 1 going from top to bottom, increasingly complex schemes are encountered. The example o3o3 considered in the following is of medium complexity and sufficient to show the principle of hexagonal discretization.

The hexagonal points are defined in relation to the center point $\mathbf{r}_{0,i,k}$. For the corner points of hexagon, $\Omega_{i,k}$. Let us define the corner points: $\mathbf{r}^c_{p,i,k}$ which are determined by the relative vectors $\mathbf{r}^{c,s_p}_{p,i,k}$ given in Eq. (2)

$$\mathbf{r}^c_{p,i,k} = \mathbf{r}_{0,i,k} + \mathbf{r}^{c,r}_{p,i,k} = \mathbf{r}_{0,i,k} + s_p(cos\varphi_p, sin\varphi_p), \tag{11}$$

where s_p is the distance of a corner point from the hexagonal center. In the case of a regular hexagon, which we treat here, $s_p = s$, the side length of the hexagon. The upper index c indicates that we use the compact grid, where each point occurs

Table 2 Transformation between redundant representation $h_{p,i,k}$ and compact representation $fl^c_{p,i,k}$. We assume the representations by corner point amplitudes and directional one-sided derivatives also at corner points. Note that for $p = 1, \ldots, 8$, the redundant representation stores the amplitudes, which are field values for $p = 1, 2$ and one-sided derivatives for $p = 3, 4, 5, 6$ relevant for the interpolation in the cell. For $p = 7, 8$, one-sided derivatives are stored, which are relevant for neighboring cells. For the redundant $fl_{p,i,k}$ representation, the 6 field values are stored for $p = 1, 2, 3, 4, 5, 6$, and for $p = 7, \ldots, 18$, one-sided derivatives are stored. The 18 amplitudes of the redundant representation are computed from the 8 amplitudes of neighboring cells, whose indices are called $i', k', p'. i', k'$, and p' are functions of p, i, k. As an example, the transformations between $fl^c_{p,i,j}$ and $fl_{p',i'.k'}$ are given. The transformations for the one-sided derivatives are done in an analog way. The transformations can be seen from Fig. 2

Redundant representation $\mathbf{r}_{p,i,k}$	Compact representation $\mathbf{r}^c_{p',i',k'}$
$h_{1,i,k}$	$h_{1,i,k}$
$h_{2,i,k}$	$h_{2,i,k}$
$h_{3,i,k}$	$h_{1,i+1,k-1}$
$h_{4,i,k}$	$h_{2,i+1,k}$
$h_{5,i,k}$	$h_{1,i+1,k}$
$h_{6,i,k}$	$h_{2,i,k+1}$

in just one hexagon. We use the one-sided derivative representation introduced in Section "Indices and Basis Functions of Hexagonal Grids in a Plane". This means that all amplitudes, field values and one-sided derivatives used as amplitudes, are defined at corner points. For $p = 1, 2$, field values are stored as amplitudes, $\varphi_1 = \frac{5}{6}\pi + \frac{p-1}{9}\pi/3$. For $p = 3, 4$, the two directional derivatives of the field at point $\mathbf{r}_{1,i,k}$ are stored, $\varphi_p = \frac{5}{6}\pi$. In an analog way, we have $\varphi_p = \frac{7}{6}pi$ for $p = 5, 6$. For $p = 7, 8$, the directional derivatives corresponding to s_3 and s_1 are stored, which means $\varphi_7 = \frac{3}{6}\pi$ for $p = 7$ and $\varphi = \frac{9}{6}\pi$ (see Figs. 1 and 2). For the meaning of the amplitudes and the resulting interpolation, see Chapter "Local-Galerkin Schemes in 1D" and Section "Indices and Basis Functions of Hexagonal Grids in a Plane". We may use values of $p > 5$ and will identify $p = 6$ with $p = 0$ for simplicity of notation (Table 2).

The spectral interpolation is used to compute the field inside the hexagon. Many assumptions for creating basis function representations of the field h are possible. Here we use a method based on dividing the hexagon into two quadrilaterals and use bilinear interpolation. There is no claim that this interpolation is the best, but it is rather simple to implement. As shown in Fig. 2, the hexagon with center point $\mathbf{r}_{0,i,k}$ is divided into two quadrilaterals along the line $\overline{\mathbf{r}_3\mathbf{r}_6}$. We construct the internal edge line, which is defined as the edge line for the definition of the two quadrilateral interpolations. Note that there is an arbitrary choice involved in choosing just this division of a hexagon into quadrilaterals. The edge parallel vector for $\overline{\mathbf{r}_3\mathbf{r}_6}$ is called $\mathbf{n}_5$ in Fig. 2. It is defined as

$$\mathbf{n}_5 = \frac{\mathbf{r}_3 - \mathbf{r}_6}{|\mathbf{r}_3 - \mathbf{r}_6|}. \tag{12}$$

The internal edge dividing the center hexagon into quadrilaterals for interpolation purposes is a diagnostically defined edge. It is indicated in Fig. 2 as a dotted line. The construction of internal points and edges is a general principle to define interpolations or functional representations for hexagons, or more general polygons. The scheme described here is just an example.

The corner points at the two ends of this diagnostic edge are already given as r_3 and r_6. We need to construct the spectral amplitudes at the internal edge $\overline{r^3 r^6}$ that are the second and third derivatives along this internal edge. The diagnostic edge $\overline{r_3 r_6}$ is defined as y-direction, meaning that n_5 is $-n^y$, with n_y being the unit vector in y-direction. For regular hexagons that we consider for simplicity, the diagnostic amplitudes belonging to this edge are $fl_{yy,0,i,k}$ and $fl_{yyy,0,i,k}$. The point of definition of these amplitudes is $r_{0,i,k}$, the center point of the hexagon. According to Section "Indices and Basis Functions of Hexagonal Grids in a Plane", the directional derivatives fl_{n_5} at points r_3 and r_6 are sufficient to derive the spectral coefficients at r_0. For the following equation, we drop the space index i, k and the pointer p and rather use the r_0 to r_6 shown in Fig. 2. For a unit vector s, fl_s means the directional derivative in direction s. Analog to Eq. (8), we obtain for fl_{s_5}

$$\begin{cases} fl_{s_{3,6}} = \alpha_1 fl_{s_{1,6}} + \alpha_2 fl_{s_{5,6}}, \\ fl_{s_{3,6}} = \alpha_3 fl_{s_{2,3}} + \alpha_4 fl_{s_{4,3}}, \end{cases} \tag{13}$$

where α_i are defined in Eq. (50) in Chapter "2D Basis Functions for Triangular and Rectangular Meshes".

The two quadrilaterals now have all amplitudes needed for a quadrilateral interpolation of third order as given in Section "Euclid's Lemma" in Chapter "Local-Galerkin Schemes in 1D" in connection with Euclid's Lemma. We can now compute values for $fl(r)$ inside the hexagon for any point r in third order. A rather simple formula is obtained for the value of fl at the center point r_0 of the hexagon. On the edge $\overline{r_3 r_6}$, we introduce the variable $y' \in (0, 2s)$ with s being the common side of the two quadrilaterals dividing the hexagon. A point r of the hexagon is then

$$r(y') = r_3 + y' s_5, \tag{14}$$

where $s_5 = r_6 - r_3$. Using Eq. (14) and y', we can apply 1D methods of Chapter "Local-Galerkin Schemes in 1D" to compute the center point amplitude $fl(r_0)$ of fl. Note that the basis function belonging to amplitude $fl_{yyy,0}$ is 0 at the center point of the hexagon:

$$fl(r_0) = fl_0 = \frac{1}{2}(fl_6 + fl_3 - s^2 fl_{yy,0}). \tag{15}$$

We suppose that $fl^{lin}(r)$ is the bilinear surface determined by the corner amplitudes fl_1, fl_2, fl_3, fl_6 for the left quadrilateral and fl_5, fl_4, fl_3, fl_6 for the right quadrilateral in Fig. 2. For its computation, Euclid's Lemma and the methods described in Section "Euclid's Lemma" in Chapter "Local-Galerkin Schemes in 1D" and Fig. 3 in Chapter "Finite Difference Schemes on Sparse and Full Grids" can be used. A number of methods are possible to compute this bilinear part $fl^{lin}(\mathbf{r})$ of $fl(\mathbf{r})$ for each of the two quadrilaterals belonging to a hexagon. These are pointed out in Chapter "2D Basis Functions for Triangular and Rectangular Meshes". The main possibilities are geometric construction following Euclid's Lemma and the introduction of local coordinates. As the whole interpolation is known, it is also possible to compute one-sided derivatives everywhere, and these can be used again to construct triangular interpolations for the 6 triangles contained in the hexagon. The triangular interpolation is not identical to the quadrilateral interpolation, but both are third-order approximations of $fl(\mathbf{r})$. The different options translate into an infinity of approaches to define numerical schemes on hexagonal grids, from which only one is described for illustration. The example is chosen for simplicity and not because it is best in any sense of the word. The high-order part $fl^{ho}(\mathbf{r})$ of $fl(\mathbf{r})$ is defined:

$$fl^{ho}(\mathbf{r}) = fl(\mathbf{r}) - fl^{lin}(\mathbf{r}). \tag{16}$$

By definition, $fl^{ho}(\mathbf{r})$ is 0 at all corners. On all four edges of a quadrilateral, $fl^{ho}(\mathbf{r})$ is a 1D function of the edge coordinate λ introduced above, being a third-degree polynomial that is 0 at the two ends of the edge. It is therefore defined by the amplitudes $fl^{ho}_{\lambda\lambda}$ and $fl^{ho}_{\lambda\lambda\lambda}$. The quadrilateral has 4 edges, meaning 8 amplitudes. Two of them, defined at the hexagon center by formula given above, are diagnostic, the other 6 prognostic. The basis functions belonging to the 8 amplitudes are different from 0 on one and only one edge. These basic functions are linear functions on lines indicated by Euclid's lemma. Analog to the situation in one dimension, the basis functions belonging to $fl^{ho}_{\lambda\lambda}$ carry mass, while $fl^{ho}_{\lambda\lambda\lambda}$ does not carry mass.

The performance of a time step is analog to the situation in one dimension. First, the time derivative of the bilinear part $fl^{lin}(\mathbf{r})$ of fl is computed by approximating the divergence at corner points. For this, any fourth-order FD equation can be used, without worrying about conservation. One-sided derivatives at corner points are dynamic variables. As the RK4 scheme used for time integration is unlikely to be stable when using it with one-sided derivatives, a way must be found to form centered differences of fourth order. Chapter "2D Basis Functions for Triangular and Rectangular Meshes" discussed a number of possibilities, including an averaging procedure of the one-sided directional derivatives. Here we suggest the use of difference stencils based on grid point values of fl. This method worked well in Steppeler et al. (2008) for a non-conserving fourth-order method on the full grid. Here we want to modify this approach to a sparse grid on hexagonal cells with conservation. As here we are not concerned with conservation issues for corner point approximation, this part of the proposed method is identical to Steppeler et al. (2008).

A stencil around the target point can be created by interpolation, and a large number of stencils is possible. This means that an infinity of different methods exists. A simple possibility is indicated in Fig. 1 in Chapter "Finite Difference

Schemes on Sparse and Full Grids". Two points on an edge representing the grid point space are used as collocation points, and two other points shown in Fig. 1 in Chapter "Finite Difference Schemes on Sparse and Full Grids" have to be interpolated inside a hexagon. There is still a more simple possibility. In Fig. 1 in Chapter "Finite Difference Schemes on Sparse and Full Grids", the center point of $\mathbf{r}_2$, $\mathbf{r}_1$ can be used in addition to $\mathbf{r}_1$ and $\mathbf{r}_2$ as stencil point, and similarly, the center point of the hexagon can be used in the stencil. Let $fl_{-2}, fl_{-1}, fl_0, fl_1, fl_2$ be the stencil points. Then we obtain for a fourth-order approximation at point $\mathbf{r}_0$ with edge coordinate λ:

$$fl_\lambda = w_{-2}fl_{-2} + w_{-1}fl_{-1} + w_0 fl_0 + w_1 fl_1 + w_2 fl_2. \tag{17}$$

Note that the grid in the stencil indicated by Fig. 1 in Chapter "Finite Difference Schemes on Sparse and Full Grids" is not equidistant. The grid distances in the stencil are $\frac{s}{2}, \frac{s}{2}, s, s$, when s is the edge length of the hexagon. This means that the values of w_i must be computed taking into account of the irregular grid in the stencil. For the one dimensional examples in this book, this was done using the Galerkin Compiler MOW_GC (Huckert and Steppeler 2021).

In order to obtain the corner point divergence, the derivative must be taken in at least two directions. In analogy to the 1D cases for squares, the second-order amplitudes $fl_{\lambda\lambda}$ at edge centers are determined by mass conservation. They are uniquely determined by requiring that the time derivative of mass associated with these amplitudes is the same as follows from flux into part of the hexagonal area. This is in analogy to the case of square cells, and there, complications coming from the no-square shape of the quadrilaterals are discussed in Chapter "2D Basis Functions for Triangular and Rectangular Meshes". For the line going through the center $\mathbf{r}_0$, λ is y, and we obtain here two fl_{yy} amplitudes, meaning that we do not have the field representation as we had at the beginning of the time step. To distribute the mass contained in these amplitudes, we may go to the one-sided derivative representation at corner points. We must then adjust the mass associated with points $\mathbf{r}_3$ and $\mathbf{r}_6$ in Fig. 1.

The properties of the schemes defined above are largely unexplored. Here just one scheme is proposed using a grid being rather easy to interpolate. The grid is shown in Fig. 1 as white points. The QUASAR formalism can be used for further efficiency. This allows to compute a third-order interpolation at a straight line within a given hexagon. The method is explained in Appendix "The Quasi-arithmetic Rendition QUASAR to Obtain a Sparse Field Representation". Let two corner points $\mathbf{r}^1$ and $\mathbf{r}^2$ of a hexagon be given, where fl^1 and fl^2 are the field values of any field fl. Let the points be connected by a straight line. Along this line, let $fl^{lin}(\mathbf{r})$ be the linearly interpolated field fl. Choose $fl(\mathbf{r}) = fl^x$ or fl^y, depending on the field to be differentiated. Let λ be the directional parameter along this line. Then the QUASAR operation computes the directional third and second derivatives to λ, by which it is possible to compute the high-order deviation along this line. QUASAR defines all sparse amplitudes of a hexagon, which are 18, some of them shared with neighboring hexagons. It derives these amplitudes using just 6 amplitudes per cell, which is a small increase of the sparseness factor. It eliminates functional forms, which are not expected to result into accurate solutions. This is the opposite of going from continuous Galerkin to discontinuous Galerkin. With QUASAR, we require fields not only to be continuous but also to be one time differentiable.

Hexagonal Options

The sparse grid described has grid points or spectral amplitudes on the surface only. The scheme is inhomogeneous, as different points of the hexagon use different formulae.

Hexagonal grids with amplitudes on the centers are in use (Skamarock et al. 2012) and are constructed in high order. The center based hexagonal schemes often use the FV method. The centered schemes can be considered as a triangular grid, where the points are at the corners of the triangles. The hexagons are then the control volumes obtained by creating the normal line for each stencil line.

Isotropy of the Hexagonal Grid in Comparison to Rhomboidal Grid

Isotropy is defined as the relation of the smallest wavelengths in two directions. Hexagonal full grids have been used (Skamarock et al. 2012) with a full atmospheric model. Sparse grid hexagonal cells were used with simple cells having one amplitude at the cell centers (Steppeler et al. 2019). The special properties of hexagonal cells compared to quadrilaterals are not intensively explored. In this section, we consider for regular grids the dependence of grid properties on the direction of wave propagation. When plane waves propagate, many grids show different accuracies of propagation for different directions. Many grids have for some directions of propagation 1D discretizations. These depend on the direction of propagation and may have different effective grid lengths. As we have discussed before, different numerical resolutions in different directions are inefficient.

For example for square grids for propagation in the directions of cell boundaries, we suppose the grid length dx, while for propagation along the diagonal, the effective grid length is $\frac{dx}{\sqrt{2}}$. This means a reduction of efficiency by a factor $\frac{1}{\sqrt{2}} = 0.7$, as because of the smaller grid length along the diagonal the time step needs to be reduced by this factor without a corresponding gain in accuracy of the simulation. The two propagation directions for the square differ by an angle of $\frac{\pi}{4}$.

Regular hexagonal grids have two propagation directions resulting in 1D difference schemes. These directions differ by $\frac{\pi}{6}$, and this can be seen geometrically from Fig. 1. One is the wave propagating along coordinate i in Fig. 1, with corner point effective grid length $\frac{\sqrt{3}s}{2} = 0.87s$, with s being the side length of the hexagons. The same grid is seen already with an angular difference of $\frac{\pi}{3}$, where for the square the grid repeats itself for $\frac{\pi}{2}$ only. The direction $\frac{\pi}{6}$ sees an irregular grid with average grid length $\frac{3s}{4} = 0.75$. This means that the difference in resolution for different directions of grid propagation for the two directions is smaller with hexagons than with squares. The same propagation error repeats itself for an angular difference of $\frac{\pi}{3}$ for hexagons and $\frac{\pi}{2}$ for squares. This means that the error of wave propagation for hexagons is more isotropic than for squares. Very little work has been done so far to investigate the different grids quantitatively and transfer this to spherical grids.

Platonic and Semi-Platonic Solids

Abstract A number of options are given for constructing Platonic and Semi-Platonic solids for spherical discretization. This chapter provides just spherical grids and a simple non-conserving toy model. However, the tools and the example provided in Chapters "Finite Difference Schemes on Sparse and Full Grids" and "Full and Sparse Hexagonal Grids in the Plane" allow to create L-Galerkin sparse grids on the sphere. Further L-Galerkin methods will be presented in Chapter "Numerical Tests."

Keywords Platonic solids · Semi-platonic solids · Solid generated grids · Spheric polygonal approximations · Spherical projections · Spherical tests

Older numerical approaches mapped the whole sphere to a plane, such as using the latitude longitude projection (Durran 2010). This generates singularities at both poles, which makes the construction of numerical schemes difficult.

An alternative is the division of the sphere into patches. The projection of icosahedrons to the sphere for discretizing has been investigated for a long time (Williamson 1968; Sadourny and Morel 1969). The icosahedron divides the sphere into 10 bilinear rhomboids. It took a long time before such grids were used for practical modeling. With hindsight, it appears that two problems were preventing the practical application of such grids. The first was that when using patches, a global coordinate mapping must not be used. The necessary singularities of such mapping would cause problems. The second was that at special lines and points, the grid is seen in applications. This phenomenon is called grid imprinting and is caused by the order of approximation not being uniform. Uniform approximation order means that at special points, such as poles or rhomboidal edge lines, the approximation order does not drop below the target value. Steppeler et al. (2008), Steppeler and Prohl (1996), and Baumgardner and Frederickson (1985) showed that icosahedral grids may be used without encountering grid imprinting. Now most new developments of global models use a division of the sphere into patches, though older methods, such as global spectral method (Lander and Hoskins 1997) on latitude longitude grids, are still in use.

© Springer Nature Switzerland AG 2022

J. Steppeler, J. Li, *Mathematics of the Weather*, Springer Atmospheric Sciences, https://doi.org/10.1007/978-3-031-07238-3_7

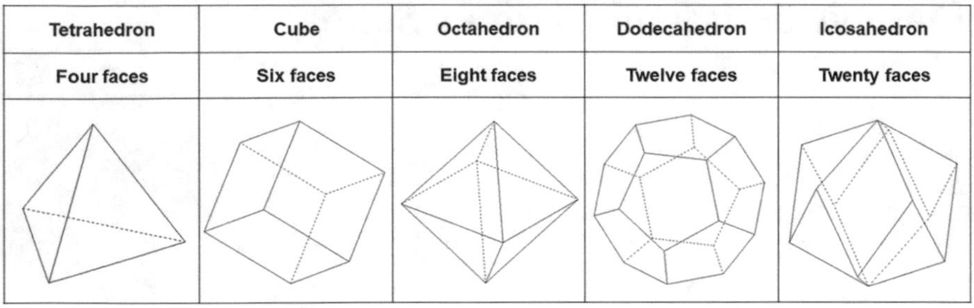

Tetrahedron	Cube	Octahedron	Dodecahedron	Icosahedron
Four faces	Six faces	Eight faces	Twelve faces	Twenty faces

Fig. 1 The 5 Platonic solids (from left to right: tetrahedron, cube, octahedron, dodecahedron, and icosahedron)

Grid imprinting occurs with the classical FD and regular o4 schemes when the resolution jumps. These classical FD schemes lose their order of approximation 2 or 4 at points where the resolution jumps. Figure 5 in Chapter "Simple Finite Difference Procedures" shows that for a 1D example, the grid imprinting disappears when the methods are modified using schemes keeping their order (2 or 4) at points where the resolution jumps.

Platonic solids are a set of equal regular polygons inscribed into the sphere. For purposes of atmospheric discretization, they are projected to the sphere. There are 5 platonic solids shown in Fig. 1: tetrahedron, cube, octahedron, dodecahedron, and icosahedron.

The grids constructed on a sphere are never platonic solids unless the coarsest grid is used and the grid is one of the five Platonic solids. Grids are created by subdividing the platonic solids. These grids are nearly regular and called semi-platonic solids. Currently the icosahedron and the cube are used to create grids (Mishra et al. 2010; Zängl et al. 2015; Müller et al. 2018). The tetrahedron can be used to create the Yin-Yang grid (Kageyama and Sato 2004; Kageyama 2005; Peng et al. 2006). However, practical realizations of the Yin-Yang grid are often done without reference to the tetrahedron.

It appears reasonable to use also Semi-Platonic solids to create the initial solid to be subdivided. *Semi-Platonic solids* are solids whose surface consists of patches which are not regular, but nearly so.

Cubed Sphere, Icosahedron, and Examples of Semi-Platonic Solids

Platonic solids are polyhedrons inscribed into the sphere which are congruent and regular. Regular means that all sides (edges) and angles are the same. There exist five Platonic solids which are shown in Fig. 1. The five choices are the only possibilities for Platonic solids. By projecting the polygonal edges onto the sphere, great circle

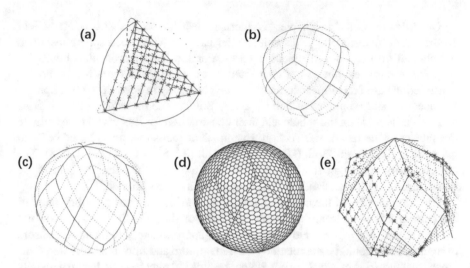

Fig. 2 Examples of Platonic and Semi-Platonic solids. (**a**) The Platonic solid tetrahedron, (**b**) Semi-Platonic solid $PL^{4,5}$, (**c**) Semi-Platonic solid $PL^{6,5}$, (**d**) the same to (**c**) carrying a hexagonal cell structure, and (**e**) the bilinear rhomboidal solid. Some points of the 10 rhomboidal grid patches are marked to illustrate the inner boundary conditions to be posed when formulation FD schemes are constructed on the grid stencils. There is no harm in storing the grid point values of fields on the inscribed bilinear grids. The FD operators, however, must be formulated on the great circle grid obtained by projection. For the efficiency of an FD formulation, the regularity of the grid projected to the sphere is essential, not the bilinear inscribed grid

solids are created, some of these can be used as a basis to create grids for global atmospheric models.

Our aim is to obtain grids for discretization by subdividing the polygonal surfaces. The tetrahedron can be subdivided into a spherical grid with just two patches. This is illustrated in Fig. 2a and investigated in Peng et al. (2006). Large grid patches result into rather irregular grids on the sphere, meaning that the smallest and largest elements of polygonal discretization differ considerably. As a rule, smaller patches result into more regular grids. Moreover, the dodecahedron is not much used for discretization as it is difficult to subdivide pentagons, even though using triangular grids this should be possible. The dodecahedron is the dual grid of the icosahedron, and grids generated by the icosahedron (see Fig. 1 in Chapter "Introduction") have a hexagonal cell structure with 12 pentagons as dual grid. This dual grid of an icosahedron-based grid may be considered as a high-resolution version of the dodecahedron.

Popular choices are grids based on the cubed sphere (Rancic et al. 1996) and the icosahedron. At this point, we limit ourselves to rhomboidal grids. Triangular grids can easily be obtained by dividing the rhomboidal cells into triangles. Hexagons are shown in Figs. 1 in Chapter "Introduction" and Fig. 2. In Fig. 2b, constructed hexagons are strongly deformed. The solid $PL^{4,5}$ appears to be more suitable for quadrilateral cells than for hexagons. In Section "The $T64$ Solid for Discretization by Quadrilateral Cells" in Chapter "Numerical Tests", the solid $T64 = PL^{4,7}$ will be shown for quadrilateral grids with smaller patches. The cubed sphere inscribed rhomboidal grid patches are plane squares. For the icosahedron, the rhomboids are bilinear surfaces, not planes connected by an edge. An alternative method of construction based on bilinear surfaces will be given later in Section "Cubed Sphere, Icosahedron, and Examples of Semi-Platonic Solids. The projection of these grids to great circle grids on the sphere and their construction for less trivial cases will be the subject of the rest of this section. Examples of realistic models based on such grids are in Skamarock et al. (2012); Zängl et al. (2015); Müller et al. (2018). The large majority of polygonal models on the sphere are based on either the cube or the icosahedron. At the moment, the Yin-Yang grid draws less interest.

For the examples of the cube and icosahedron grids obtained by subdivision of the cube and icosahedron, the rhomboidal cells of the grids in Fig. 2b–d are not congruent and not regular. However, they are approximately regular. A measure of homogeneity could be the relation of the maximum and minimum cell size. This means that the grids of Fig. 2 are not Platonic solids but may be called approximately Platonic. Such solids being nearly regular are called Semi-Platonic solids.

In the following classes of Semi-Platonic solids, $PL^{m,n}$ will be defined, dividing the sphere into rhomboidal patches of similar but not identical shape. For the terminology of the Semi-Platonic solid $PL^{m,n}$ (Fig. 3), m represents the number of rhomboids meeting at the pole and n is the number of circles with corner points between the north and south poles. These are circles on the sphere in planes vertical to the axis connecting north and south poles. This means that except for the equator, m circles defined by such corner points (the points on one blue circle in Fig. 3b) are not great circles. For the purpose of defining n, we consider the poles as degenerated circles carrying m corners, which are all identical. Note that n is also the number of circles with corner points in one great circle connecting the north and south poles. For example, the cube sphere can be represented as $PL^{3,4}$, while the icosahedron is $PL^{5,4}$. The Semi-Platonic solid in Figs. 2c–e is $PL^{6,5}$. Note that this does not mean that solids for larger m are impossible or not useful.

Due to the division of the sphere into rhomboids/triangles, we can also classify the Semi-Platonic solids by the number of triangles or rhomboids it contains. In general, the Semi-Platonic solids $PL^{m,n}$ contain $m \cdot (n - 2)$ bilinear rhomboids and $2m \cdot (n - 2)$ triangles. For example, the cubed sphere is $PL^{3,4} = R6 = T12$ meaning that it contains 12 triangles or 6 bilinear rhomboids. The icosahedron is called $PL^{5,4} = R10 = T20$.

In order to construct the Semi-Platonic solids, we use the Earth centered Cartesian coordinate system with unit orthogonal vectors $\mathbf{n}^x, \mathbf{n}^y, \mathbf{n}^z$, as shown in Fig. 4a. For the spatial vector $\mathbf{r}$, we have the coordinate representation:

$$\mathbf{r}(x, y, z) = r_x \mathbf{n}^x + r_y \mathbf{n}^y + r_z \mathbf{n}^z. \tag{1}$$

Then, we introduce the definition of longitude θ and complementary latitude ϕ for points on the unit sphere:

$$\mathbf{r} = sin\phi cos\theta \mathbf{n}^x + sin\phi sin\theta \mathbf{n}^y + cos\phi \mathbf{n}^z. \tag{2}$$

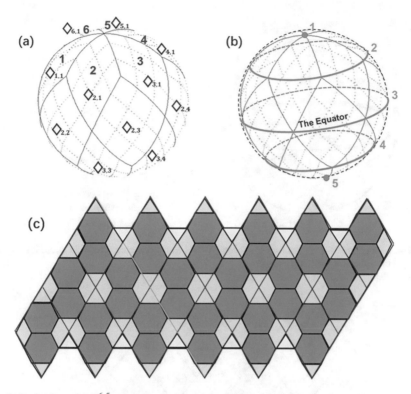

Fig. 3 Definition of $PL^{6,5}$. (**a**) The number of rhomboids aggregating at the north pole and (**b**) the blue circles carrying 6 corner points (this is illustrated in Fig. 5a). Circle 3 is the equator, while Circles 1, 2, 4, and 5 are circles being vertical to the earth axis. Note that Circles 1 and 5 are degenerated to the south and north poles. (**c**) is the division of hexagons on the patches. Note that the six triangles on the top and the six on the bottom form the hexagon at the poles on the sphere. Polygons that belong to more than one rhomboidal patch are shown in yellow. These exist on all corners of the 18 rhomboidal patches: two hexagons at the poles, 2 times 6 pentagons at mid-latitudes, and 6 hexagons at the equator

Therefore, points on the surface of the unit sphere satisfy

$$|\mathbf{r}| = \sqrt{r_x^2 + r_y^2 + r_z^2} = \sqrt{(cos\phi sin\theta)^2 + (sin\phi sin\theta)^2 + (cos\phi)^2} = 1. \qquad (3)$$

In the following, we discuss the Semi-Platonic solid $PL^{m,n} = PL^{6,5}$ (Fig. 2c), which is defined by the corners of the constituting rhomboids/triangles in Fig. 5. There are $m \cdot (n - 2) + 2 = 20$ corner points for $PL^{6,5}$, which are indexed as $\mathbf{r}_{i,j}$ where $i = 1, 2, ..., m$, $j = 1, 2, ..., n$. We note that for any j, $\mathbf{r}_{i+m,j} = \mathbf{r}_{i,j}$. When $j = 1$, $\mathbf{r}_{i,j} = \mathbf{r}_{i,1}$ while when $j = n$, $\mathbf{r}_{i,j} = \mathbf{r}_{i,n}$. The definition of $\mathbf{r}_{i,j}$ for $PL^{6,5}$ is as follows:

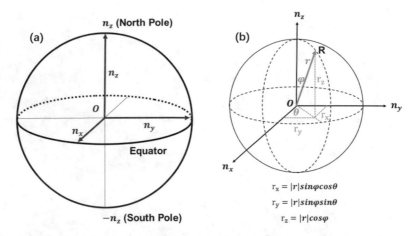

Fig. 4 Two coordinate system for the sphere. (**a**) The Earth centered Cartesian coordinate system and (**b**) the spherical coordinate system

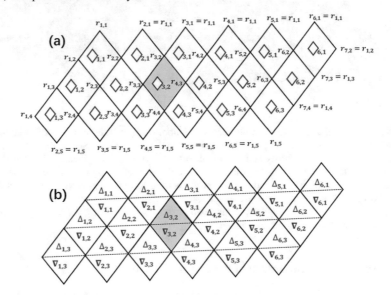

Fig. 5 The rhomboidal and triangular patches rolled out into plane for $PL^{6,5}$. The corner points $r_{i,j}$ and the patches $\diamond_{i,j}/\triangle_{i,j}/\nabla_{i,j}$ are shown

$$
\begin{cases}
\mathbf{r}_{1,j} = \mathbf{n}^z, \\
\mathbf{r}_{2,j} = \cos\left(\theta_i + \dfrac{2\pi}{12}\right) \sin\phi_0 \mathbf{n}^x + \sin\left(\theta_i + \dfrac{2\pi}{12}\right) \sin\phi_0 \mathbf{n}^y + \cos\phi_0 \mathbf{n}^z, \\
\mathbf{r}_{3,j} = \cos\theta_i \mathbf{n}^x + \sin\theta_i \mathbf{n}^y, \\
\mathbf{r}_{4,j} = \cos\left(\theta_i - \dfrac{2\pi}{12}\right) \sin(\pi - \phi_0)\mathbf{n}^x + \cos\left(\theta_i - \dfrac{2\pi}{12}\right) \sin(\pi - \phi_0)\mathbf{n}^y + \cos(\pi - \phi_0)\mathbf{n}^z, \\
\mathbf{r}_{5,j} = -\mathbf{n}^z,
\end{cases}
$$

$$(4)$$

where $\theta_i = \frac{2(i-1)\pi}{6}$ $(i = 1, 2, \ldots, n)$. For ϕ_0, we have the following equation, which is solved numerically:

$$|\mathbf{r}_{1,1} - \mathbf{r}_{1,2}| = |\mathbf{r}_{1,2} - \mathbf{r}_{2,3}|. \tag{5}$$

Note that Eq. (5) is one way to choose ϕ_0. The analog equation for the icosahedron and cubed sphere follows from the requirement that these are Platonic solids. For a semi-Platonic solid as treated here, other options than Eq. (5) are possible. The edges of the rhomboidal patches have equal length. The family of $PL^{m,4}$ solids, to which the cube $(m = 3)$ and icosahedron $(m = 5)$ belong, is constructed in the same way.

There are 3 rows of rhomboids or 6 rows of triangles. For rhomboids, one row intersects at the north pole, another at the south pole, and one centered at the equator. For a rhomboid numbered $\diamond_{i,j}$, we provide another index system for the four corner points $\mathbf{r}_1^{\diamond i,j}, \mathbf{r}_2^{\diamond i,j}, \mathbf{r}_3^{\diamond i,j}, \mathbf{r}_4^{\diamond i,j}$ $(i = 1, 2, \ldots, m, j = 1, 2, \ldots, n)$ shown in Fig. 6:

$$\begin{cases} \mathbf{r}_1^{\diamond i,j} = \mathbf{r}_{i,j}, \\ \mathbf{r}_2^{\diamond i,j} = \mathbf{r}_{i,j+1}, \\ \mathbf{r}_3^{\diamond i,j} = \mathbf{r}_{i+1,j+1}, \\ \mathbf{r}_4^{\diamond i,j} = \mathbf{r}_{i+1,j+2}. \end{cases} \tag{6}$$

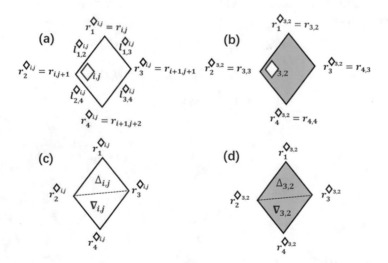

Fig. 6 Relations of index systems for rhomboids/triangles and the sphere. (**a**) The general case for rhomboid, (**b**) an example for $\diamond_{3,2}$, (**c**) the general case for triangle, and (**d**) an example for $\triangle_{3,2}/\nabla_{3,2}$

For two triangles numbered $\triangle_{i,j}/\nabla_{i,j}$, the index system for the corner points is shared with Eq. (6) (Fig. 6). Note that according to Eq. (6), the rhomboids $\diamond_{i,j}$ can be indexed by $\mathbf{r}^{i,j}$. The rhomboid index is defined by its top corner point.

For the inscribed polyhedron in Fig. 2c, n corner points $\mathbf{r}_{i,j} (j = 1, 2, \ldots, n)$ are connected by straight lines, the so-called edge lines, Fig. 5. These lines can be projected to the sphere to obtain the great circle grid, shown in Fig. 2c. The projection of a polygonal edge to the spherical surface is a great circle segment. For each special system $\mathbf{r}^{\diamond i,j}$, the edges of the grid are generated by the following steps:

1. Connect $\mathbf{r}_1^{\diamond i,j}$ and $\mathbf{r}_2^{\diamond i,j}$, $\mathbf{r}_3^{\diamond i,j}$ and $\mathbf{r}_4^{\diamond i,j}$, $\mathbf{r}_2^{\diamond i,j}$ and $\mathbf{r}_4^{\diamond i,j}$, $\mathbf{r}_1^{\diamond i,j}$ and $\mathbf{r}_3^{\diamond i,j}$ in turns.
2. Define for each rhomboid $\diamond_{i,j}$ the vectors on the edge lines: $\mathbf{l}_{1,2}^{\diamond i,j}$, $\mathbf{l}_{3,4}^{\diamond i,j}$, $\mathbf{l}_{1,3}^{\diamond i,j}$, $\mathbf{l}_{2,4}^{\diamond i,j}$
   using Eq. (6) shown in Fig. 7b. For example, $\mathbf{l}_{1,3}^{\diamond i,j} = \mathbf{r}_3^{\diamond i,j} - \mathbf{r}_1^{\diamond i,j} = \mathbf{r}_{i+1,j+1} - \mathbf{r}_{i,j}$.
3. Divide the edge into a 1D grid using a division of parameter $\lambda_0 = 0 < \lambda_1 < \ldots < \lambda_i < \lambda_{i+1} < \ldots < \lambda_{pe} = 1$ which can be chosen in such a way that the projection to the sphere is a regular grid. p is the number of points in the edge lines. Alternatively the two parallel edges $\mathbf{l}_{1,2}^{\diamond i,j}$, $\mathbf{l}_{3,4}^{\diamond i,j}$ can be divided, creating $p_e + 1$ edge points. According to Euclid's Lemma, this leads to the same grid in $\diamond_{i,j}$.
4. Define the grid for the four edge lines.
5. Construct the 2D grid on the bilinear surface associated with the points $\mathbf{r}_k^{\diamond i,j}$ ($k = 1, 2, 3, 4$) by connecting opposite edge points.

Note that in Step 4 for example, on the edge line $\mathbf{l}_{1,3}^{\diamond i,j}$, the high-resolution grid points are defined as

$$\mathbf{r}_{1,3,p}^{\diamond i,j} = \mathbf{r}_1^{\diamond i,j} + \lambda_p(\mathbf{r}_3^{\diamond i,j} - \mathbf{r}_1^{\diamond i,j})$$
$$= \mathbf{r}_{i,j} + \lambda_p(\mathbf{r}_{i+1,j+1} - \mathbf{r}_{i,j}) \qquad (7)$$
$$= \mathbf{r}_1^{\diamond i,j} + \lambda_p \mathbf{l}_{1,3}^{\diamond i,j}.$$

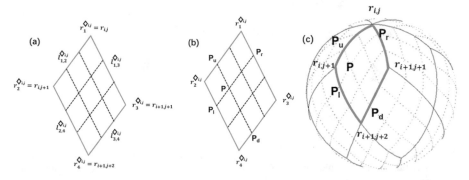

Fig. 7 Divisions for the rhomboids on the plane and the sphere. The bilinear grid in 3D space is illustrated in Fig. 2e. The bilinear grid patches are curved surfaces in 3D space. The grid patches shown in (**a**) and (**b**) are projections of the curved grid patches to the plane. Comparison with (**c**) shows that a regular grid patch on the plane (**b**) becomes irregular when projected to the sphere (**c**)

Using the earth centered coordinate system $\mathbf{n}^x, \mathbf{n}^y, \mathbf{n}^z$ (Fig. 4a), the bilinear function $z(x, y)$ is given as

$$z(x, y) = a_0 + a_x x + a_y y + a_{x,y} xy,\tag{8}$$

in which x, y are coordinates in the plane vertical to $\mathbf{r}_1^{\diamond i,j}$. This is the interpolating function to be used when describing the inscribed polygon consisting of rhomboids for the case of Semi-Platonic solids, as opposed to the great circle rhomboids on the sphere. The triangular version of the inscribed polyhedron consists of plane triangles. The cube–sphere is special as the rhomboidal surfaces are planes, where in general inscribed rhomboids are bilinear surfaces. The bilinear grid can be constructed using Euclid's Lemma, illustrated in Fig. 2d for the tetrahedral case. It is a generalization of Euclid's law of parallels.

Now we discuss how to generate 2D grids on the bilinear surface associated with the points $\mathbf{r}_k^{\diamond i,j}$ ($k = 1, 2, 3, 4$) in Step 5 (Fig. 7, for an illustration of the grid generation, see Fig. 2e). Note that in the index system $\mathbf{r}_k^{\diamond i,j}$ ($k = 1, 2, 3, 4$), the vectors of edge lines are expressed as $\mathbf{l}_{1,2}^{\diamond i,j}, \mathbf{l}_{3,4}^{\diamond i,j}, \mathbf{l}_{1,3}^{\diamond i,j}$, and $\mathbf{l}_{2,4}^{\diamond i,j}$. In the following, we omit the subscripts i, j of diamonds in $\mathbf{r}_k^{\diamond}$ ($k = 1, 2, 3, 4$). Euclid's Lemma (see Chapter "Local-Galerkin Schemes in 1D" and Fig. 2) can be applied to the polygonal approximations on the sphere. Bilinear surfaces are used before all rhomboids are projected to the sphere. Since all rhomboids can be divided into triangles, we can use triangular interpolation, which is done on planes. Let $\mathbf{r}_k^{\diamond}$ ($k = 1, 2, 3, 4$) be the points in one rhomboid of the sphere. We divide the opposite edges $\mathbf{l}_{1,3}^{\diamond}$ and $\mathbf{l}_{2,4}^{\diamond}$ by the factor λ and the points on $\mathbf{l}_{1,3}^{\diamond}$ and $\mathbf{l}_{2,4}^{\diamond}$ can be defined as $\mathbf{r}_{1,3,p}^{\diamond}$ and $\mathbf{r}_{2,4,p}^{\diamond}$ by Eq. (7) (marked as P_l and P_r in Fig. 7b). Then we connect the division points P_l and P_r by a straight line. In the same way, we follow the operations for the edges $\mathbf{l}_{1,2}^{\diamond}$ and $\mathbf{l}_{3,4}^{\diamond}$ with division factor μ and connect the division points P_u and P_d by a straight line again. The two connection lines meet at a point P and P divides the two straight lines in the same division relations λ. Two directions λ are taken at two different points. The grid indices for the grid inside $\diamond_{i,j}$ will be called I, J. The vector at P is expressed as

$$\mathbf{r}_{I,J}^{\diamond} = \mathbf{r}_2^{\diamond} + \lambda_I \mathbf{l}_{4,2}^{\diamond} + \lambda_J (\mathbf{r}_1^{\diamond} + \lambda_{p'} \mathbf{l}_{3,1}^{\diamond} - \mathbf{r}_2^{\diamond} - \lambda_{p'} \mathbf{l}_{4,2}^{\diamond}).\tag{9}$$

In the following, $I (= 0, 1, 2, \ldots, I_e)$ and $J (= 0, 1, 2, \ldots, J_e)$ are the indices for the division on every patch. By connecting the 1D grids on the edges, the bilinear grid is created and shown in Fig. 7. Note that the corner points of the rhomboids are on the sphere, while the grid points $P = \mathbf{r}_{I,J}^{\diamond}$ defined within the rhomboids are not. The storage of grid points can be done on the inscribed rhomboidal grid. It is essential that the numerical procedures to solve the dynamic equations are solved on the sphere in Fig. 7c. The projection to the unit sphere using Eq. (4) to obtain the great circle grid shown in Fig. 7c is

$$\mathbf{r}_{i,j}^{\odot} = \frac{r_P^{\diamond i,j}}{|r_P^{\diamond i,j}|} = \frac{r_{I,J}^{\diamond i,j}}{|r_{I,J}^{\diamond i,j}|}. \tag{10}$$

For the corner points belonging to more than one rhomboid, an extended grid can be created by interpolation. This extended grid can be used to create FD schemes which allows to compute time derivatives for edge and corner points.

Geometric Properties of Spherical Grids

A common method to create FD equations on the sphere is to project the spherical surface to a plane and solve the transformed equations (see Durran 2010). Here the equations are directly solved on the sphere. Geometric measures are computed using the 3D grid vectors $\mathbf{r}_{i,j}$. The spherical great circle difference $\delta\phi$ between two points $\mathbf{r}_a$ and $\mathbf{r}_b$ on the sphere with radius R is obtained by

$$\delta\phi(\mathbf{r}_a, \mathbf{r}_b) = arccos\left(\frac{\mathbf{r}_a \cdot \mathbf{r}_b}{|\mathbf{r}_a| \cdot |\mathbf{r}_b|}\right) = arccos\left(\frac{\mathbf{r}_a \cdot \mathbf{r}_b}{R^2}\right). \tag{11}$$

Equation (11) can used for neighboring grid points to compute grid length of different spherical solids on the sphere.

Let us consider any patch $\mathbf{r}^{\diamond i,j}$ ($i = 1, 2, \ldots, m, j = 1, 2, \ldots, n$) for $PL^{m,n}$ in Fig. 8. Please note that in this section we still omit the subscript i, j in $\mathbf{r}^{\diamond i,j}$. Even though the patch index is omitted, the result and the grid structure depend on the patch. For Platonic solids, such as the icosahedron, the grids are identical. $T64$ is a Semi-Platonic solid. Neither all grid sizes nor all angles of coordinate lines are identical for different patches of $T64$. For $T64$ the rhomboidal patches around the poles and those centered around the equator differ from each other but among themselves have identical grids.

The points $\mathbf{r}_k^{\diamond}$ ($k = 1, 2, 3, 4$) defining the corner points of a rhomboidal patch are defined in Eq. (6). For each patch in Eq. (9), a grid $\mathbf{r}_{I,J}^{\diamond}$ ($I = 0, 1, 2, \ldots, I_e, J = 0, 1, 2, \ldots, J_e$) is defined. The origin of the I, J indices is at $\mathbf{r}_2^{\diamond}$. We have $\mathbf{r}_{0,0}^{\diamond} = \mathbf{r}_2^{\diamond}, \mathbf{r}_{I_e,0}^{\diamond} = \mathbf{r}_1^{\diamond}, \mathbf{r}_{I_e,J_e}^{\diamond} = \mathbf{r}_3^{\diamond}, \mathbf{r}_{0,J_e}^{\diamond} = \mathbf{r}_4^{\diamond}$. For any two points $\mathbf{r}_{I_1,J_1}^{\diamond}, \mathbf{r}_{I_2,J_2}^{\diamond}$ in I-direction we call $d\phi(\mathbf{r}_{I_1,J_1}^{\diamond}, \mathbf{r}_{I_2,J_2}^{\diamond})$ the spherical distance of these points. For the spherical grid length $d\phi(\mathbf{r}_{I_1,J_1}^{\diamond}, \mathbf{r}_{I_2,J_2}^{\diamond})$ and the spherical edge length $\mathbf{l}_{1,2}^{\diamond}$ we obtain:

$$\begin{cases} d\phi(\mathbf{r}_{I+1,J}^{\diamond}, \mathbf{r}_{I,J}^{\diamond}) = R\delta\phi(\mathbf{r}_{I+1,J}^{\diamond}, \mathbf{r}_{I,J}^{\diamond}), \\ d\phi(\mathbf{l}_{1,2}^{\diamond}) = R\delta\phi(\mathbf{r}_{0,J}^{\diamond}, \mathbf{r}_{I_e,J}^{\diamond}), \end{cases} \tag{12}$$

where R is the radius of the sphere. Before the rhomboid is projected to the plane, the distance between the edge lines in $\mathbf{r}^\circ$ is defined as

$$\begin{cases} dx_{1,2}^\circ = |\mathbf{l}_{1,2}^\circ|, \\ dy_{2,4}^\circ = |\mathbf{l}_{2,4}^\circ|, \\ dx_{3,4}^\circ = |\mathbf{l}_{3,4}^\circ|, \\ dy_{1,3}^\circ = |\mathbf{l}_{1,3}^\circ|. \end{cases} \tag{13}$$

Note that a regular 1D grid on an edge in straight line norm may not be regular in the great circle norm and vice versa. For a particular 1D segment of a spherical grid, it can always be achieved that this 1D part is regular. It will, however, not be possible to achieve this for all 1D grid segments. This means that there are many alternative ways to define grids within a rhomboidal or triangular patch of the sphere. So there is the possibility of grid optimization. If a grid used great circle grid lines, this offers advantages for the numerical efficiency of the model. Obviously, grids can be constructed where in two directions grid points with indices I and J lying on great circles. This offers advantages for the construction of FD schemes, as neighboring points do not have to be obtained by interpolation. The maximum and minimum grid lengths are then defined as

$$\begin{cases} d\phi_{max\{I,J\}}^\circ = max\{d\phi(\mathbf{r}_{I,J}^\circ, \mathbf{r}_{I,J+1}^\circ)\}, \\ d\phi_{min\{I,J\}}^\circ = min\{d\phi(\mathbf{r}_{I,J}^\circ, \mathbf{r}_{I,J+1}^\circ)\}, \end{cases} \tag{14}$$

which allows to compute maximum and minimum grid lengths for grids on semi-Platonic solids ($I = 0, 1, 2, \ldots, I_e, J = 0, 1, 2, \ldots, J_e$). Unfortunately currently no systematic evaluation of such grid parameters is available. As already visible in the plotted results of Fig. 1, the results of T36 show the smallest difference between the smallest and largest grid lengths. The properties of the grids in dependence of the resolution are not well investigated. Rather than using statistics derived from the above formula, model developers rely on the optical impression of the plotted grids. Apart from the grid length, the angle of coordinate lines is an important information and its computation will be described in the following and this will allow the computation of vectors being important for the evaluation of the dynamic equations. First the tangential vector in the direction of the arc between two vectors $\mathbf{r}_1$ and $\mathbf{r}_2$ will be computed.

For any point $\mathbf{r}$ on the unit sphere, a unit vector $\mathbf{n}$ is called tangential to the sphere and perpendicular to $\mathbf{r}$, if $\mathbf{r} \cdot \mathbf{n} = 0$. A system of tangential vectors will be constructed for every grid point. The tangential unit vector $\mathbf{n}_{I,J}$ at point $\mathbf{r}_{I,J}$ in the direction of a line segment between the two points $\mathbf{r}_{I,J}$, $\mathbf{r}$ on the unit sphere is given as

$$\mathbf{n}_{I,J} = \mathbf{r} - \mathbf{r}_{I,J} \frac{\mathbf{r} \cdot \mathbf{r}_{I,J}}{|\mathbf{r}_{I,J}|^2} \tag{15}$$

for $|\mathbf{r}_{I,J}| = 1$. Now consider the three points $\mathbf{r}_{I,J}, \mathbf{r}_{I+1,J}, \mathbf{r}_{I,J+1}$ and define the tangential vector as $\mathbf{n}_{i,j}^I$ along I-direction for points $\mathbf{r}_{I,J}$ and $\mathbf{r}_{I+1,J}$. Let similar $\mathbf{n}_{i,j}^J$ be the tangential vector in J-direction, defined by Eq. (15) for points $\mathbf{r}_{I,J}$ and $\mathbf{r}_{I,J+1}$. The tangential vectors with superscript i being tangential to the same great circle are called *line parallel*. As we have seen above that grids can be constructed on great circles systems of line parallel tangential vectors and these will simplify solution procedures. In principle, each grid point may have tangential vectors in arbitrary directions. However, this would create a large effort for transformations in the solution procedure. Using the two tangential vectors of point $\mathbf{r}_{i,j}$, we can define the coordinate angle ϕ^{coord} according to Eq. (11) as

$$\phi_{i,j}^{coord} = \delta\phi(\mathbf{n}_{i,j}^I, \mathbf{n}_{i,j}^J). \tag{16}$$

The orthogonal unit vector system $\mathbf{e}^I$ and $\mathbf{e}^J$ is formed from the tangential vectors $\mathbf{n}^I$ and $\mathbf{n}^J$:

$$\begin{cases} \mathbf{e}^I = \dfrac{\mathbf{n}^I}{|\mathbf{n}^I|} \\ \mathbf{e}^J = \dfrac{\mathbf{n}^J - \frac{\mathbf{n}^I \cdot \mathbf{n}^J}{|\mathbf{n}^I|}\mathbf{e}^I}{|\mathbf{n}^J - \frac{\mathbf{n}^I \cdot \mathbf{n}^J}{|\mathbf{n}^I|}\mathbf{e}^I|} \end{cases}. \tag{17}$$

The orthogonal vectors belonging to a grid point are shown in Fig. 8. For any vector $\mathbf{r}_{i,j}$ on the unit sphere, any tangential unit vector $\mathbf{n}$ at this point can be obtained from the orthogonal system defined in Eq. (17):

$$\mathbf{n} = \mathbf{e}^I \cos\alpha + \mathbf{e}^J \sin\alpha, \tag{18}$$

with $\alpha = arccos(\mathbf{n} \cdot \mathbf{e}^I)$.

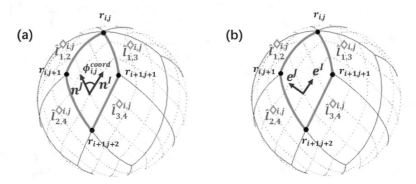

Fig. 8 Tangential vectors and unit orthogonal vector system on the sphere. (a) Tangential vectors and (b) unit orthogonal vector system for any point $\mathbf{r}_{i,j}$. Note that $\mathbf{e}_{i,j}^I$, $\mathbf{e}_{i,j}^J$ are different at different point $\mathbf{r}_{i,j}$, and so do $\mathbf{n}_{i,j}^I$, $\mathbf{n}_{i,j}^J$. $\mathbf{n}^I$ and $\mathbf{n}^J$ are along the dotted-curves, respectively, and only $\mathbf{e}^I$ is along the dotted-curve. $\mathbf{e}^J$ is perpendicular to $\mathbf{e}^I$ which is not along the dotted-curve. For the extension into the halo area, see Fig. 11

For any tangential unit vector $\mathbf{n}$ at unit point $\mathbf{r}$, we can define the directional derivative h_t of a field $h(\mathbf{r})$. A spheric function $h(\mathbf{r})$ is a function based on the 3D vector $\mathbf{r}$ on the unit sphere. For any tangential vector $\mathbf{n}$, the directional derivative of a function $h(\mathbf{r})$ is defined as

$$h_{\mathbf{n}}(\mathbf{r}) = \frac{\partial h(\mathbf{r})}{\partial \mathbf{n}} = \frac{\partial h(\mathbf{r} + t\mathbf{n})}{\partial t} \tag{19}$$

in which h is defined for $\mathbf{r}$ on the unit sphere. For other values of $\mathbf{r}$, as occur in Eq. (19), we can define $h(r) = h\left(\frac{\mathbf{r}}{|\mathbf{r}|}\right)$. Equation (19) is the analog on the sphere of the directional derivative on a plane given in Section "Alternative Methods to Compute Derivatives" in Chapter "Finite Difference Schemes on Sparse and Full Grids". In this book we do not use the projection of large areas to a plane. For small areas, such as the area of a computational cell the projection to the sphere may be useful to obtain approximations for geometric quantities such as cell surface. Let a cell be given by four spherical points $\mathbf{r}_i^s$ with $|\mathbf{r}_i^s| = 1 (i = 1, 2, 3, 4)$. Note that the four points are the corner points of the cell. Let the projection plane be determined by $\mathbf{r}_1^s$ and assume without loss of generality $\mathbf{r}_1^s$ to be in z-direction, $\mathbf{r}_1^s = \mathbf{n}^z$. Then, the projection plane $P(x, y, z)$ is defined as

$$P(x, y, z) = x\mathbf{n}^x + y\mathbf{n}^y + z\mathbf{n}^z = x\mathbf{n}^x + y\mathbf{n}^y + \lambda\mathbf{r}_1^s. \tag{20}$$

The tangential plane to the unit sphere is obtained for $\lambda = 1$. To obtain the projected vector, define the earth center angles β_i as

$$\beta_i = arccos(\mathbf{r}_i^s \cdot \mathbf{r}_1^s), \text{ for } i = 1, 2, 3, 4. \tag{21}$$

The four points on the projection plane $\mathbf{r}_i^p (i = 1, 2, 3, 4)$ are

$$\begin{cases} \mathbf{r}_1^p = \lambda\mathbf{r}_1^s, \\ \mathbf{r}_i^p = \dfrac{\lambda\mathbf{r}_i^s}{\mathbf{r}_1^s \cdot \mathbf{r}_i^s}, \end{cases} \tag{22}$$

which is the 3D version on the projection plane.

Let $\mathbf{r}_i^p (i = 1, 2, 3, 4)$ be plane vectors as shown in Fig. 9. The intersection angles are

$$\begin{cases} \xi_{1,2} = arccos\left(\dfrac{\mathbf{r}_1^p \cdot \mathbf{r}_2^p}{|\mathbf{r}_1^p| \cdot |\mathbf{r}_2^p|}\right), \\ \xi_{1,3} = arccos\left(\dfrac{\mathbf{r}_1^p \cdot \mathbf{r}_3^p}{|\mathbf{r}_1^p| \cdot |\mathbf{r}_3^p|}\right). \end{cases} \tag{23}$$

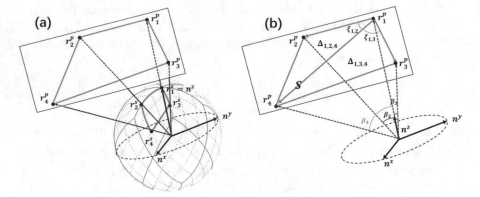

Fig. 9 A schematic of one projection from the sphere to the plane determined by $\mathbf{r}_i^s = \mathbf{n}^z$

The area of quadrilateral S is defined as

$$S = \triangle_{1,2,4} + \triangle_{1,3,4}$$
$$= \frac{1}{2}|\mathbf{r}_2^p - \mathbf{r}_1^p||\mathbf{r}_4^p - \mathbf{r}_1^p|sin(\xi_{1,2}) + \frac{1}{2}|\mathbf{r}_3^p - \mathbf{r}_1^p||\mathbf{r}_4^p - \mathbf{r}_1^p|sin(\xi_{1,3}). \tag{24}$$

Equations of Motion on the Spherical Grid and Non-conserving Finite Difference Schemes

In this section we consider 2D problems on the surface of the sphere. The superscript s of the vector $\mathbf{r}^s$ on the sphere will be omitted in the following. Let $h(\mathbf{r})(|\mathbf{r}| = 1)$ be any function on the unit sphere. For each point $\mathbf{r}$, we assume two liner independent tangential vectors $\mathbf{n}^I(\mathbf{r})$ and $\mathbf{n}^J(\mathbf{r})$ forming a system of tangential basis function at point $\mathbf{r}$:

$$\begin{cases} \mathbf{n}^I \cdot \mathbf{r} = 0, \\ \mathbf{n}^J \cdot \mathbf{r} = 0, \end{cases} \tag{25}$$

where $\mathbf{n}^I \cdot \mathbf{n}^J \neq 1$ and $|\mathbf{n}^I| = |\mathbf{n}^J| = 1$. We assume that $\mathbf{r}_{I,J}$ is one of the grid points in a patch. For the semi-Platonic solid we have in Section "Cubed Sphere, Icosahedron, and Examples of Semi-Platonic Solids" constructed the grid such that in the two main directions I, J the grid points neighboring to I, J are on great circles. Then the orthogonal tangential vector unit system $\mathbf{e}^I, \mathbf{e}^J$ is defined following Eq. (17).

Now let $\mathbf{r}^{\odot}(\phi)$ be the great circle determined by $\mathbf{e}^I$ and $\mathbf{r}$ which is given by

$$\mathbf{r}^{\odot}(\phi) = \mathbf{r}\cos\phi + \mathbf{e}^I \sin\phi, \tag{26}$$

where ϕ is the differentiation parameter. For ϕ small compared to $\frac{\pi}{2}$, $\sin\phi$ can be approximated by ϕ and $\cos\phi$ by one. Under this approximation, the sphere near $\mathbf{r}$ is approximately a plane and Eq. (26) becomes the directional derivative in direction $\mathbf{e}^I$ according to Eq. (17).

The directional derivative of h at the point $\mathbf{r}$ in direction $\mathbf{e}^I$ or $\mathbf{e}^J$ can be obtained according to Eq. (19) as

$$\frac{\partial h}{\partial \mathbf{e}} = \frac{\partial h(\mathbf{r}^{\odot}(\phi))}{\partial \phi}, \tag{27}$$

where the great circle Eq. (26) is used. This is the analog to the directional differentiation in planes, where the differentiation parameter for directional derivatives was called s.

The spherical grids $\mathbf{r}_{I,J}$ obtained in Sections "Cubed Sphere, Icosahedron, and Examples of Semi-Platonic Solids" as projections form bilinear grids have rows and columns of grid points which are on great circles. We describe as an example the partial differentiation along line $\mathbf{e}^I$. Choose $\mathbf{r}_I^{\odot} = \mathbf{r}^{\odot}(\phi_I)(I = -2, -1, 0, 1, 2)$ for fourth order or $i = -1, 0, 1$ for second approximation order, we define

$$h_{I,J} = h(\mathbf{r}_{I,J}^{\odot}). \tag{28}$$

Let $\delta\phi$ be the angular distance of $\mathbf{r}_{I,J}$ and $\mathbf{r}_{I+1,J}$. When $\delta\phi$ is independent of I, the grid is called regular in $\mathbf{e}^I$ direction. Then from Eqs. (14), (15) in Chapter "Simple Finite Difference Procedures" and (27), we obtain the differential approximation:

$$\frac{\partial h}{\partial \mathbf{e}^I} = w_{-2}h_{I-2,J} + w_{-1}h_{I-1,J} + w_0 h_{I,J} + w_1 h_{I+1,J} + w_2 h_{I+2,J}, \tag{29}$$

in fourth order and

$$\frac{\partial h}{\partial \mathbf{e}^I} = w_{-1}h_{I-1,J} + w_0 h_{I,J} + w_1 h_{I+1,J}, \tag{30}$$

for approximation order 2. For the tangential vectors $\mathbf{e}^I$ and $\mathbf{e}^J$, we omit the lower index $I < J$ indicating the grid point. Also the dependence of w on I, J is not indicated by an index. Equations (29) and (30) are the analog on the spherical surface to Eq. (29) in Chapter "2D Basis Functions for Triangular and Rectangular Meshes", valid on the plane. For the regular grid $\mathbf{r}_i^{\odot}$, the weights are given in Eq. (51) in Chapter "Simple Finite Difference Procedures". Note that the same values for w as on the plane can be used when the distance of grid points on the plane is replaced by great circle angular distances. For the irregular grid, the weights w_i can be obtained by the Galerkin compiler MOW_GC. The situation is analog to the differentiation on the sphere. For the most general case the w_i can be computed using the program MOW_GC (see Chapter "Simple Finite Difference Procedures").

The corner point differentiation Eqs. (29) and (30) are the analog to the differentiation in a plane, when differentiation along a straight line is replaced by differentiation along a great circle. The procedure at corner points is identical to the solution method described in Section "Third-Order Differencing for Corner Points with a Second-Degree Polynomial Representation" in Chapter "Finite Difference Schemes on Sparse and Full Grids" for the non-conserving method. At the boundary points, points in a neighboring grid patch are needed. Such points in neighboring patches can be obtained by interpolation. The situation is illustrated in Fig. 11.

When this method to compute derivatives is used at all points of a full grid, as illustrated in Fig. 1c in Chapter "2D Basis Functions for Triangular and Rectangular Meshes" and used to form the spatial derivatives for all fields, we obtain a non-conserving FD approximation on the sphere, which is discussed in Section "Conserving L-Galerkin Schemes on the Sphere". In analogy to discretizations on the plane, conserving schemes can be obtained by introducing amplitudes on the edges. These can be second and third derivatives on the edges. On the plane an example is the o3o3 scheme. On the inscribed bilinear grid, these amplitudes can be introduced as for the plane. The second and third derivatives must be with respect to the angular coordinate. We have also sparseness as in the plane. The definition of basis functions can be taken over from the plane case in the bilinear grid. The spherical grid is obtained by projecting points on the bilinear surface to the sphere by the transformation:

$$\mathbf{r}^s = \frac{\mathbf{r}}{|\mathbf{r}|}, \tag{31}$$

where $\mathbf{r}$ is a point on the bilinear surface and $\mathbf{r}^s$ the corresponding point on the sphere. All basis functions defined on the plane for a plane grid $\mathbf{r}_{I,J}$ are in the same way defined on the bilinear surface, when a bilinear grid $\mathbf{r}_{I,J}$ is given. By the transformation Eq. (31), the grid is transformed to the sphere, where the grid $\mathbf{r}^s_{I,J}$ is induced by the grid $\mathbf{r}_{I,J}$. The same is true for edge grid points, such as $\mathbf{r}_{I+\frac{1}{2},J}$. Please note that in the system of $\mathbf{r}^\diamond$, $\mathbf{r}_{I+\frac{1}{2},J}$ can be notated as $\mathbf{r}^\diamond_{1,2,\frac{1}{2}}$ and similar for the other edges. All basis functions $b(\mathbf{r})$ defined on the plane in Chapter "2D Basis Functions for Triangular and Rectangular Meshes" are defined also on bilinear surfaces for the bilinear grids. Examples are the linear hat functions and quadratic high-order functions introduced in Chapter "2D Basis Functions for Triangular and Rectangular Meshes". $b(\mathbf{r})$ is defined on one or more grid cells. We consider the case of quadrilateral cells as an example $\mathbf{r}(\lambda, \mu)$, where the point $\mathbf{r}$ depends on two coordinates. For a vector $\mathbf{r}'$ outside the bilinear surface (Fig. 10), there is a vector of the surface defined by the equation $\mathbf{r}' a^{red} = \mathbf{r}(\lambda, \mu)$. $\mathbf{r}(\lambda, \mu)$ is the vector on the bilinear surface obtained by a simple geometric construction where λ and μ are the

Fig. 10 A schematic of the relationship between **r** and **r'**

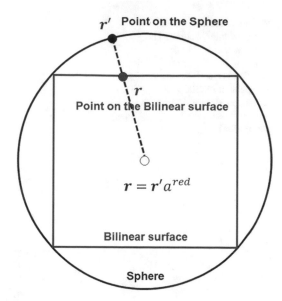

bilinear coordinates. $a^{red}(\mathbf{r}')$ is called the reduction factor of $\mathbf{r}'$. It is obtained by a simple geometric construction.

On the sphere for every basis function b on the bilinear surface, we have defined the corresponding function $b^s(\mathbf{r}^s)$ on the corresponding great circle surface. The corresponding great circle surface is the projective image of the bilinear surface. The connection between b and b^s is

$$b^s(\mathbf{r}^s) = b(\mathbf{r}^s a^{red}), \tag{32}$$

where a^{red} is the reduction factor to bring a point on the sphere to the bilinear grid, as explained in Fig. 10.

Many of the operations used in Chapter "2D Basis Functions for Triangular and Rectangular Meshes" for the definition of the o3o3 scheme, such as the formation of the collocation point amplitudes, can be done in the bilinear grid and then transformed by Eq. (31) to the surface of the sphere. However, the differentiation and integrals of functions must be done on the sphere. For differentiation this means that for example in Eq. (29) the differentiation is with respect to the spherical distance along the great circle. The weights $w_{i'}$ in Eq. (29) are computed from the spherical distances rather than the bilinear distances.

An example for the necessity of great circle arithmetic is the flux through a great circle line between two points $\mathbf{r}_1$ and $\mathbf{r}_2$ in λ-direction on the unit circle. Let $fl(\alpha)$ be the flux vertical to the great circle line $\mathbf{l}_{1,2}$. The definition of the flux can be done on the bilinear line: $\mathbf{r} = \mathbf{r}_1 + t(\mathbf{r}_2 - \mathbf{r}_1) = \mathbf{r}_1 + t\mathbf{l}_{1,2}$ with $t \in (0, 1)$. If the flux

Fig. 11 Obtaining stencil points by interpolation. The numbers 0,1,2,3 are the values of indices i and j for a rhomboid (Reproduced from Steppeler et al. 2008, © American Meteorological Society, used with permission)

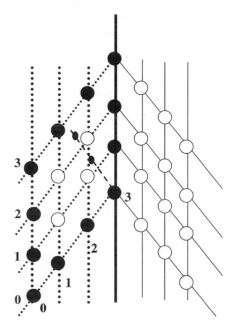

through the great circle arch between $\mathbf{r}_1$ and $\mathbf{r}_2$ is needed, the differential must not be dt. We rather have for the flux F:

$$F = \int_0^1 fl(t)d\omega(t), \tag{33}$$

where the great circle $d\omega$ is a function of t.

Spherical angles between great circles are quite different than the angles of the lines on the bilinear surfaces from which they were generated. For example for the cube after projection three great circles meet at the pole (Fig. 1 in Chapter "Introduction"). This means that the great circle angle between two great circles from the cubed sphere are $\frac{1}{3} \cdot 360° = 120°$. The angles of the plane surfaces of the cube, however, have angles of $90°$.

Inner points of a rhomboidal patch are those, where the neighbors to be used in Eq. (29) are grid points in the same patch. For points on the boundary of a patch this is not the case and points for the difference stencil Eq. (29) have to be obtained from another patch. As this grid has another orientation, the required points must be obtained by interpolation. The situation is shown in Fig. 11 and further explanations are given in Section "Numerical Methods of Hexagonal Grids on the Plane".

For the interpolation process, the coordinates of the point are obtained by great circle extrapolation and a cloud of points around it must be used to interpolate. A popular way of doing the interpolation is the second-order Baumgardner cloud method of Baumgardner and Frederickson (1985) described in Section "Baumgardner's Cloud Derivative Method" in Chapter "Finite Difference Schemes on Sparse and Full Grids". In this method, the cloud of stencil points for differentiation is chosen to consist of all near neighbors, which in a rhomboidal grid is often a number of about 6. Then the grid point values are approximated by a least square method using a second-order polynomial method. This polynomial is then used to interpolate to the target point. Baumgardner's cloud interpolation in Section "Equations of Motion on the Spherical Grid and Non-conserving Finite Difference Schemes" can also be used to compute derivatives and was used to create an atmospheric model based on the icosahedral grid. Unfortunately it is difficult to generalize this cloud method to third order. To do this, it would be necessary to use points beyond next neighbors. This extended cloud of points on the full grid is too large to make this method practical and efficient. This method is illustrated in Fig. 1c in Chapter "2D Basis Functions for Triangular and Rectangular Meshes" but not used here, as we want third-order approximations as a minimum.

Here the serendipity interpolation described in Appendix "The Serendipity Interpolation on the Sphere" is used. The serendipity method from a regular grid forms a larger cell of 4×4 points. If the grid has a constant cell length of dx, the enlarged cell is of approximate size $3dx \times 3dx$. From the enlarged cell, only the 12 points on its outer edges are used for interpolation, not the four points in the inside of the enlarged cell. The interpolation is shown in Fig. 11. A simple test case is obtained by solid body rotation.

Another method to achieve numerical efficiency is to approximate the divergence $fl = (uh, vh)$ by derivatives of h:

$$\nabla \cdot \mathbf{fl} = uh_x + vh_y + u_x h + v_y h. \tag{34}$$

For an approximation of the mass contained in a pixel defined by $\mathbf{r}_{I,J}, \mathbf{r}_{I+1,J}$, $\mathbf{r}_{I+1,J+1}, \mathbf{r}_{I,J+1}$, we have in a simple approximation for the center cell amplitude:

$$\overline{h}_{I,J} = \frac{1}{4}(h_{I+1,J} + h_{I,J} + h_{I+1,J+1} + h_{I,J+1}), \tag{35}$$

and the mass associated with density $h(\mathbf{r})$ is

$$M = \sum_{(I,J)} \overline{h}_{I,J} S(\mathbf{r}_{I,J}), \tag{36}$$

where $S(\mathbf{r}_{I,J})$ is the area of the pixel defined by $\mathbf{r}_{I,J}, \mathbf{r}_{I+1,J}, \mathbf{r}_{I+1,J+1}, \mathbf{r}_{I,J+1}$. We speak of mass conservation, if

$$M_t = \frac{\partial M}{\partial t} = 0. \tag{37}$$

A numerical scheme is said to be conserving, if Eq. (37) is satisfied. There is no reason to expect the different versions of dynamic approximations we have discussed in connection with Eq. (29) are conserving. So the schemes discussed so far may be considered as non-conserving.

For L-Galerkin methods on the sphere, $S(\mathbf{r}_{I,J})$ in Eq. (36) is to be computed as the integral of the basis function corresponding to $h_{I,J}$ or other amplitudes. The property of being conserving depends on the form of the approximation for mass. So a scheme to be conserving, is a suitable alternative to Eq. (35). Then the mass element equation exists, which leads to Eq. (37). There are two non-conserving alternatives given in Sections "Shallow Water Tests on the Sphere: Solid Body Rotation, Solid Body Flow, Advection, and Williamson Test No. 6 and Test of the o3o3 Scheme on the Cubed Sphere Grid Using the Shallow Water Version of the HOMME Model" in Chapter "Numerical Tests" of fourth or second order. An example of a non-conservative toy model including predictions of the solid body rotation and other tests will be given in Section "A Simple Non-conserving Homogeneous Order Discretization on the Sphere" and Section "Shallow Water Tests on the Sphere: Solid Body Rotation, Solid Body Flow, Advection, and Williamson Test No. 6" and Fig. 5 in Chapter "Numerical Tests". 3D test cases being rather near to the real atmosphere have been suggested (Jablonowski and Williamson 2006), but this book is limited to 2D tests for the sphere.

Further Spherical Test Problems

In this book we stick to 2D tests for the surface of the sphere. A more exhaustive set of 2D shallow water tests has been suggested (Jablonowski and Williamson 2006) but only a limited selection is described here. Here we use three shallow water tests.

Let $h(\mathbf{r}), u(\mathbf{r}), v(\mathbf{r})$ be the density field h and two velocity components. As described in Section "Equations of Motion on the Spherical Grid and Non-conserving Finite Difference Schemes", $u(\mathbf{r})$ and $v(\mathbf{r})$ are defined referring to tangential vectors $\mathbf{e}^I$ and $\mathbf{e}^J$, which again are defined in relation to coordinate lines by spherical patches, such as 10 rhomboidal patches for the icosahedron or 12 rhomboidal patches with T24. Therefore, the moments in $u(\mathbf{r})$ and $v(\mathbf{r})$ directions can be defined as

$$\begin{cases} fl^u(\mathbf{r}) = h(\mathbf{r})u(\mathbf{r}), \\ fl^v(\mathbf{r}) = h(\mathbf{r})v(\mathbf{r}), \end{cases} \tag{38}$$

where $\mathbf{u} = u(\mathbf{r})\mathbf{e}^I + v(\mathbf{r})\mathbf{e}^J$. We use the conservation form of dynamic equations:

$$\begin{cases} fl^u_t = -fl^u u_x - fl^v u_y h_x + f v \mathbf{r} \cdot \mathbf{n}^{c,z}, \\ fl^v_t = -fl^u v_x - fl^v v_y h_y - f u \mathbf{r} \cdot \mathbf{n}^{c,z}, \\ h_t = -fl^u_x - fl^v_y, \end{cases} \tag{39}$$

where $\mathbf{n}^{c,z} = \frac{\mathbf{r}}{|\mathbf{r}|}$ is the unit vector in direction of $\mathbf{r}$ and f is the Coriolis parameter.

The discretization using the RK4 time scheme was described in Section "Equations of Motion on the Spherical Grid and Non-conserving Finite Difference Schemes". The discretization using the o2 and o4 schemas of Section "Homogeneous and Inhomogeneous Difference Schemes" in Chapter "Simple Finite Difference Procedures". The last term in Eq. (39) is the horizontal component of the Coriolis term. The first two tests are without Coriolis term: $f = 0$. The solid body rotation is a test for maintaining the large scale flow. Define u, v as tangential vectors and h = constant.

A gravitational wave with $f = 0$ is a good test of the isotropy of the scheme. When the wave is excited in one point, a circular symmetric wave arises if initial values for h constitute a point excitation. Different wave velocities in different directions will result in a distortion of the circular symmetry of the wave.

Shallow water models are often used with staggered grids, as for the plane it is exemplified in Section "The $T64$ Solid for Discretization by Quadrilateral Cells" in Chapter "Numerical Tests". Grid staggering can be done also on the sphere. The schemes described here do not principally prevent staggering. However, the main aim of this section is to go toward high order of approximation, meaning order 3 or 4. For such high order and for spectral elements the advantages of staggering are less important. In this section, we stick to unstaggered schemes.

Conserving L-Galerkin Schemes on the Sphere

L-Galerkin schemes were treated in Chapters "2D Basis Functions for Triangular and Rectangular Meshes and Finite Difference Schemes on Sparse and Full Grids" for planes. To create similar L-Galerkin schemes on the sphere, we notice that the bilinear grids are formal generalizations of plane quadrilateral grids. For example, the o3o3 method in two dimensions needs six steps, which in the same way can be performed on bilinear grids:

1. Formation of a sparse grid. This consists of the corner grid using the corners of the computational cells. The high-order or edge grid is created by subdividing each connection of two corners $\mathbf{r}_1$ and $\mathbf{r}_2$ by creating equally spaced point on the connection line. The grid of corner points is called $\mathbf{r}_{I,J}$. Equivalently the midpoint $\mathbf{r}_m = \frac{1}{2}(\mathbf{r}_1 + \mathbf{r}_2)$ can be used to store a field h with the amplitudes h_{ss} and h_{sss}, the second- and third-order directional derivatives in the direction of $\mathbf{r}_2 - \mathbf{r}_1$. This grid can be projected to the sphere. Projections of rows of grid points are on great circles if the points are on a line in the bilinear grid. Some of the operations for the approximation procedure need to be performed on the great circle grid.
2. Definition of amplitudes $h_{I,J}$, $h_{ss,I+\frac{1}{2},J}$, $h_{s's',I,J+\frac{1}{2}}$, $h_{sss,I+\frac{1}{2},J}$ and $h_{s's's',I,J+\frac{1}{2}}$. The field inside a computational cell is obtained by the basis function representation using these amplitudes.

3. Time derivatives at corner points are obtained by creating a stencil and taking the FD schemes on this stencil. For the special simple case of a line stencil one dimensional numerical differentiation can be used.
4. The edge amplitude $h_{t,ss,I+\frac{1}{2},J}$ for example is obtained by applying the principle of conservation to a computational cell and the relevant amplitudes.
5. The third-order amplitudes $h_{t,sss,I+\frac{1}{2},J}$ are obtained by numerical differentiation of the $h_{t,ss,I+\frac{1}{2},J}$ field.
6. Using the amplitudes of time derivatives RK4 or any other time differencing method can be used to make a time step.

Steps 1 to 6 can be performed on bilinear/spherical grids in the same way as on planes. The grid construction, Step 1 can be done in the same way as for the plane. After projection to the sphere, a spherical grid is obtained. The only difference to a grid in a plane is that the resulting spherical grid is not regular, even when the grid is bilinear. This needs to be taken into account with Step 3, where differencing formula Eq. (15) in Chapter "Simple Finite Difference Procedures" corresponding to irregular grids need to be employed. While it can be achieved that some rows or columns of grid points are regular on the sphere, a totally regular grid cannot be obtained. For Step 2, it is easiest to assume that the second- and third-order derivatives used as amplitudes are derivatives with respect to the spherical angle. With Step 3, there is no difficulty to perform it on great circles, when difference formula for irregular grids are taken. In Step 4, flux integrals must be taken. The difference to the grid on a plane is that integrals on the sphere must be taken. With Step 5, the special case that the bilinear grid consists of straight lines is easy to treat. The spherical grid obtained by projection is a grid on great circle lines. This is treated by adjusting the weights used for differentiation Eq. (15) in Chapter "Simple Finite Difference Procedures". To perform Step 6, there is no difference between a model on a plane and on a spherical grid. This means that Steps 1 to 6 can be performed in a similar way as a model on a plane if the bilinear grid is a regular or irregular structured grid where the bilinear grid points are on straight lines. The only difference is that for the sphere the formula Eq. (15) in Chapter "Simple Finite Difference Procedures" for differentiation on irregular grids always need to be taken, where on a plane the regular grid case exists. For limited area models formulated on a quadrilateral, this means that we have all tools to construct a spherical model. When the model area is small the resulting grid is rather regular while for large quadrilaterals the grid may become irregular, which is a computational price to be paid for going toward the spherical geometry. Figure 1 in Chapter "Introduction" illustrates the grid irregularity caused by the spherical geometry.

With global models on the sphere, we have the difficulty that the grid will consist of several patches which create internal boundaries. One way of dealing with such internal boundaries is to pose boundary values. For the case of a non-conserving scheme, this will be treated in Section "A Simple Cut-Cell System Based on the Staggered Low-Order Basis Functions" in Chapter "Finite Difference Schemes on Sparse and Full Grids". As corner point differentiation with the o3o3 method is done with the same stencil as used in Section "A Simple Cut-Cell System Based on the Staggered Low-Order Basis Functions" in Chapter "Finite Difference Schemes on Sparse and Full Grids", the method can be transferred to the o3o3 case on the sphere. However, no sparse L-Galerkin methods on spherical grids have been explored until now. The same is true for sparse hexagonal grids on the sphere, even though the tools provided above would suggest such models. Existing quadrilateral and hexagonal models are not sparse.

The method described above is for small patches, a rather regular grid, on semi Platonic solids. Examples are shown in Fig. 1 in Chapter "Introduction". We have discussed above that grid refinement in global forecast modes is target oriented. This means that the grid is refined in the area where the forecast is of interest for the owners of the model. For some models, such as MPAS (Park et al. 2014) this grid refinement is within the global model using two way interaction. If just the medium range forecast of about 10 days is of interest, a regular grid for the global model is appropriate. For the longer forecast, all areas of the world are important. Resolving some areas of low meteorological interest in low resolution saves little computer time that does not justify the overhead of adaptive resolution. This is totally different for pollution modeling, where areas near the origin of pollution and along the path require an increased resolution. This may be obtained by adaptive meshes. Climate simulation where the increased resolution of critical areas is explored is another area of adaptive global modeling.

For global weather forecasting, except for resolution increase in a target area, the currently used models are based on regular resolution. For operational weather forecasts less than 10 weather services worldwide run a global model. A lot of countries use limited area models based on one-way interaction on the global models. This is the reason why for global modeling the regular grids investigated in this section may have an application. As o3o3 is suitable for sudden increase of resolution (Chapter "Local-Galerkin Schemes in 1D") two way nesting is possible with o3o3. A totally irregular mesh with o3o3 will need more development work and probably would need more computer power.

A Simple Non-conserving Homogeneous Order Discretization on the Sphere

For the formal definition of the grid see Section "Further Spherical Test Problems". Test equations are the shallow water equations and the advection equation which for our tests must be taken in two dimensions.

The grid is shown in Fig. 1c in Chapter "Introduction". For each grid point different orthogonal basis vectors are formed, as shown in this figure. So each grid-point has its own orthogonal coordinates x, y and the shallow water equation (39) is used. So for every grid point a different coordinate is used. The grid consists of great circles. Let $x - y'$ be the coordinates in the direction of the great circles. Then the directional derivatives are computed using the neighboring grid points. Inside one of the great circle rhomboids seen in Fig. 1a, field f can be indexed $f_{i,j}$. For example, in the direction of the index i let $\eta^{i'}$ be $|\mathbf{r}_{i,j} - \mathbf{r}_{i-i',j}|$, where $|\cdot|$ is the great circle distance, $i \in (-2, -1, 0, 1, 2)$ using the $\eta^{i'}$ this allows to compute weights $w_{i'}$ and can compute the directional derivative applying:

$$f_{x,i,j} = w_{-2}f_{i-2,j} + w_{-1}f_{i-1,j} + w_0 f_{i,j} + w_{+1}f_{i+1,j} + w_{+2}f_{i+2,j}. \qquad (40)$$

The derivative to y' is formed analogously and from this the derivatives to x and y can be formed and using Eq. (39) the time derivative of f can be formed.

This procedure is unproblematic for inner points of the rhomboid. For points near the boundary some of the points are obtained by interpolation, as indicated in Fig. 11. The interpolation is done by the serendipity interpolation described in Appendix "The Serendipity Interpolation on the Sphere".

For input–output purposes and plotting normally the latitude–longitude grid with coordinates (λ, ϕ) is used. The 3D vector $\mathbf{r} = (x, y, z)$ on the earth surface is

$$\begin{cases} X = R\cos\lambda\cos\phi, \\ Y = R\cos\lambda\sin\phi, \\ Z = R\sin\phi, \end{cases} \qquad (41)$$

with R being the radius of the earth. A transformation between the latitude longitude used for output and plotting and the icosahedral grid used for model calculations will require interpolation.

A few results of this shallow water model will be shown in Chapter "Numerical Tests".

Hexagonal Grids on the Sphere

The grid is illustrated in Fig. 2d. The grid construction and labeling of the amplitude points can be done in analogy to the plane. Amplitudes are defined at corner points and not at cell centers. As in the plane, it is convenient to use cell centers to organize the amplitudes where each cell stores all amplitudes. When this is done, the representation is redundant as corner points may belong to more than one cell. For a compact representation in analogy to the plane case, a subset of corner amplitudes is stored in order to obtain a compact storage. The full representation can be obtained by message passing from the compact representation. The center points of hexagons can be subsets of the rhomboidal corner grid, where it must be observed that center points may belong to more than one rhomboidal patch and the corners belonging to a hexagon center may belong to different patches. Not all semi-Platonic solids will result in nearly regular hexagons. Figure 2d is based on the $T36$ solid and the icosahedron also appears to be a good choice.

Numerical Tests

Abstract A few tests are proposed being based on toy models.

Keywords Advection tests · Open boundary condition · Shallow water spherical tests · Mountain waves · Numerical efficiency

1D Homogeneous Advection Test for *onom* Methods, SEM2 and SEM3

The test example is the homogeneous 1D advection equation:

$$h_t(x, t) = -u_0 h_x(x, t), \tag{1}$$

where $h(x, t)$ is the field, $u_0 = 1$ is the constant velocity field, and the lower indices t or x mean differentiations to time or space. We use an interval of 601 points ($i = 0, 1, 2, \ldots, 600$) with $dx_i = 1$ for regular grid and periodic boundary conditions. To investigate the mass-conservation property of *onom*, an irregular grid is also used where

$$\begin{cases} dx_i = 2, & \text{for } i = 180, 181, \ldots, 210, \\ dx_i = 1, & \text{for otherwise.} \end{cases} \tag{2}$$

There are three setups of the initial conditions that are called the peak initial condition, the No. 4 smooth initial condition, and the No. 8 smooth initial condition centered at x_{150}. The peak condition is defined as

$$
\begin{cases}
h_{(x_{148}, t_0)} = h(x_{152}, t_0) = \dfrac{4}{3}, \\
h_{(x_{149}, t_0)} = h(x_{151}, t_0) = \dfrac{8}{3}, \\
h(x_{150}, t_0) = 4,
\end{cases}
\tag{3}
$$

and $h(x_i, t_0) = 0$ for all other i in the experiments.

The smooth initial condition is defined as

$$
h(x_i, t_0) = 4 exp \left[-\frac{1}{l} (x_1 - x_{150})^2 \right],
\tag{4}
$$

where $l = 4$ is No. 4 smooth initial condition and $l = 8$ is No. 8 smooth initial condition. A time step $dt = 1.0$ is used by the fourth-order Runge–Kutta integration, while the number of steps is 30 000. We set the comparison of standard o3o3, spectral o3o3, SEM3, and o4 as an example to show the advection results in regular grids with peak initial condition (Fig. 1a). The codes of all methods (including o2o3, standard o3o3, spectral o3o3, SEM2, SEM3, o4, and the physical interface) are listed in Chapter "Examples of Program".

Figure 1 shows the initial values representing the peak solution, as defined above over the distance of 300 and 30 000 for standard non-conserving o4 (Fig. 1b, c), standard o3o3 (Fig. 1d, e), spectral o3o3 (Fig. 1f, g), and SEM3 (Fig. 1h, i). The transport over the larger distance of 30 000 shows an increased dispersion error. Figure 2 is the same as Fig. 1 for the No. 4 smooth solution.

For the transport over 30 000 with $dt = 1$, the maximum of $h(x)$ is given in Table 1 for the four schemes investigated. SEM3 is the best by a small margin, and the two versions of o3o3 are rather similar in accuracy to the classical o4 scheme. As a quantitative measure of accuracy for the transport over the distance of 30 000, the forecasted maximum of $h(x)$ divided by four with solutions using $dt = 1, dx = 1$ is shown in Table 1. The higher the maximum is, the more accurate the scheme is. The accuracies for the four schemes are rather similar. For o4 and o3o3, the results are also shown for the larger time steps $dt = 2$ and $dt = 2.5$. The rather strong dependence of the results on the time step is remarkable. The spread of the results with dt is such that we cannot conclude more than that the fourth-order schemes are similar in accuracy.

The FORTRAN codes of all methods (including o2o3, standard o3o3, spectral o3o3, SEM2, SEM3, o4, and the physical interface) are listed in Section "Dispersion Analysis of o2o3 and o3o3 Methods".

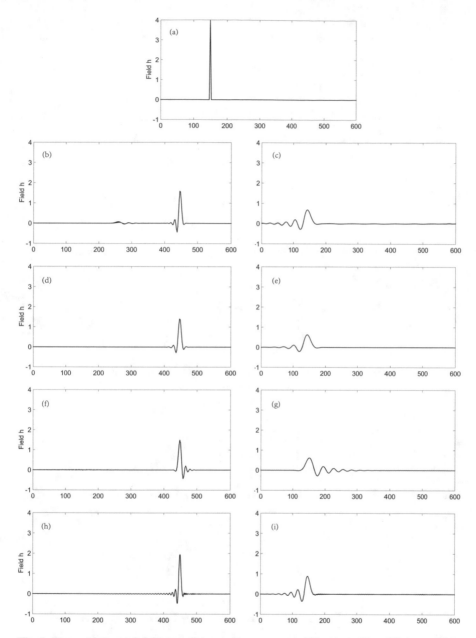

Fig. 1 Runs with the peak initial condition and transport over 300 points with a different spacing for dx = 1 and dt = 1: (**a**) initial values, (**b**) standard o4 spatial difference, (**d**) o3o3 standard difference, (**f**) o3o3 spectral difference, and (**h**) SEM3; transport over 30 000 points: (**c**) standard o4 spatial difference, (**e**) o3o3 standard difference, (**g**) o3o3 spectral difference, and (**i**) SEM3 at all points (reproduced from Steppeler et al. (2019))

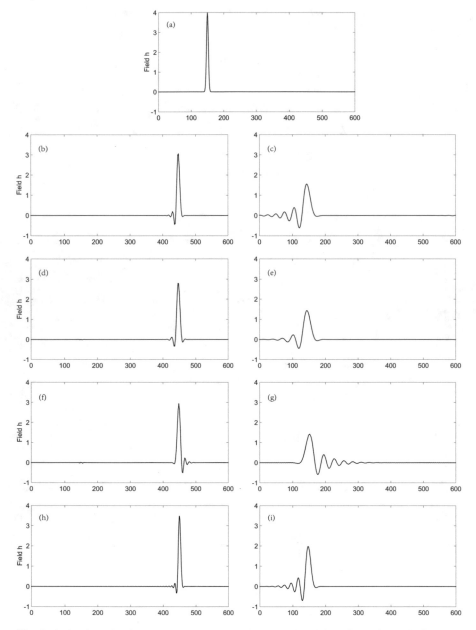

Fig. 2 As in Fig. 1, but for the No. 4 smooth solution (reproduced from Steppeler et al. (2019))

Table 1 Maximum of field $h(x)$ after a transport over 30 000. The higher the value, the more accurate is a scheme

Scheme	Standard o4	Standard o4	Standard o3o3	Standard o3o3	Spectral o3o3	SEM3
Time step (s)	1.0	2.0	1.0	2.5	1.0	1.0
Peak	0.135	0.146	0.1	0.139	0.079	0.209
NO 4	0.279	0.313	0.214	0.309	0.149	0.402
NO 8	0.364	0.488	0.311	0.501	0.208	0.419

A Numerical Example of Open Boundary Condition for a Fast Wave

As seen in Section "Open Boundary Condition" in Chapter "Local-Galerkin Schemes in 1D", the formulation of open boundaries is a matter of trial and error. Attempts to obtain an open boundary by discretizing an analytic boundary condition have failed. An open boundary for the advection process was given in Chapter "Local-Galerkin Schemes in 1D".

For the fast waves, we consider the linearized shallow water wave equation, obtained from a 1D version of Eq. (58) by linearizing around the state $u = 0$, $h = h_0$. We obtain

$$
\begin{cases}
u_t = -h_x, \\
h_t = -h_0 u_x.
\end{cases}
\tag{5}
$$

In Eq. (5), the lower indices mean differentiation. As always in this book, we use the RK4 scheme for time discretization (see Section "The Runge–Kutta and Other Time Discretization Schemes" in Chapter "Simple Finite Difference Procedures"). We assume a regular grid x_i with amplitudes h_i and u_i. The spatial derivatives in Eq. (5) are done by standard (non-conserving) o4 differentiation given in chapter "Simple Finite Difference Procedures". We assume dx and $h_0 = 1$, which results into a wave velocity of 1. This simple non-conserving model is sufficient to show the principle of open boundary discretization. The model area contains 301 points, $i \in \{0, 1, \ldots, 300\}$. In order to apply FD schemes at points $i = 299$ and $i = 300$, we need the amplitudes $u_{301}, u_{302}, h_{301}, h_{302}$ outside the domain, which we obtain by posing boundary conditions. While with homogeneous advection Eq. (2) in Chapter "Simple Finite Difference Procedures" waves move in just one direction, the fast gravitational waves from Eq. (5) are moving to the right and to the left.

An open boundary is obtained by

$$
\begin{cases}
h_{301} = h_{302} = h_{300}, \\
u_{301} = u_{302} = u_{300}.
\end{cases}
\tag{6}
$$

Fig. 3 The fast waves at different times (initial and after being swallowed by the boundary at $i = 300$). Corresponding to (d), form boundary condition (8**4). With this boundary condition, the computational mode has a smaller amplitude

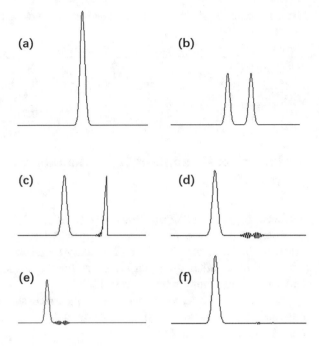

For initial values, we chose

$$\begin{cases} u_i = 0, \\ h_i = exp\left[-\left(\dfrac{x_i - x_{250}}{8}\right)^2\right]. \end{cases} \tag{7}$$

The results are shown in Fig. 3 with $dt = 1$. Equation (6) shows the wave at different times, and the part of the wave moving to the right is swallowed by the boundary. The boundary creates a small amplitude computational mode that is of wavelength $2dx$ and therefore can be filtered by hyper-diffusion. The computational mode moves faster than the wave moving to the left and will eventually overtake it. The subject of open boundaries is largely un-researched, and Eq. (6) is intended as an example to encourage further work.

Equation (6) can be replaced by boundary values from linear extrapolation:

$$\begin{cases} h_{301} = h_{300}(h_{300} + (h_{300} - h_{299})), \\ h_{301} = 2h_{300}(h_{300} + (h_{300} - h_{299})). \end{cases} \tag{8}$$

Similar for u. Compared to Eq. (6), Eq. (8) produces much less noise. This result is also shown in Fig. 3. The generation of computational modes is common with internal or external boundaries or physical processes when these create small-scale fields.

The *T* 64 Solid for Discretization by Quadrilateral Cells

The two patch systems for the sphere (see Chapter "Platonic and Semi-Platonic Solids") in widely practical use are the cubed sphere and the icosahedron. Plots of the grids are given in Fig. 1 in Chapter "Introduction". It appears that the icosahedron is suitable for hexahedral cell structures and was used for this purpose (Skamarock and Klemp 2008). The cubed sphere is suitable for quadrilateral grid patches, and with practical applications, it was used for this purpose (Taylor et al. 1997). The icosahedron is called *T* 20 as it consists of 20 triangular patches. The cubed sphere is *T* 12. An improved version for hexagonal grids is the *T* 36 solid introduced in Chapter "Platonic and Semi-Platonic Solids". In this section, we look for an improved spherical patch systems on the sphere for quadrilaterals as a potential alternative to the cubed sphere. As the cubed sphere covers the sphere by 6 rhomboidal patches, these rather large areas show strongly deformed grids. The patches of the cubed sphere deviate strongly from an orthogonal grid. Because at patch corners, three edges meet. The angle between these edges is 120° for all patch corners, meaning a strong deviation from a 90° grid. The rather deformed character of the cubes sphere grid is seen in Fig. 1 in Chapter "Introduction".

The *T* 64 solid introduced here has right angle patches at the poles. As at all corners 4 edges meet. For this reason and because of the smaller patch sizes, *T* 64 may be expected to be better suitable for quadrilateral cells than *T* 12 (cubed sphere). Unfortunately, such information must remain qualitative as no systematic evaluation of grid cell regularity for patches on the sphere is available. It is common to obtain such information by visualizing the grids.

The *T* 64 patch system is shown in Fig. 4. For the northern hemisphere, the rhomboidal bilinear system is shown. The spherical grid is obtained by projecting this to the sphere. The system has corners of the rhomboidal patches at the two poles and the equator on 4 circles being not great circles. Using the terminology introduced in Chapter "Platonic and Semi-Platonic Solids", this grid is classified as $PL^{4,7}$. The equator and the two corner carrying circles next to it support 8 corners, while the other circles and the poles carry 4 corners each. It can also be called *R*32, as it consists of 32 rhomboidal patches. A discretization would use 32 patches with quadrilateral cell systems. Comparing Figs. 1 in Chapter "Introduction", a more regular grid system may be expected with *T* 64 than with *T* 12.

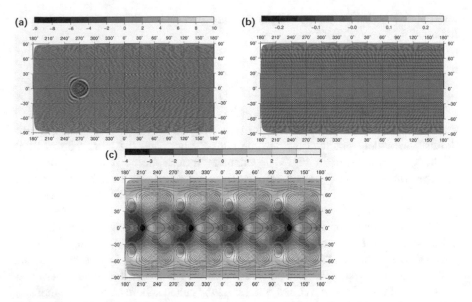

Fig. 4 The *T*64 solid for the northern hemisphere. The bilinear rhomboids are shown, which must be projected to the sphere to get the great circle rhomboids. Solid lines are the rhomboidal edges for the front surface. For the back surface, the rhomboidal edges are shown as dashed lines. Dotted lines show the division of the rhomboids into triangles. This is not a platonic solid but after projection shows more regularity for the grid (not shown) in the patches than the cubed sphere (*T*12)

Shallow Water Tests on the Sphere: Solid Body Rotation, Solid Body Flow, Advection, and Williamson Test No. 6

For shallow water modeling on the sphere, a set of tests were proposed in Williamson (1968). Some of these tests will be reported here: the first test is the transport of a structure in a flow representing a solid body rotation. The flow components are given in the (λ, ϕ) coordinates. u and v are the components of velocity corresponding to the coordinates λ and ϕ.

The velocity fields are $v = 0$ and $u = \cos\phi$. The solution transports the initial values along the equator, and after one rotation, the solution is shown in Fig. 5a. The error consists of some spectral shown in Fig. 5.

The same solid body rotation can also be used with the shallow water equations. When using Coriolis parameter $f = 0$ in the dynamic equation, this is a stationary solution. It can be used to see the effect of internal boundaries. In this grid, the internal boundaries are the boundaries of rhomboids. In normal predictions, these internal boundaries must not be seen. For the solid boundary solution, which is large scale and stationary, the error can be made visible by plotting a small error level. The solution of a 5-day prediction is shown in Fig. 5b, and the error is smaller than 0.2 geopotential meters. For comparison, Fig. 5c shows the solution of a Rossby wave with realistic Coriolis parameter, and the isoline spacing is 100 geopotential meters. So with the spacing used for the physical solution of meteorological waves, the inhomogeneity caused by the internal boundaries of the grid is not seen. This absence of a disturbance caused by the internal boundaries is caused by the uniform fourth-order approximation order. The approximation order is called uniform, when it is present at all grid points and at the points of an internal boundary.

When introducing a numerical diffusion, the error of the solid body shallow water solution is further reduced to 0.001 geopotential meters. The solution using the meteorological wave can be compared to the correct solution. Note that this high resolution was obtained around 2000, and using today's computer, even higher resolution could be used. The latter has been determined by very high-resolution simulation. This solution, called Williamson test No. 6, is convergent for the 3-day prediction and from Day 5 on it, shows unpredictability as small disturbances amplify to the error which a randomly chosen solution would have.

The Williamson tests (Williamson 1968) are still used to test approximations on the sphere. Recently, 3D test solutions have also been proposed (Williamson 1968). In comparison, for the effect of order one at the internal boundaries, see Thuburn et al. (2001).

2D Mountain Wave Test

In this section, we want to answer an unsolved question: whether continuous Galerkin method with unstructured adaptive mesh technique can accurately represent the underlying terrain and simulate mountain waves, which have a dominant effect on atmospheric motions as the horizontal resolution approaches or exceeds

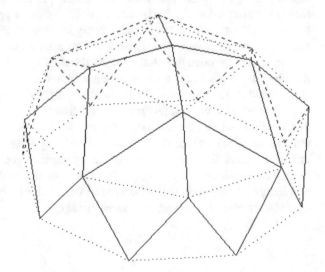

Fig. 5 (**a**) Advection by a constant velocity along the equator, prediction after one rotation, (**b**) solid body rotation without Coriolis force: The error is smaller than 0.2, (**c**) prediction of the meteorological relevant Rossby wave. The line spacing is 200 geopotential meters, so that the error shown in (**b**) is not visible (reproduced from Steppeler et al. (2008), © American Meteorological Society, used with permission)

$O(10^1)$ km (Gallus and Klemp 2000). We use the benchmark 2D test of Lock et al. (2012) for flow over a bell-shaped mountain with steady boundary conditions to form a stable upward-propagating mountain wave in a stratified atmosphere. The computational domain is 60 *km* wide in the horizontal direction and 16 *km* deep in the vertical direction, with a simulation time of 50 000 *s*. The time step is set to 5.0 *s*, and mesh adaptation is performed every 10 time steps. The resolution of the adaptive meshes varies from 0.2 *km* to 2 *km* with respect to the solution of the state variables (the velocity vector here). For comparison purposes, the control run is conducted in a fixed mesh with horizontal and vertical resolutions of $dx = dz = 0.2$ *km*. For spatial discretization, first-order CG method is applied. For temporal discretization, we utilize the semi-implicit Crank–Nicolson scheme with $\lambda = 0.5$. The underlying 2D bell-shaped mountain is defined as

$$h(x) = \frac{h_0}{1 + \frac{x^2}{a^2}}, \tag{9}$$

where $h_0 = 400$ *m* is the maximum height of the mountain, and the half-width of the mountain is $a = 1000$ *m*. We use a constant Brunt–Väisälä frequency of $N = 0.01 s^{-1}$ to define the stratified background, and the bottom potential temperature is 293.15 *K*. The initial velocity of the flow is $\mathbf{u} = (10.0, 0.0)$ *m/s*. We apply no flux boundary conditions along the bottom surface. Open lateral boundary conditions are used at the inflow and outflow boundaries. To prevent the oscillation of the waves reflected at the top and the lateral boundaries, an absorbing layer is applied. For stability, we define two continuous diffusions for the lateral boundaries.

Figure 6 illustrates the contours of two components of the velocity until a steady-state velocity is achieved by (a) theory for vertical velocity (Gallus and Klemp 2000); (b) simulation for vertical velocity by a fixed mesh; (c) simulation for vertical velocity by an adaptive mesh; (d) simulation for horizontal velocity by a fixed mesh; (e) simulation for horizontal velocity by an adaptive mesh. All the flow patterns using the CG method show good agreement with the analytic solution, and the contours of the vertical velocity are stacked vertically above the terrain. A comparison between the results for the fixed mesh and the adaptive mesh reveals that the adaptive mesh is able to simulate mountain waves with a similar quality as the fixed mesh. Moreover, the results for the fixed mesh exhibit smooth vertical velocity contours, while a few numerical artifacts can be detected at the periphery of the contours (e.g., the outermost contour) for all the adaptive mesh cases. This may be seen as a small error arising from adaptivity. Although an increase in diffusion makes the numerical noise disappear, it is accompanied by a reduction in the magnitude of the velocity. Therefore, partial node locking at the bottom boundary can eliminate the noise around the peak of the mountain and maintain the magnitude of the velocity (not shown here).

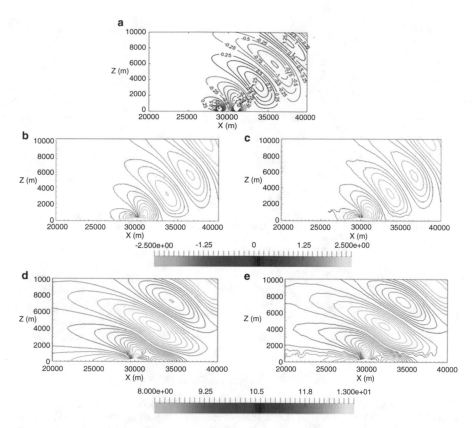

Fig. 6 Two components of the velocity solution for the mountain wave simulation over a 2D bell-shaped terrain with a contour interval of 0.25 m/s. (**a**) The analytic solution reproduced from Gallus and Klemp (2000); (**b**) simulation for vertical velocity by a fixed mesh; (**c**) simulation for vertical velocity by an adaptive mesh; (**d**) simulation for horizontal velocity by a fixed mesh; (**e**) simulation for horizontal velocity by an adaptive mesh. Used from Li et al. (2021)

To evaluate the impact of the horizontal (vertical) resolution, we keep the minimum vertical (horizontal) mesh size at 200 m, while the minimum horizontal (vertical) mesh sizes are 1600, 800, 400, 200, 100, and 50 m. The results with variations in the horizontal mesh size are shown in Fig. 7. These results (the accuracy and location of the wave contours) agree well with the analytic solution except for those for the coarse mesh scheme. According to the increase in the horizontal mesh resolution, the amplitudes of the vertical velocity are increased somewhat positively at the peak of the mountain, and the contour of the vertical velocity becomes smooth, although there is slight noise on the bottom boundary. However, when the mesh size is less than 200 m, the effect of increasing the horizontal resolution is not obvious in terms of smoothness, and the continuity of the solution is different from the results in Gallus and Klemp (2000). With the increase in vertical resolution, the vertical velocity contour is near the analytic solution and is similar to the result in Fig. 7a.

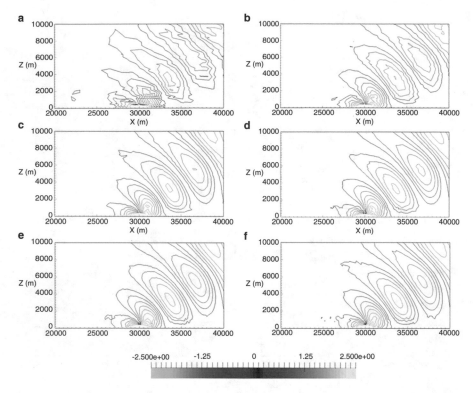

Fig. 7 Vertical velocity solution of the mountain wave simulation over a 2D bell-shaped terrain with different horizontal mesh sizes. The maximum mesh size in both the horizontal and vertical directions is 2000 m, and the minimum vertical mesh size is 200 m. The minimum horizontal mesh sizes are (**a**) 1600 m, (**b**) 800 m, (**c**) 400 m, (**d**) 200 m, (**e**) 100 m, and (**f**) 50 m. The contour interval is 0.25 m/s. Used from Li et al. (2021)

The Kalman Filter Data Analysis

As described in section "Data Assimilation", data analysis is the art of creating an initial state for a discretized model using observations. In this section, an example for the Kalman filter assimilation method is given, which is simple enough to experiment with it on a PC without needing specialized software. This section has the purpose to generate interest in this subject, and the interested reader has to use the specialized literature (Nakamura and Potthast 2015).

As it is un-practical to create observations for every grid point, the observations are typically incomplete. An example is the vertical velocity, which is difficult to measure and few direct observations exist. So the observed vertical velocities as shown in Fig. 4 in Chapter "Introduction" are obtained from other observations.

The Kalman filter analysis allows to analyze dynamic amplitudes for which no observations exist. It also accounts for the fact that the model itself is not 100%

accurate but rather has errors. Compared to other methods, such as 4D variational analysis, it is a simple way to add data analysis while observing the dynamic constraints represented by the model.

As in this book data analysis is a side topic, we give a very simple example with the intention to raise interest into this subject and allow some simple analysis experiments. For the dynamic equations, we use the Lorenz model as described in Section "The Lorenz Paradigmatic Model" in Chapter "Simple Finite Difference Procedures". It has the dynamical variables C, L, M, which are considered time dependent. As the time can go to infinity and we are interested only in intervals where there is predictability, we have for the different initial times an infinity of cases.

Let t_1 be the initial time and (t_1, t_3) be the time interval for which we would like to do forecasts. This time interval must be such that for the accuracy of our model and observations we have some predictability. The data analysis time interval is (t_1, t_2), with $t_2 < t_3$. Let us consider the time trajectory in C, L, M space with parameter $q = 2$. For simplicity, we assume that C or other parameters are directly observable. We could employ an observation operator that creates observations as a function of C, L, M. For example, we may assume that C stands for a velocity, and we observe a transported field, which is a function (observation operator) of C. While observation operators are a common feature with data analysis systems, we do not include this difficulty here.

We assume to have a model of the dynamic system, for example the Lorenz model, with state vector $\phi = (C, L, M)$ with parameter $q = 1$. The reader can easily investigate what predictability this model has by assuming the model error $(q = 2/q = 1)$ is the only error of the system. This model error may depend on the initial time of the forecast, and in this respect, the simple model is similar as the real atmosphere except that for the real atmosphere the model error is more difficult to investigate as the true atmospheric state is not available.

We assume to have a first guess ϕ_0 for the initial state, and ϕ_0 could be a forecast from the previous analysis period. Now we disturb the initial state to get an ensemble of i_e initial states, which could be chosen by adding a small vector to ϕ_0.

For maximum simplicity, we assume that just component C of ϕ is observed. We can assume that in phase space of ϕ we disturb equally in all three directions. This will sometimes bring us out of the attractor described in Section "The Lorenz Paradigmatic Model" in Chapter "Simple Finite Difference Procedures". A more refined proceeding would be to make the disturbance within the attractor. The disturbance of the initial value allows by time integration to obtain the trajectories first time (t_1, t_2).

To change the trajectory to a more likely one, we need a cost function:

$$J(\phi) = \frac{1}{w_b}(\phi_b - \phi)^2 + \frac{1}{w_{obs}}(y - \phi)^2, \tag{10}$$

where ϕ is the optimal trajectory to be determined, ϕ_b is the approximated trajectory with $q = 1$ (to be determined by an ensemble of forecasts), y are the observations, w_b is the error variance of the trajectories, and w_{obs} are the error variances of observations.

In more refined versions of Eq. (10), error variances toward the unobserved amplitudes L, M can be introduced.

Test of the o3o3 Scheme on the Cubed Sphere Grid Using the Shallow Water Version of the HOMME Model

The o3o3 scheme was implemented with the shallow water version of the HOMME model. The cubed sphere quadrilateral grid is shown in Fig. 1 in Chapter "Introduction", and the Williamson test case No. 6 is described in Section "The $T64$ Solid for Discretization by Quadrilateral Cells". The error is shown in Fig. 8 as a function of grid length. The SE3 scheme shown for comparison uses the full grid, while the o3o3 scheme in two dimensions is sparse with sparseness factor $\frac{5}{9}$. The results shown in Fig. 8 are of comparable accuracy.

Projections of Semi-platonic Solids to Triangular Surfaces

Quadrilateral grid patches can be obtained by projecting bilinear surfaces to the sphere, as shown in Chapter "Platonic and Semi-Platonic Solids". Such a bilinear surface cannot be realized on a paper sheet. A deformable paper such as a rubber sheet would be necessary. However, when triangular patches are formed by dividing a rhomboidal path into two triangles, these can be realized on a plane, and the plane triangle is projected to the sphere to obtain the great circle grid. When the triangular patches are arranged in a way that the triangles are connected, we call this a rollout into the plane of the inscribed triangular mesh of the sphere.

This section gives a number of rollouts for some semi-Platonic solids treated in this book, and these are shown in Fig. 9. Rollouts can be used to illustrate the internal boundaries of a patch system for the sphere. A simple way of creating boundary amplitudes is to compute the one-sided derivatives and average. Rollouts can also be used to create 3D models of the solid. To do this, cut out the solid from Fig. 9 along the solid lines, if they are not internal. At the internal solid lines and the dotted line, fold the paper and glue together.

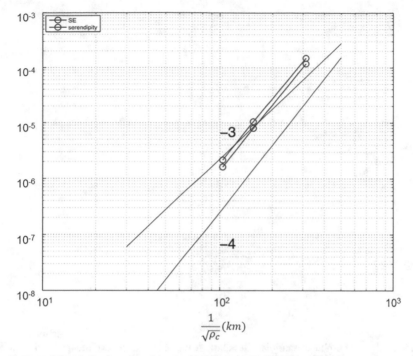

Fig. 8 Test of the o3o3 scheme on the cubed sphere grid. The Williamson test case 6 is performed (see Section "The $T64$ Solid for Discretization by Quadrilateral Cells"). The error is shown for the SE3 scheme (blue, standard discretization for HOMME) and the o3o3 scheme (red) as a function of grid length. Even though both schemes are third order by construction, a third-order convergence happens due to super-convergence. SE3 used the full grid, while o3o3 is sparse. The test was performed with the help of Dr. Marc Taylor

Figure 9a shows the rollout of $T4$, the tetrahedron. The $T4$ has the special feature that when combining triangles to rhomboids, these overlap. We have 3 rhomboids. This means that $T4$ can also be called $R3$. Figure 9b shows the cube, which has the special property that folding along the dotted line is not necessary. The folding angle is π.

Figure 9c shows the octahedron. No meteorological models based on the octahedron have been done. Note that on the bilinear and triangular grids, the two edges meeting at a corner are not on the same line. However, after projection to the sphere, they are on the same great circle. Figure 9d shows the icosahedron. Figure 9e shows $T24$. All the 12 rhomboids have equal sides. The 8 triangles surrounding the north pole, and the 8 surrounding the south pole are identical. The 8 triangles having a side on the equator are identical among themselves but not to the polar triangles.

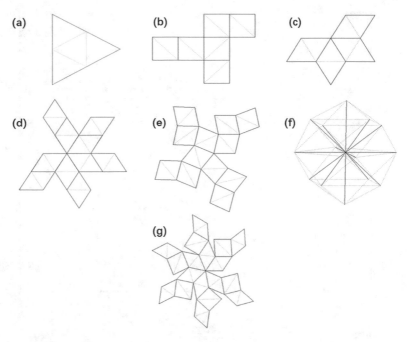

Fig. 9 Rollouts of different semi-platonic solids. Along dotted lines, two triangles are combined to form a rhomboid. Solid lines cell boundaries. (**a**) $T4$ (tetrahedron), (**b**) $T12$ (cube), (**c**) $T8$ (octahedron), (**d**) icosahedron, (**e**) $T24$, and (**f**) the hedgehog diagram corresponding to (**e**). The hedgehog representation shows the corner points of (**e**) as vectors pointing from the Earth's center to the point. From such 3D representations of grid points, geometric relations between points, such as angular distances, can be computed with ease. (**f**) Hedgehog diagram for $T24$. (**g**) Rollout of T36

CPU Time Used with a 3D Version of o3o3 Scheme

Chapter "2D Basis Functions for Triangular and Rectangular Meshes" suggested that by using sparse grids a reduction of computer time to run a model is possible. The largest sparseness factors and the largest saving were suggested for 3D models and hexagonal grids. In this section, we want to explore this in practice. A 3D cell structure with cubes as cells is assumed. The full grid is used with $96 \times 96 \times 96$ points. As a control experiment, the standard fourth-order FD scheme o4 defined in Chapter "Simple Finite Difference Procedures" is used. Figure 10b shows the full grid. The sparse grid used with the o3o3 representation is shown in Fig. 10c. For the full grid used with o4, a standard structured 3D grid is used. For the sparse grid Fig. 10c, a compact storage was devised. All black points indicated in Fig. 10c are stored together with the point indicated as P in Fig. 10c. P has three integer indices $i, j, k = 0, 1, 2, 3, \ldots$. The 7 points indicated as black in Fig. 10c are then indicated by a pointer index $p = 0, 1, 2, 3, 4, 5, 6$. This means that an amplitude of the field h is $h_{i,j,k,p}$, where i, j, k run through the Cartesian grid and p through 7 values. For

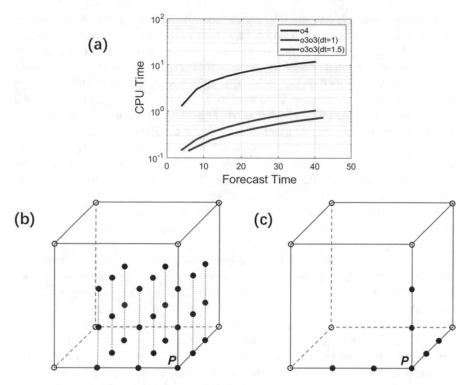

Fig. 10 (**a**) CPU times as functions of the forecast time for the o4 full grid scheme with $dt = 1$, the sparse grid o3o3 scheme o3o3 with $dt = 1$ and $dt = 1.5$. A cubic cell for the full grid is shown in (**b**), and the sparse grid cell used with o3o3 is shown in (**c**)

the spectral representation, the amplitudes are

$$\begin{cases} h_{i,j,k,0} = h(\mathbf{r}_{i,j,k}), \\ h_{i,j,k,1} = h_{xx}(\mathbf{r}_{i+\frac{1}{2},j,k}), \\ h_{i,j,k,2} = h_{xxx}(\mathbf{r}_{i+\frac{1}{2},j,k}), \\ h_{i,j,k,3} = h_{yy}(\mathbf{r}_{i,j+\frac{1}{2},k}), \\ h_{i,j,k,4} = h_{yyy}(\mathbf{r}_{i,j+\frac{1}{2},k}), \\ h_{i,j,k,5} = h_{zz}(\mathbf{r}_{i,j,k+\frac{1}{2}}), \\ h_{i,j,k,6} = h_{zzz}(\mathbf{r}_{i,j,k+\frac{1}{2}}). \end{cases} \tag{11}$$

The 3D version of o3o3 is analog to the 2D version described in Chapter "2D Basis Functions for Triangular and Rectangular Meshes".

The required CPU time as a function of forecast time is shown in Fig. 10a. Using the same time step $dt = 1$ for both methods, o3o3 is more than 10 times faster than o4. Our example has the sparseness factor 4. So the speedup is larger than explained by sparseness. The o3o3 computations are done completely in spectral space, without ever transforming to grid point space. Looking at the algebra, the third derivative amplitude (see Chapter "Local-Galerkin Schemes in 1D") requires much less computations than the second derivative and the corner point amplitudes. This would bring the expected CPU time to 7, still much lower than the 12 CPU seconds used with the full grid scheme o4. This can be explained by the fact that the second derivative amplitudes are computed in a simpler way as the corner point time derivatives. Many of the terms used for the second derivative amplitude are computed in second, rather than third or fourth order. Only corner point amplitudes are computed with a similar computational effort as a point in the o4 scheme. However, for corner points, the sparseness factor is 27.

For the 3D computations, we used an alternative version of o3o3 than that used in Chapter "Local-Galerkin Schemes in 1D". The differencing stencil for the computation of corner point derivatives was changed to use only corner points and interval mid-points. In this way, the third derivative amplitude becomes diagnostic, meaning that the 0-space problem encountered in Chapter "Local-Galerkin Schemes in 1D" with o3o3 disappeared. While already the standard o3o3 of Chapter "Local-Galerkin Schemes in 1D" had a higher CFL than o4. With the new version, CFP increased to 3 for the one dimensional case, where with RK4 o4 has a CFL of 2. The CPU curve with the increased time step for o3o3 is also shown in Fig. 10a. This brings the increase of efficiency up to a factor of 16, compared to o4. The computer economy obtained with hexagonal cells and the physical interface shown in Section "The Interface to Physics in High-Order L-Galerkin Schemes" in Chapter "Local-Galerkin Schemes in 1D" could speculatively bring the efficiency factor into the order of a factor 40 for this very simple example.

The simple example given is suitable to demonstrate substantial savings to be obtained with sparse L-Galerkin methods. The saving should be larger than the factor 3 of saving normally obtained with the semi-Lagrange method. The numbers obtained, however, would need confirmation by less trivial examples. In our simple example, the effect of code optimization was not considered. We used computation in spectral space with a constant velocity field, where the multiplication of fields requires 1 multiplication per amplitude. If instead we have curved velocity fields, this would require 3 multiplications per field. Also, the savings must be investigated using less trivial examples than done so far.

1D Example for the QUASAR System

The QUASAR methods use piecewise polynomial fields that are assumed differentiable at corner points, and corner node points as well as corner derivatives are assumed as dynamic variables. So using this representation, all amplitudes are defined at corner points. Assuming differentiability of fields at corner points heuristically implies an increased resolution, the neglected fields (meaning those with strong differences of derivatives at corner points) may be assumed to be not predicted accurately enough to be worth representing. So this means that we have grid sparseness even in one dimension. When the cell length is dx, we may consider a collocation grid of $dx' = \frac{dx}{3}$. The representation is equivalent to a collocation grid of $dx'' = \frac{dx}{2}$.

Note that this sparseness in one dimension is just a heuristic assumption. The performance of a QUASAR method is comparable to a method with the dx' grid using methods explained in Chapters "Simple Finite Difference Procedures" and "Local-Galerkin Schemes in 1D". This has not yet been done. The reader interested in using such methods must be warned that nothing about QUASAR has been explored so far and this book did not do this. Therefore, instead of long explanations of an unexplored method, we just show the performance of a QUASAR method for the $4dx''$ or $6dx'$ wave.

Figure 11 shows the initial field and the field shifted by dx''. The represented fields may be the piecewise polynomial representation. Note that in discretization mode (a) is given by third-order polynomials, using basis functions e_i, $b^2_{i+\frac{1}{2}}$ and $b^3_{i+\frac{1}{2}}$. (b) is represented by quadratic polynomials using $b^2_{i+\frac{1}{2}}$. The $4dx''$ wave of any phase can be described as a linear combination of (a) and (b). Both (a) and (b) show

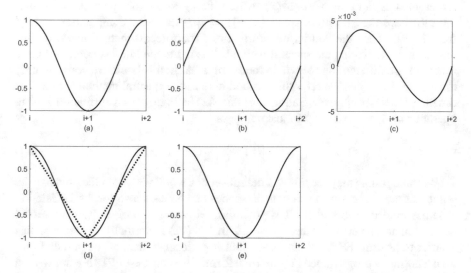

Fig. 11 Prediction of the $4dx'$ wave by a QUASAR method: (**a**) and (**b**) show the waves as the initial field and shifted by $4dx'$. This is the representation by cos and sin functions. For the numerical treatment, (**c**) gives the representation by piecewise polynomials; in this case, second-order polynomials given by b^2 (x) basis functions (see Chapter "Local-Galerkin Schemes in 1D"). The two representations are hard to distinguish, which speaks for the accuracy of such representations. Due to the small difference of the two representations, (**a**) and (**b**) may also be taken for the polynomial representations. The differentiation of the initial state (**a**) is the polynomial representation (**b**). The differentiation of (**b**) gives the dotted line in (**d**), where the solid line in (**d**) is the regularized version of this function. So the regularized x-derivative of (**b**) is (**e**)

functions being continuous and differentiable, which means that they are QUASAR representations. The derivative of (a) is (b) multiplied by a factor. This means in particular that a time step performed on (a) leads to a continuous and differentiable function and no regularization is necessary. This means that the time step performed on (a) is exact. By differentiation, the third-degree representation (a) is transformed into the second-degree polynomial representation (b). When differentiating (b), the result is a continuous but not differentiable representation by a piecewise linear function given as the dotted line in (d). The regularization process described in Section "An Example of a Regularization Operator" in Chapter "2D Basis Functions for Triangular and Rectangular Meshes" will result into the solid line of Fig. 11d.

Operations of Linear Spaces

The solution of prognostic equations is classically reduced to operations on numbers. For Galerkin and L-Galerkin methods, the concept of discretization space appears, and operations are considered to be operations on spaces. The oldest Galerkin method in meteorology is the sin/cos wave method, where the basis functions are $b^k(x) = e^{Ikx}$, with I being the imaginary unit and a field $h(x)$ is represented as $h(x) = \sum_k a_k b^k(x)$, where the a_k forms the spectral space $\{a_k\}$. We also have the grid point space $\{h(x_i)\} = \{h_i\}$, and the computation of $\{a_k\}$ from $\{h_i\}$ is called the Fourier transformation. The forecast method based on this representation is called the spectral method. It is not treated in this book. All basis functions are different on the whole modeling area. In this book, we consider only local basis functions. We refer this method here as the spectral method for the first time used the concept of operations on spaces. The Fourier transformation $\mathbb{F}$ is an operation on spaces, rather than just numbers:

$$\mathbb{F} : \{h_i\} => \{a_k\}. \tag{12}$$

While some people may see this as a definition of a word, we would like to make the point that the concept of operations on spaces can be used also for the L-Galerkin schemes treated in this book. It is potentially a powerful tool to organize model development. The code for the operator $\mathbb{F}$ in Eq. (12) is normally written as a fast Fourier transform (FFT), which means that it performs orders of magnitude faster than a normally programmed Fourier transform. Often the best FFTs are not written by the model developers themselves but rather come from another institute. The model developer just needs to know the properties of FFT, not how it is done. This leads to the effect that FFT is not part of the model code, and so the model can be rather short and easy to understand. So the concept of operation on spaces leads to a simplification of model code and a better organization of the work of model development. The authors believe that the concept of operations on spaces can be useful in the development of complex models.

Hexagonal spherical models often are rather lengthy programs with a handbook of several hundred pages. The use of operators on spaces brings this to a size that it is possible to be printed in a book. For the operations used, it is not necessary to know how they are coded, but rather their properties.

Summary and Outlook

Abstract This section deals with potential applications and important research areas of numerical models of the atmosphere.

Keywords Parallel processing · Computer requirements · Atmospheric contaminants · Elliptical earth · Linear algebra

Since the first successful numerical weather prediction in 1950 (Charney et al. 1950), NWP models have become quite realistic in representing many of the relevant physical processes of the atmosphere (Zängl et al. 2015). This book deals only with the numerical aspect of this undertaking. It is amazing that in spite of 70 years of development, this particular aspect of NWP is still full of problems to be solved. New numerical methods are invented and big efforts take place to implement them in realistic models. Some of these methods, such as classic Galerkin, are already more than 100 years old and some forms of them were implemented. However, for use in atmospheric models, some methods are new and some important aspects did not find their way to atmospheric models. There are many versions and aspects of known numerical schemes that have not found attention and are of potential key importance for applications. While a particular numerical approach may be formulated on a few pages, its implementation in a realistic model may easily take a decade as we are not only concerned with the numerical procedure itself, but rather how the final forecast is affected. Often trivial aspects have to get a high attention, such as numerical parameters. Such parameters may be the amount of numerical diffusion necessary or the filtering of orography. In particular for weather forecasting, the forecast must be ready in a given time as a forecast after the target date is not useful. This requires compromises and not all numerical aspects can be treated with optimal accuracy. This again needs a lot of investigations on the impact of approximations of different accuracies on the forecast, even principally new numerical approaches were developed in recent years and continue to be investigated. In this section, we want to point to some areas where numerical developments offer scientific potentials.

J. Steppeler, J. Li, *Mathematics of the Weather*, Springer Atmospheric Sciences,
https://doi.org/10.1007/978-3-031-07238-3_9

The reader should not be discouraged by the numerical paradoxon that an increased accuracy by high-order schemes as proven by toy models did not materialize in the final forecast product. Numerical developments were necessary to implement higher-resolution models and increased resolution does improve forecast scores. As another example, the introduction of the nonhydrostatic equations leads to new forecast elements, such as meso-scale weather, materializing in localized precipitation forecasts. In other words, this is the weather that is experienced directly every day. This has been appreciated not only by the specialist, but also by the general public.

Computer Aspects of Parallel Processing

The processing speed of computers, measured in FLOPS (FLOating Point operations per Second), has increased enormously (Adams et al. 2019) and this development is shown in Fig. 1. For example from Fig. 1, it follows that a graphic card with 30 flops would be the top performing computer in 2000. Models running on supercomputers in 1990 can now run on PCs. Considering the models ran operationally in 1990, this means that now a PC can run a global model of about $100\,km$ grid size. A possible performance in the order of exaflops (10^{19} operations per second) is expected to be realized soon. Currently, global models for forecasting are possible in the meso-scale, meaning a mesh length of about a few kilometers. Such high resolutions are possible, but the cost is still considered rather high and

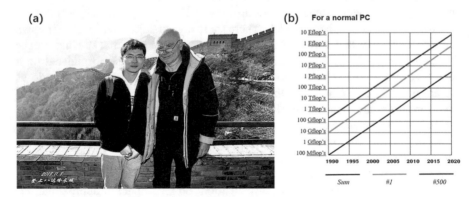

Fig. 1 (**a**) The collaboration task to build a long wall. Big engineering efforts, such as building a huge wall, cause problems of communication, transport, and work coordination in a similar way as occur with multiprocessing computer architectures. (**b**) The exponential increase in the speed of computers (TOP500 2019). The red curve is the sum of the performances of all top500 supercomputers. The green curve is the best performing supercomputer, and the blue curve is the performance level at the particular time to be in the top500 supercomputer family. For small deviations from this linear trend, see TOP500 (2019)

therefore, some modern model developments use adaptive or variable resolution, to limit the high resolution to areas most important for the target region (Skamarock et al. 2012; Savre et al. 2016). This use of adaptive variable resolution is relatively new, even though methods suitable for variable resolution in the form of classical Galerkin methods (see Chapter "Local-Galerkin Schemes in 1D") are known for more than 100 years, even before the arrival of computers. Many numerical methods used in the past, such as centered FD schemes, are not very suitable for variable resolution. This is the reason why some of these models use very smoothly varying meshes (Thuburn et al. 2001). So the introduction of variable and adaptive resolution into atmospheric models means the introduction of numerical techniques that are new in the sense that they have not been used a lot before for atmospheric modeling. Such methods are Voronoi grids (Skamarock et al. 2012), classical Galerkin methods (Savre et al. 2016), or L-Galerkin methods, in particular SEs (Taylor et al. 1997; Giraldo 2001).

The new high performance computers achieve their power not just by increasing the speed of the single processors, but rather by a very high degree of parallel processing. Even current computers have millions of processors. In order to use such high performance, one way is to go to rather high resolution. For example, a regular grid on the sphere on a 10-*km* grid has about 10^8 horizontal grid points for a model with 10 dynamic fields, and it is usual to store the whole vertical column belonging to one horizontal point in one processor. The use of many millions of processors poses new requests on numerical procedures (Adams et al. 2019). For example, Gaussian elimination, though efficient with just a few processors, is heavy on communication and thus less suitable for distributed memory. It can be used only if the data are not distributed between too many processors.

This is easy to understand and the problem does not only occur with informatics problems. Consider the construction of a long wall (see Fig. 1). When the wall is 2100 *km* long and has 25,000 towers, and you have 25,000 teams of 10 bricklayers each, it would obviously be difficult to keep all workers busy when they are all working on the same tower. A much better idea would be to assign each of the 25,000 teams a different tower to build, even then the communication between the teams cannot be too extended. For example, when building with natural stones of irregular shape, one method could be to transport each stone between all teams until a place is found where it fits optimal. This is also not a good way of proceeding, as easily more time could be spent with communication and transport than with bricklaying.

This latter point has a complete analogy to solution procedures with the classical Galerkin procedure (see Chapter "Local-Galerkin Schemes in 1D") and occurs even in 1D. Consider a row of 40,000 grid points along the equator with a $dx = 1$ grid. Consider the test problem of homogeneous advection with velocity $h_t = -h_x$. Classical Galerkin approximation with linear hat functions leads to the following mass matrix equations for the grid point time derivatives:

$$\frac{1}{6}h_{t,i-1} + \frac{2}{3}h_{t,i} + \frac{1}{6}h_{t,i+1} = -(h_{i+1} - h_{i-1}) = rs_i, \tag{1}$$

which is called compact FD scheme, and it is fourth-order accurate, while taking just the right-hand side rs_i of Eq. (1) as an approximation for second order.

There is a computationally efficient solution of Eq. (1), as a recursive solution by Gauss elimination exists:

$$
\begin{cases}
rs_i' = rs_i - rs_{i-\frac{1}{4}}, \\
b_i = \dfrac{1}{6}, \\
c_1 = \dfrac{2}{3}, \\
c_i = c_i - \dfrac{1}{6c_1 - 1}, \\
h_t = \dfrac{rs_i' - b_i h_{t,i+1}}{c_i}.
\end{cases}
\tag{2}
$$

As b_i and c_i can be pre-computed, Eq. (2) requires only 3 multiplications per index i. So the Gauss elimination is computationally efficient and was applied to small and medium sized problems. In many current models, it is still employed for the vertical direction. Consider that each of the grid point values h_i resides on a different computer, and there are 40,000 points along the equator. We need message passing with 40,000 computers to obtain one amplitude. So this problem is so heavy in message passing that it is probably inefficient. A program is called optimal scalable, when the use of computer time per grid point does not depend on the size of the problem and the number of processors involved. Let n be the size of the field (the number of grid points). When the CPU time of a program for a range of n is linear dependent on a value $n = n_0$, the program is scalable up to n_0. For many numerical methods, this curve lets out after some value of n_0. As pointed out, classical Galerkin is not a good candidate for scalability. However, L-Galerkin methods, such as SEs, have been found to be scalable up to high processor numbers.

This should be enough to exemplify the fascinating tasks in the area of mathematics/computer science to be used in numerical prediction (https://www.top500.org/).

Numerical Weather Prediction for Small Research Groups and Owners of Private PCs

As we have seen in Section "Computer Aspects of Parallel Processing", many readers of this book will own a PC of a computational power that would be a supercomputer as recent as in the year 2000. The numerical resolution of an NWP model determines the scale of atmospheric motion that can be simulated. The different scales, from macro α to micro γ, describe quite different atmospheric phenomena. The scales are often described according to Orlanski (1975) and are listed in Table 1, together with the atmospheric phenomena they describe. For all

Table 1 Scales of atmospheric motion following Orlanski (1975)

Horizontal scale	Atmospheric phenomenon	Names
$(10,000\,km, +\infty)$	Long atmospheric wave	Macro α
$(2500\,km, 10,000\,km)$	Baroclinic waves, lows, highs	Macro β
$(250\,km, 2500\,km)$	Fronts	Meso α
$(25\,km, 250\,km)$	Cloud clusters	Meso β
$(2.5\,km, 25\,km)$	Heat Islands, thunderstorms	Meso γ
$(250\,m, 2.5\,km)$	Shallow convection	Micro α
$(25\,m, 250\,m)$	Thermal turbulence	Micro β
$(0, 25\,m)$	LES, turbulence, flow inside rooms, flow inside private gardens	Micro γ

scales, research by single researchers with just a PC is possible. In the year 2000, meteorologists were interested in numerically modeling meso-scale phenomena. This is interesting as the forecasted meso-scale features translate immediately into every day weather, while to verify lows and highs one must have a barometer. The transition to meso-scale forecasting required new dynamic equations that were identical to those used in CFD. These developments were discussed in a number of meetings (SRNWP, MOW, AMCA). These numerical forecasting developments were quite sophisticated, and the authors of this book are fascinated by the fact that now private PCs have the computer power available. This already indicates the potential big strides a private person is able to make just using his/her private PC. This section wants to give some advice concerning the possibilities for small research groups.

Models without the full forecast infrastructure of full physics and data assimilation are called toy models. Such models can test the numerical procedures (including large eddy simulations in 2D) described in this book just using a normal PC. Also 3D modeling is possible, but the resolution is limited. The software requirements are a compiler and a plot program, such as FORTRAN, MATLAB, and NCAR graphics.

When the ambition is to go to real-life predictions, this is also possible and will require further software. For real-life forecasts and climate prediction, one possibility is to work together with one of the big computer centers, such as given in Table 1 in Chapter "Introduction". For the private researcher, data are available in the Internet (ECMWF 2016). Data as measured by a large number of thermometers, satellites, etc., are transformed into gridded data by a process called data assimilation. The ERA5 data are already gridded. The reader having such ambitions is warned that a lot of software needs to be installed. An example of a model having up-to-date physics and a whole infrastructure to do global and local forecasts is MPAS, and this model is freely available for many people. For limited area modeling, the WRF model (Skamarock et al. 2007) is a popular choice because it is easily available and has an up-to-date scientific standard. It is possible to do private forecasts on small computers, and the authors of this book know one person having done this as early as 2005. However, the reader is advised to the possibility of working together with one of the big forecast centers and concentrate on certain aspects of such work. Now it is possible to do things in small research groups that were the domain of the big centers 20 years ago.

For persons interested in real forecasts, it is now in principle possible to do forecasts of climate or environmental studies for a private garden, for example. The purpose could be climate evaluation or environmental protection for atmospheric contaminants.

New Applications for NWP

The first numerical weather forecast (Charney et al. 1950) was done mainly with weather forecasting in mind. However, the prospects for climate evaluation were already seen then. Climate evaluation is done by long integrations of atmospheric models. When localization of climate evaluation is required, small-scale models nested into global models are used.

New application areas for atmospheric models have appeared. Technically, these are made possible by the increase in available computer power, which translates into high resolution.

Environmental models are mainly concerned with the transport of pollutants. After the accident at Chernobyl, weather services made transport models for radiative constituents of the atmosphere. As trajectories can be computed backward, it is possible to go from measured pollutants to the origin. Models are now available with hundreds of chemical constituents that allow to tackle a new range of climate questions. For the purpose of clean air in cities, pollution emissions can be followed by modeling, and the results feed into measures for cleaner target areas. Pollution transport models are now available down to the scale of streets.

Being different from the NWP model using a structured grid, environmental protection for atmospheric contaminants intends to utilize the dynamic adaptive mesh in accordance with the spreading and evolving of the atmospheric contaminants. For example, the 3D dynamically adaptive CFD model Fluidity, developed by Imperial College London, uses the classic Galerkin method with a triangular grid structure to predict the atmospheric transport processes (Savre et al. 2016; Pain et al. 2005, 2001; Zheng et al. 2021, 2015). Zheng et al. (2015) utilized Fluidity to simulate the 3D realistic dispersion of plumes in Shanxi–Hebei–Shandong–Henan area. Figure 2 shows the evolution of the concentration of SO_2 including 3D pollutant plumes on the surface for the concentration greater than $100 \, \mu gm^{-3}$. The adaptive mesh is denser in the areas with higher gradients of SO_2 and the steep slope of the terrain and relatively sparser in the remaining domain while maintaining promising accuracy and reducing the computational cost.

Another example is the application of Fluidity to the COVID-19 simulation in London (Zheng et al. 2021). With the outbreak of the COVID-19 pandemic, hundreds of millions of people were infected throughout the world. Zheng et al. (2021) investigated the effects of virus spreading and their survival time with certain meteorological conditions in London (Fig. 3) using adaptive mesh in Fluidity.

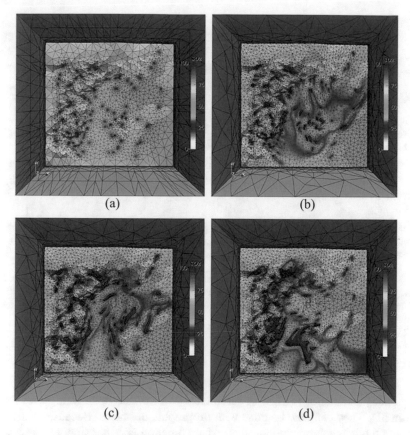

Fig. 2 Power plant plumes prediction for the SO$_2$ concentration in Shanxi–Hebei–Shandong–Henan area at four different times. The horizontal computational domain is 1090 × 1060 km with the resolution of 5 × 5 km. The simulation period is from 00:00 UTC on 10 January 2013–15 January 2013. The meteorological fields are provided by WRF model with 20 sigma layers. There are about 100 power plants in this area. The results were obtained using the Fluidity model of Imperial College London (Figure reused from Zheng et al. 2015). (**a**) 20130110 00:00(UTC). (**b**) 20130111 12:00(UTC). (**c**) 20130113 00:00(UTC). (**d**) 20130114 12:00(UTC)

They demonstrated that the aerosolized coronavirus particles can be transported to even hundreds of meters away from the source location under the meteorological conditions. Furthermore, the Fluidity developer now focuses on generating urban canopy algorithm to quantify the impact of urban green environment on reducing the coronavirus spreading and the urban heat island phenomena that are exacerbating the effects of climate changes resulting in more frequent and extreme heatwaves.

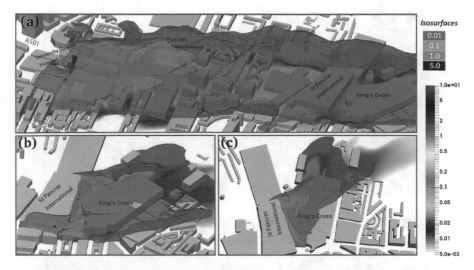

Fig. 3 3D simulation of the evolution of virus concentrations ($copies/m^3$) in London: (**a**) University College Hospital, (**b**) St Pancras International Station, and (**c**) King's Cross station. The results were obtained using the Fluidity model of Imperial College London (Figure reused from Zheng et al. 2021)

Large Eddy Simulation

Large eddy simulations (LES) of turbulence can now be done in high resolution and in 3D. They can now be done with orographic obstacles. Because of the high resolution of such computations, the surface function is rather smooth, as opposed to weather forecast models where the resolution of the orography is usually not good due to operational compromises. Because of this good resolution of surface topography with LES, the coordinate transformation method for orographic representation is valid and often used. Alternative methods of orographic representation have been presented in Chapter "2D Basis Functions for Triangular and Rectangular Meshes".

For LES calculations, the molecular physical friction term is of key importance. This is a term involving second derivatives and is similar to the second-order Laplace operator. From this, it is clear that second-order FD scheme, such as centered differences, is not suitable for this purpose. For error terms, it is necessary that they are much smaller than the physical diffusion term being of importance for this solution.

Therefore, high-order methods, in particular spectral methods based on trigonometric functions, are a popular choice for LES. As in connection with the new computer generation, alternative highly accurate L-Galerkin methods (for example, SEs) are developed. It is an open question if such methods are also useful for LES.

In particular, high-order cut cells may be an alternative to the coordinate transformation method. As with LES, an accurate representation of the surface is often considered essential, the representation of mountains by linear splines, as used in Chapter "Local-Galerkin Schemes in 1D".

The Elliptical and the Potato Shaped Earth

As we have seen above, modeling has become much more accurate since 1950. In this connection, it may be asked if approximations used today in many current numerical models should be further developed in future. In this book, we assume the shape of the Earth to be a sphere. Approximations on the sphere are not much more difficult than on a plane. Spherical geometry is known for a long time, and a great circle is obtained by cutting the spherical surface with a plane going through the Earth's center. This makes the construction of grids or cells as well as the formulation of FD schemes easy. The two main axes of the Earth's ellipsoid differs by about 21 km. This makes it clear why the spherical Earth makes a set of other approximations necessary, called the shallow atmosphere approximation. Among these approximations is the neglection of the centrifugal force. With the difference of the half axes being 20 km, this would mean that very thin air would appear at the poles. This means that the neglection of the centrifugal force goes together with the assumption of a spherical Earth. Deep atmospheric circulations are now the object of research.

In addition, we have the difficulty of formulating grids on the ellipsoid. The formulation of such schemes on elliptical surfaces goes beyond the scope of this book. Also, in the opinion of the authors, the development of such schemes should go together with practical computations.

Another problem is that the geopotential does not have the shape of a sphere or an ellipsoid. Due to an inhomogeneous distribution of mass in the Earth's interior, there are deviations of about 100 m to the reference ellipsoid. Also this interesting subject must be omitted in this book.

Data Assimilation

Data assimilation is the method to create the model field on a given mesh, which for weather forecasting is often nearly regular.

Current research of data assimilation goes far beyond simple interpolation of data and involves versions of the model. The principle is that the model trajectory in phase space is chosen to fit observed data optimally. At the same time, observation and even model errors are tried to take into account.

Data assimilation is not the subject of this book, but rather there are specialized books on this subject (Doms and Schättler 2002). Because of the importance of this subject, however, it is possible that a reader taking on modeling professionally will end up in this subject, and many assimilation schemes require important changes to the model program. There are two approaches:

1. The 4D Var method obtains a best fit of the model trajectory by using the adjoint model, which is the adjoint of a linearized version of the forecast model A difficulty is that this adjoint is specific to a model and must be reprogrammed every time when the model is changed.
2. The Kalman filter model uses an ensemble of forecasts. The principle is to find the best forecasts from the past and from these forming the data assimilated initial state of the model. This would be very accurate with a huge size of the ensemble, but in practice the size of the ensemble is in the order of 100 (Nakamura and Potthast 2015).

Global Models for Forecasting and Climate Research

It is characteristic for atmospheric models that convergence experiments are difficult to do. As seen from Table 1 when refining the grid length, new atmospheric phenomena occur. This means that while refining the grid, we do not necessarily get the same phenomenon more accurate, but rather create new phenomena, such as going from the macro- to the meso-scale. Forecasts of more than 4 days or climate simulations depend on the interaction of all parts of the atmosphere; in other words, they require a global model. Very often, around the target area, the grid is refined to go in this area to the meso-scale when the global model is in the macro-scale. This can be done using one-way interaction, when the global model runs independent of the high-resolution model and is steered by data (Doms and Schättler 2002; Skamarock et al. 2007). Also models with two-way interaction exist (Doms and Schättler 2002; Skamarock et al. 2007). In both cases, the aim is to improve the shorter range forecasts in the target area rather than to extend the global prediction. Such models have a global-resolution part and a fine-mesh part for the target region.

It should be noted that with current computer technology it is not possible to increase the resolution beyond the resolution of current forecast models systematically for the purpose of convergence investigations such as shown in Fig. 2 in Chapter "Finite Difference Schemes on Sparse and Full Grids" for the case of a simple toy model. The quality of global forecasts depends critically on the resolution and the increase of resolution has been one of the major factors for progress in forecasting. This means that global forecast centers tend to use top of the line computers and use them exclusively to do the forecast. A 10-day forecast typically needs a few hours. Currently, meso-scale global models are feasible in the order of $5\,km$ mesh length. By using several days for a forecast and combining several big computers, it is theoretically possible to do the forecast experimentally by an increase of resolutions. Considering that a doubling of resolution of a model takes 16 times the computer time, the potential increase of resolution is limited and this approach is rather un-practical.

In order to reach an increased resolution for the global part of the forecast, one could think of using adaptive models. The technology of adaptivity exists for atmospheric models (AMCG 2014; Savre et al. 2016; Yamazaki and Satomura 2010). It allows to resolve only those parts of the model area highly where the fields change considerably. Such adaptivity has an obvious application for environmental predictions where for example the origin of a pollution is resolved higher than the rest of the model. For examples of atmospheric adaptive solution, see Sections "Computer Aspects of Parallel Processing–Numerical Weather Prediction for Small Research Groups and Owners of Private PCs". While this is in principle also possible to do for global models, current global forecast models do this only to increase the resolution in the target area. One could think of increasing the resolution in the boundary or resolve the higher parts of the atmosphere less. When considering that extensive high pressure areas would need less resolution, there is in principle a potential of saving. When looking at a global weather map, such high pressure areas take only about $\frac{1}{3}$ of the area of the globe, and the question arises if the overhead of adaptive modeling offsets the benefits of fewer grid points. Currently, this question seems to have been decided against adaptivity in the field of global forecasting. None of the global models used for practical forecasting uses adaptivity for the purpose of improving the global forecast, rather than the forecast in a target region.

Linear Algebra

Even in a time, when the Internet should provide good information on any potential field of study, it comes as a surprise to many meteorology students that they share basic mathematical courses in calculus and linear algebra with mathematicians and engineers. Normally students with a great interest in mathematics do not study meteorology, but rather mathematics, theoretical physics, or engineering. It would be nice if this book could help to change this situation.

Eventually, most students gain a good knowledge of calculus, as this is the basis for courses in theoretical meteorology. For linear algebra, this is true to a lesser extent. However, with global spectral models, expressions such as phase space, grid point, or spectral space are used frequently. This section wants to repeat some basic notions of linear algebra in connection with the subjects covered here and thus contribute to the precise use of such terms.

Let λ be a real or complex number; then we call the set of such numbers C. C may represent real or complex numbers. A linear space H is any set where the operations $\lambda\mathbf{x}$ and $\mathbf{x}+\mathbf{y}$ are defined and result into vectors of H for any 2 members $\mathbf{x}$ and $\mathbf{y}$ of this set:

$$\lambda \in C, \mathbf{x}, \mathbf{y} \in H \implies \lambda\mathbf{x} \in H, \mathbf{x}+\mathbf{y} \in H, \lambda(\mathbf{x}+\mathbf{y}) = \lambda\mathbf{x}+\lambda\mathbf{y}. \tag{3}$$

Elements of H are called vectors. There is the $\mathbf{0}$ vector with the property:

$$\mathbf{x} + \mathbf{0} = \mathbf{x}. \tag{4}$$

Operations such as division through a vector or multiplication of two vectors are not necessarily possible within H. Further operations may be defined Depending on the existence of such operations, H may have different names, such as normed space, Hilbert space, etc. A frequently used extra operation is the scalar product "·" of two vectors:

$$\begin{cases} \mathbf{x}, \mathbf{y} \in H, \\ \lambda \in C, \\ \mathbf{x} \cdot \mathbf{y} = \overline{(\mathbf{y} \cdot x)}. \\ (\lambda \mathbf{x}) \cdot \mathbf{y} = \lambda(\mathbf{x} \cdot \mathbf{y}), \end{cases} \tag{5}$$

where $^-$ indicates the operation of forming the conjugate complex. A finite set of vectors $\mathbf{x}_i (i \in \{1, 2, 3, \ldots, i_e\})$ is called linear independent if

$$\sum_i \lambda_i \mathbf{x}_i = 0 \tag{6}$$

implies $\lambda_i = 0$ for all i.

The space H is called n dimensional if n is the maximum number of linear independent vectors in H. A set of linear independent vectors $\mathbf{x}_i (i \in \{1, 2, \ldots, n\})$ is called a basis of space H.

In spaces, when a scalar product is defined, $\mathbf{x}_i$ can be an orthonormal basis by the property:

$$\mathbf{x}_i \cdot \mathbf{x}_j = \delta_{i,j} = \begin{cases} 1, & \text{for } i = j, \\ 0, & \text{for } i \neq j, . \end{cases} \tag{7}$$

Examples of spaces are the grid point values h_i of a field h for grid distance dx:

$$H_{dx}^g = \{h_i\}, \tag{8}$$

if the discretization area is finite, H^g, and the grid point space is finite dimensional. The space H^{ana} consists of all analytic functions of the spatial vector $\mathbf{r}$. This is infinite dimensional. H^{sp-o3} is the space of all piecewise cubic functions as defined in Eq. (15) in Chapter "Local-Galerkin Schemes in 1D". This is called the spectral space. A space H^1 is contained in H^2 ($H^1 \subset H^2$), when each element of H^1 is an element of H^2.

For two spaces H^1 and H^2, a linear mapping f is a function from H^1 to H^2:

$$f : H^1 => H^2, \tag{9}$$

such that $\mathbf{x}_1 \in H^1, \mathbf{x}_2 \in H^2$:

$$\begin{cases} f(\mathbf{x}_1 + \mathbf{x}_2) = f(\mathbf{x}_1) + f(\mathbf{x}_2), \\ f(\lambda \mathbf{x}_1) = \lambda f(\mathbf{x}_1). \end{cases} \tag{10}$$

f is called a one-to-one mapping if different $\mathbf{x}_1$ and $\mathbf{x}_2$ imply that the images $f(\mathbf{x}_1)$ and $f(\mathbf{x}_2)$ are different. The image of H^1, $f(H^1)$ is the set of all $y \in H^2$ being the $Im\, y = f(\mathbf{x})$ of an $\mathbf{x} \in H^1$. If $H^2 = f(H^1)$, f is called an isomorphism. An $\mathbf{x}$ from H^1 and its image $f(\mathbf{x})$ from H^2 are considered identical. So H^1 is considered imbedded in H^2 if an isomorphism exists. For example, it may happen that grid point and spectral spaces have the same dimension. Then there is a one-to-one mapping between grid point and spectral space. They may be considered identical due to the existence of an isomorphism. However, it is also possible to have these spaces under different names, even though they are different manifestations of the same mathematical object.

As an example, consider in H^{sp-o3}, which is a space of functions, the functions $g_i(\mathbf{x})$ with the properties $g_i(\mathbf{x}_i) = 1$ and $g(\mathbf{x}_{i'}) = 0$ for all $i \neq i'$. Then we can construct an isomorphism between H^{sp-o3} and H^g_{dx} by mapping the grid point values $\{h_i\}$ to the spectral function by

$$f(\{h_i\}) = \sum_i h_i g_i(\mathbf{x}), \tag{11}$$

in which the element $\{h_i\}$ of the grid point space is mapped by f to the member of the function space. The brackets in Eq. (11) indicate that the element of the grid point space is the ensemble of grid point values h_i. The isomorphism allows to identify the set of grid point values h_i with the function on the right-hand side of Eq. (11). The grid point space is therefore a part of the spectral space:

$$f(H^g_{dx}) = H^{sp-o3}, \tag{12}$$

where the $g_i(\mathbf{x})$ in Eq. (11) are the grid point basis in the spectral space. The basis function representation Eq. (12) forms the spectral basis. In Eq. (17) in Chapter "Local-Galerkin Schemes in 1D", the basis is described by multiple indices.

Consider that the basis functions are renumbered, and we have a basis of spectral representation $s_{i'}(\mathbf{x})$. According to Eq. (12), the grid point space is mapped identical to the grid point space. The spectral transforms between grid and spectral representations become just coordinate transforms between the two basis function representations:

$$\sum_i h_i g_i(\mathbf{x}) = \sum_{i'} h^s_{i'} s_{i'}(\mathbf{x}). \tag{13}$$

The spectral coefficients in Eq. (13) are denoted by $h^s_{i'}$.

When we have a space with a scalar product, Eq. (13) can be written as a matrix equation by scalar multiplied with $s_i(\mathbf{x})$:

$$\{h_i\}\mathbf{A} = \{h_i^s\}\mathbf{B}, \tag{14}$$

where $\{\}$ indicate the vectors formed of h_i or h_i^s, and the matrices $\mathbf{A}$ and $\mathbf{B}$ are defined as

$$\begin{cases} \mathbf{A} = \{g_i \cdot s_{i'}\}, \\ \mathbf{B} = \{s_i \cdot s_{i'}\}, \end{cases} \tag{15}$$

with $\cdot$ indicating the scalar product.

The solution of Eq. (14) is the spectral transform. Directly forming the inverse of $\mathbf{A}$ or $\mathbf{B}$ is a way to perform the transform, though being very inefficient. The phase spaces tend to be rather high dimensional, and the matrix operations Eq. (14) are then rather inefficient. The whole point of L-Galerkin methods as opposed to the classical Galerkin method is to split the operations Eq. (14) into smaller dimensional sub-problems.

Small dimension sub-spaces are often convenient when investigating local interpolation problems. For example, for the third-order differencing with the tripod in Fig. 3 in Chapter "Finite Difference Schemes on Sparse and Full Grids", we have to consider the third-order correction to the second-order representation in the triangle. A convenient basis chosen for this correction allows that one basis function is 0 where we want to have an amplitude.

Examples of Program

Abstract This chapter provides programs in Matlab to perform some of the tests described in previous chapters.

Keywords Computer programs · Von Neumann method · Advections methods · Irregular grid differentiation weights · o3o3

In this chapter, we will show examples of how to use the code of o*nom*.

Dispersion Analysis of o2o3 and o3o3 Methods

We set the dispersion analysis of standard o3o3 and spectral o3o3 as an example Fig. 1 in Chapter "Simple Finite Difference Procedures". The codes of o3o3 and o2o3 methods for dispersion analysis are listed below (Steppeler et al. 2019).
The MATLAB code for o3o3 dispersion analysis:

```
% Compute evolution matrices for o3o3
[A1,A2,A3,A4,A5,A6] = o3o3_stand_mat();
[B1,B2,B3,B4,B5,B6] = o3o3_spect_mat();
% Compute eigenvalues
count = 4; vectorX = [0:0.001:1]*2*pi;
evalvalue = zeros(length(vectorX), count-1);
for j = 1:length(vectorX)
   M=A1*exp(-2i*vectorX(j))+A2*exp(-1i*vectorX(j))+A3
      +A4*exp(1i*vectorX(j))+A5*exp(2i*vectorX(j))+A6*exp(3i*vectorX(j));
   N=B1*exp(-2i*vectorX(j))+B2*exp(-1i*vectorX(j))+B3
      +B4*exp(1i*vectorX(j))+B5*exp(2i*vectorX(j))+B6*exp(3i*vectorX(j));
   [V,D] = eig(M);
   evalvalue(j,:) = diag(D);
   [VN,DN] = eig(N);
```

© Springer Nature Switzerland AG 2022
J. Steppeler, J. Li, *Mathematics of the Weather*, Springer Atmospheric Sciences,
https://doi.org/10.1007/978-3-031-07238-3_10

```
    evalvalueN(j,:) = diag(DN);
end
function [A1,A2,A3,A4,A5,A6] = o3o3_stand_mat()
    % define evolution matrix.
    A1 = zeros(3); A2 = zeros(3); A3 = zeros(3);
    A4 = zeros(3); A5 = zeros(3); A6 = zeros(3);
    % element length, distance between collocation points, half length of the
element
    dx_element = 1;
    dx_collocation = 1/3*dx_element;
    dx_half = 3/2*dx_collocation;
    % Value Updates
    A2(1,1) = (1/12/dx_collocation)*(2/3) +(-2/3/dx_collocation)*(1/3);
    A2(1,2) = (1/12/dx_collocation)*(-1/9) +(-2/3/dx_collocation)*(-1/9);
    A2(1,3) = (1/12/dx_collocation)*(1/162)+(-2/3/dx_collocation)*(-1/162);
    A3(1,1) = (1/12/dx_collocation)*(1/3) +(-2/3/dx_collocation)*(2/3) ...
                    + (2/3/dx_collocation)*(2/3) +(-1/12/dx_collocation)*(1/3);
    A3(1,2) = (2/3/dx_collocation)*(-1/9) +(-1/12/dx_collocation)*(-1/9);
    A3(1,3) = (2/3/dx_collocation)*(1/162)+(-1/12/dx_collocation)*(-1/162);
    A4(1,1) = (2/3/dx_collocation)*(1/3) +(-1/12/dx_collocation)*(2/3);
    A3(2,1) = 3/2/dx_half^3;
    A4(2,1) = -3/2/dx_half^3;
    A2(2,:) = A2(2,:) + A2(1,:)*(3/2/dx_half^2);
    A3(2,:) = A3(2,:) + A3(1,:)*(3/2/dx_half^2);
    A4(2,:) = A4(2,:) + A4(1,:)*(3/2/dx_half^2);
    A3(2,:) = A3(2,:) + A2(1,:)*(3/2/dx_half^2);
    A4(2,:) = A4(2,:) + A3(1,:)*(3/2/dx_half^2);
    A5(2,:) = A5(2,:) + A4(1,:)*(3/2/dx_half^2);
    A3(3,:) = A3(3,:) + (1/2/dx_element)*A2(2,:);
    A4(3,:) = A4(3,:) + (1/2/dx_element)*A3(2,:);
    A5(3,:) = A5(3,:) + (1/2/dx_element)*A4(2,:);
    A6(3,:) = A6(3,:) + (1/2/dx_element)*A5(2,:);
    A1(3,:) = A1(3,:) - (1/2/dx_element)*A2(2,:);
    A2(3,:) = A2(3,:) - (1/2/dx_element)*A3(2,:);
    A3(3,:) = A3(3,:) - (1/2/dx_element)*A4(2,:);
    A4(3,:) = A4(3,:) - (1/2/dx_element)*A5(2,:);
```

The function calling for the standard o2o3 evolution matrices is listed below:

```
function [C1,C2,C3,C4] = o2o3_standard_mat()
    % define evolution matrix
    C1 = zeros(2); C2 = zeros(2); C3 = zeros(2); C4 = zeros(2);
    % element length, distance between collocation points, half length of the
element
```

```
    dx_element = 1.0;
    dx_collocation = dx_element / 2.0;
    dx_half = 1.0 * dx_collocation;
    % Value Updates
    coeff_h_t_i_lef_2 = + 1 / (12 * dx_collocation);
    coeff_h_t_i_lef_1 = - 2 / ( 3 * dx_collocation);
    coeff_h_t_i_cen = 0.0;
    coeff_h_t_i_rig_1 = + 2 / ( 3 * dx_collocation);
    coeff_h_t_i_rig_2 = - 1 / (12 * dx_collocation);
    bf = basis_function(dx_half, 0.0 * dx_half);
    C1(1,1) = coeff_h_t_i_lef_2 * 1.0 + coeff_h_t_i_lef_1 * bf(1);
    C1(1,2) =                        coeff_h_t_i_lef_1 * bf(2);
    C2(1,1) =                        coeff_h_t_i_lef_1 * bf(3) ...
            + coeff_h_t_i_cen * 1.0 + coeff_h_t_i_rig_1 * bf(1);
    C2(1,2) = coeff_h_t_i_rig_1 * bf(2);
    C3(1,1) = coeff_h_t_i_rig_1 * bf(3) + coeff_h_t_i_rig_2 * 1.0;
    C2(2,1) = + 3 / 2 / dx_half^3;
    C3(2,1) = - 3 / 2 / dx_half^3;
    C1(2,:) = C1(2,:) + C1(1,:)*(3/2/dx_half^2);
    C2(2,:) = C2(2,:) + C2(1,:)*(3/2/dx_half^2);
    C3(2,:) = C3(2,:) + C3(1,:)*(3/2/dx_half^2);
    C2(2,:) = C2(2,:) + C1(1,:)*(3/2/dx_half^2);
    C3(2,:) = C3(2,:) + C2(1,:)*(3/2/dx_half^2);
    C4(2,:) = C4(2,:) + C3(1,:)*(3/2/dx_half^2);
end


function bf = basis_function(dx, collocation)
    e_plus = 1/2 + collocation / (2 * dx);
    e_minu = 1/2 - collocation / (2 * dx);
    b_2 = 1/2 * (collocation^2.0 - dx^2.0);
    bf = [e_minu, b_2, e_plus];
```

1D Homogeneous Advection Test

The FORTRAN codes of all methods (including o2o3, standard o3o3, spectral o3o3, SEM2, SEM3, o4, and the physical interface) are listed here. The results of Fig. 2 in Chapter "Simple Finite Difference Procedures" are produced with the parameters (imod = 1 and 2, initialconditionh = 0), while Fig. 2 in Chapter "Numerical Tests" are with the parameters (imod = 2 and 3, initialconditionh = 4). The program can be run for regular resolution. For irregular resolution, the user must define the grid x_1 and compute the weights $w_m^2(i)$, $w_m^1(i)$, $w_m^0(i)$, $w_p^1(i)$, $w_p^2(i)$. For the irregular example presented in this book, this was done using the weights given in Chapter "Local-Galerkin Schemes in 1D", and further results using an arbitrary irregular 1D grid can be done using the Galerkin compiler MOW_GC (Huckert and Steppeler 2021).

! program onom_SEM.f90 (This is the correct version, created by J. Steppeler, J. Li modified 25 Feb 2019)

! ————————————————————————————————————

```
MODULE const
PARAMETER (ie=600,ib=6,it=7)
real dx,dt,f,u00,h00,h00y
real wm2,wm1,w0,wp1,wp2
real wwm2(0:ie),wwm1(0:ie),ww0(0:ie)
REAL wwp1(0:ie),wwp2(0:ie)
real x1(-ib:ie+ib)
real xp(-ib:ie+ib)
real a21 ,a22 ,a31 ,a32 ! coefficients for spectral to gp transform
real a2x1,a2x2,a3x1,a3x2 ! grid point values for derivatives of h2 and h3
real a2x0,a2x3,a3x0,a3x3 ! grid point values for derivatives of h2 and h3
real aglinp,aglinm ! Galerkin coefficient linear part, lower and upper interval
real a2xg0,a2xg3 ! Galerkin coefficient o2 part
real a3xg0,a3xg3 ! Galerkin coefficient o3 part
real aa2,aa3 ! coefficients for gp to spec transform
real fl1,fl2 ! coefficients for linear interpolation
real dx3 ! large grid interval
!!!!!!!!!!!!!!!!!!!!!!!!!!!!!!!!!!!!!!!!!!!!!!!!!!!!!!!
! FOR NUMERICAL METHOD
parameter ( imod = 1 )
! Choice for the numerical method
! 1–control run, standard o4 differenceing
! 2–o3o3 dt = 2.5
! 3–o3o3 completely in spectral spave: not available
! 4– not available
! 5–SEM3 dt = 1.5
! 6–SEM2
!!!!!!!!!!!!!!!!!!!!!!!!!!!!!!!!!!!!!!!!!!!!!!!!!!!!!!!!!!!!
!! FOR PLOTTING
parameter ( line_number = 5 )
parameter ( plot_together = 1 )
parameter ( matlab = 0 ) ! 0–close matlab files, 1–open matlab files
INTEGER :: unit1Matlab, unit2Matlab, unit3Matlab
PARAMETER (unit1Matlab = 0711)
PARAMETER (unit2Matlab = 0712)
PARAMETER (unit3Matlab = 0713)
!! FOR INITIAL CONDITION
parameter ( initialconditionh = 8 ) ! default: 8; 0–peak solution, 4–smooth 4,
8–smooth 8, 1–constant = 4
```

parameter (hconstant = 0) ! if hconstant = 1 then initialconditionh = 1
line_number = 1
  parameter (condensation = 0)
  ! CLOSE CONDENSATION
  ! 0–no condensation
  ! OPEN CONDENSATION
  ! 11–naive condensation method at third point
  ! 12–spline condensation method at third point
  ! 21–naive condensation method at every point
  ! 22–spline condensation method at every point
  END MODULE const
  ! ——————————————————

  MODULE field
  USE const
  real h (-ib:ie+ib,it) ! density
  real ha2 (-ib:ie+ib,it) ! coefficient of h2
  real ha3 (-ib:ie+ib,it) ! coefficient of h3
  real hxx (-ib:ie+ib) ! coefficient of h2
  real hta2(-ib:ie+ib) ! coefficient of h2 t
  real hta3(-ib:ie+ib) ! coefficient of h2 t
  real hlin(-ib:ie+ib) ! h linearly interpolated
  real hho (-ib:ie+ib) ! high order part of h
  real h2mid(0:ie) , htmid(0:ie), htxmid(0:ie), ht3(-ib:ie+ib)
  real hmid (0:ie)
  real hxmid(0:ie)
  real hdiff(0:ie)
  real hxxx(-ib:ie+ib) ! coefficient of h3 for SEM3
  real hxxSE2 (-ib:ie+ib) ! coefficient of h2 for SEM2
  END MODULE field
  ! ——————————————————

  SUBROUTINE initialize( )
  USE const
  USE field
  #ifdef NOGKS
  USE gkssubstitute
  *PRINT *, 'Note: symbol NOGKS is defined - GKS/NCAR will not be used !'*
  #else
  *PRINT *, 'Note: symbol NOGKS not defined - GKS/NCAR will be used !'*
  #endif
  ! characteristic polynomials
  ! initialise fields, low resolution and high resolution
  *dx3 = 3. * dx*
  ! weights for 4rth order differentiation
  i = 6
  *x1(-ib) = -ib * dx* !x1 is for u-positions

```
do i=-ib+1,ie+ib
x1(i) = x1(i-1) + dx
enddo
! MOW_GC routine to compute
! weights for regular resolution
wm2 = 8.33333358E-02
wm1 = -0.666666687
w0 = 0.00000000
wp1 = 0.666666687
wp2 = -8.33333358E-02
#ifdef MOREOUTPUT
print*,wm2,wm1,w0,wp1,wp2,',wm1,wm2,w0,wp1,wp2,'
print*,wwm2(5),wwm1(5),ww0(5),wwp1(5),wwp2(5),',wwm1,wwm2,ww0,
wwp1,wwp2,'
    #endif
if (imod .eq. 1) then ! control run
dt = 2.0 ! stable stability limit control imod = 1
else if (imod .eq. 2) then
dt = 2.5 ! stable imod = 2
else if (imod .eq. 4) then
dt = 2.8 ! stable for Centred FD (imod = 4) unstable 2.9
else if (imod .eq. 5) then
dt = 1.5
else if (imod .eq. 3) then
dt = 0.5 !! smaller timesteps for scheme
else if (imod .eq. 6) then
dt = 0.5 !!
else if (imod .eq. 7) then
dt = 0.5 !!
end if
dt = 1.
u00 = 1.
h00 = 10.
! the following is for imod=5, comment otherwise
if (imod == 5) then
do i = -ib, ie+ib, 3
xxmid = (x1(i)+x1(i+3)) * .5
xx11 = -.447214 * 1.5
xx22 = -xx11 ! Gauss Lobtto points
x1(i+1) = xxmid - .447214 * 1.5
x1(i+2) = xxmid + .447214 * 1.5
end do
else if (imod == 6) then
do i = -ib, ie+ib, 2
xxmidSE2 = (x1(i)+x1(i+2))*.5
```

```
xx11SE2 = 0.0
x1(i+1) = xxmidSE2 - xx11SE2
end do
end if
xp(-ib:ie+ib) = x1(-ib:ie+ib) * 10. / real(ie) - 5.
pi = atan(1.) * 4.
h = 0.
if (initialconditionh == 0) then
h(:,1) = 0.
else if (initialconditionh == 4) then
do i = -ib, ie+ib
h(i,1) = 4.*exp(-(x1(i)*dx/4.-x1(150)/4.*dx)**2)
enddo
else if (initialconditionh == 8) then
do i = -ib, ie+ib
h(i,1) = 4.*exp(-(x1(i)*dx/8.-x1(150)/8.*dx)**2)
enddo
else if (initialconditionh == 1) then
do i = -ib, ie+ib
h(i,1) = 4.0
enddo
end if
! periodic boundary conditions
h(- ib:0,1) = h(ie- ib:ie,1);
h(ie:ie+ ib,1) = h(0: ib,1)
! Open GKS
CALL OPNGKS()
END SUBROUTINE initialize
SUBROUTINE finalize()
#ifdef NOGKS
USE gkssubstitute
#endif
!
CALL CLSGKS()
END SUBROUTINE finalize
PROGRAM o3o3_2_vers_book
use const
use field
#ifdef NOGKS
USE gkssubstitute
#endif
character(len = 80) :: filename_new_banyin1, form_banyin1
character(len = 80) :: filename_new_banyin2, form_banyin2
character(len = 80) :: filename_new_banyin3, form_banyin3
! Definitions for SEM3
```

```
eem(x) = .5-.5*x/1.5;eep(x)=.5+.5*x/1.5
eemx(x) = -.5/1.5;eepx(x)=.5/1.5
hh2(x) = (x**2-1.5**2)/2. ; hh3(x)=(x**3-x*1.5**2)/6.
hh2x(x) = (2.*x)/2. ; hh3x(x)=(3.*x**2-1.5**2)/6.
! Definitions for SEM2
eemSE2(x) = .5-.5*x/1.0;eepSE2(x)=.5+.5*x/1.0
eemSE2x(x) = -.5/1.0;eepSE2x(x)=.5/1.0
hh2SE2(x) = (x**2-1.0**2)/2.
hh2SE2x(x) = (2.*x)/2.
call initialize()
h(:,1) = 0.
h(150,1) = 4.
do i=0,ie
h(i,1) = 4. * exp(-(x1(i) * dx/4. - x1(250)/4. * dx) ** 2)
enddo
! periodic boundary conditions
h(- ib:0,1) = h(ie- ib:ie,1);
h(ie:ie+ ib,1) = h(0: ib,1)
CALL SET( 0. , 1. , 0. , 1. , -5.9, 5.9, -5. , 5. , 1 )
call GSLN(3)
if (matlab == 1) then
select case (ie+ib)
case (0:9)
write(form_banyin1,'(i1)') ie+ib
case (10:99)
write(form_banyin1,'(i2)') ie+ib
case (100:999)
write(form_banyin1,'(i3)') ie+ib
case (1000:9999)
write(form_banyin1,'(i4)') ie+ib
case (10000:99999)
write(form_banyin1,'(i5)') ie+ib
end select
write(filename_new_banyin1,*) "initial",trim(form_banyin1),".txt"
open(unit1Matlab,file=filename_new_banyin1)
do i = -ib, ie+ib
write(unit1Matlab, *) i, h(i,1), h(i,2), h(i,3), h(i,4), h(i,5), h(i,6), h(i,7), max-
val(abs(h(:,1) - h(:,6)))
end do
close(unit1Matlab)
else if (matlab == 0) then
end if
! advection over a distance of 3000 dx
if (hconstant == 0) then
nnend = 150 ! Integration steps
```

else if (hconstant == 1) then
nnend = 1
end if
iplot = 0
iplotrep = 0
do ntime = 1, nnend
! mass
fmass=0.
do i=0,ie-2
*fmass = fmass+h(i,1) * (x1(i+1) - x1(i))*
enddo
h(:,2) = h(:,1)
h(:,4) = 0.
do irk = 1, 4
h(:,3) = 0.
u00 = 1.
h(-1,1) = h(1,1)
h(ie+1,1) = h(ie-1,1)
! periodic boundary conditions
h(- ib:0,1) = h(ie- ib:ie,1);
h(ie:ie+ ib,1) = h(0: ib,1)
a21 = -1. / 2.
if (imod.eq.1) then ! control run
do i = -ib+2, ie+ib-3, 3
h(i,3) = -u00(wm2*h(i-2,1)+wm1*h(i-1,1) + wp1*h(i+1,1)+wp2*h(i+2,1))*
h(i+1,3)   =   -u00(wm2*h(i-2+1,1)+wm1*h(i-1+1,1)+wp1*h(i+1+1,1)   +*
*wp2*h(i+2+1,1))*
h(i+2,3)   =   -u00(wm2*h(i-2+2,1)+wm1*h(i-1+2,1)+wp1*h(i+1+2,1)   +*
*wp2*h(i+2+2,1))*
enddo
else if (imod.eq.2) then ! o3o3 run
do i = -ib+3, ie+ib-3,3
! interchange the two statements to get the old method (suitable for reg grid only)
*h(i,3) = -u00 * (wm2*h(i-2,1) + wm1*h(i-1,1) + wp1*h(i+1,1) + wp2*h(i+2,1))*
end do
do i = -ib+3, ie+ib-3,3
ha2(i,1) = -(-3.((h(i+3,1)-h(i,1))/ 3. +(h(i+3,3)+h(i,3))/2.))*4./9.*
end do
do i = -ib+3, ie+ib-3,3
hhhm2 = (h(i,3)+h(i-3,3)).5+ha2(i-3,1)*(-1.5**2)*
hhhm1 = h(i,3);hhhp1=h(i+3,3);
hhhp2 = (h(i+3,3)+h(i+6,3).5+ha2(i+3,1)*(-1.5**2))*
hhhm3 = h(i-3,3)
hhhp3 = h(i+6,3);
hhh0 = (h(i,3)+h(i+3,3)).5*

$hhhx = (wm2*hhhm2+wm1*hhhm1+wp1*hhhp1+wp2*hhhp2)/1.5 - (h(i+3,3)-h(i,3))/3.$

$hhhhxm = (wm2*hhhm3+wm1*hhhm2+wp1*hhhm1+wp2*hhhp1)/1.5 - (h(i+3,3)-h(i,3))/3.$

$ha3(i,1) = hhhx / (-1.5 ** 2)$!new version of o3o3 V2

$ha3(i,1) = (ha2(i+3,1) - ha2(i-3,1))/6.$!standard o3o3 V1

end do

do i = -ib+3, ie+ib-3, 3

$h(i+1,3) = h(i,3)*2./3.+h(i+3,3) / 3.+ha2(i,1)*.5*(.5**2-1.5**2)-ha3(i,1) * .5*(.5**2-1.5**2)*.5/3.$

$h(i+2,3) = h(i,3)/3.+h(i+3,3)*2./ 3.+ha2(i,1)*.5*(.5**2-1.5**2)+ha3(i,1)* .5*(.5**2-1.5**2)*.5/3.$

end do

else if (imod.eq.3) then ! spectral differencing at main nodes

print*,'not available'

stop

else if (imod.eq.5) then ! SEM3

! linear interpolated h

$xx11 = -.447214*1.5 ; xx22=-xx11$! Gauss lobtto points

$hlin = h(:,1)$

do i = -ib, ie+ib-3, 3

$hlin(i+1) = (hlin(i)*eem(xx11) + hlin(i+3)*eep(xx11))$

$hlin(i+2) = (hlin(i)*eem(xx22) + hlin(i+3)*eep(xx22))$

end do

$h(:,3) = 0.$

$h(:,6) = 0.$

$hxx(i) = 0.$

do i = -ib+3, ie+ib-3, 3

$del1 = h(i+1,1)-hlin(i+1)$

$del2 = h(i+2,1)-hlin(i+2)$

$hxx(i) = (del1+del2)/(hh2(xx11)+hh2(xx22))$

$hxxx(i) = (del1-del2)/(hh3(xx11)-hh3(xx22))$

end do

do i = -ib+3, ie+ib-3, 3

$hxl = (h(i+3,1)-h(i,1))/3.$

$hxlm = (h(i,1)-h(i-3,1))/3.$

$h(i,3) = -.5*(hxl-hxx(i)*1.5+hxxx(i)*hh3x(-1.5))-.5*(hxlm+hxx(i-3)*1.5+hxxx(i-3)*hh3x(-1.5))$

end do

do i = -ib+3, ie+ib-3, 3

!SE3

$h(i+1,3) = -(h(i,1)*eemx(xx11)+h(i+3,1)*eepx(xx11)+hxx(i)*hh2x(xx11)+hxxx(i)*hh3x(xx11))$

$h(i+2,3) = -(h(i,1)*eemx(xx22)+h(i+3,1)*eepx(xx22)+hxx(i)*hh2x(xx22)+hxxx(i)*hh3x(xx22))$

```
h(i+1,6) = -(h(i,1)*eemx(xx11)+h(i+3,1)*eepx(xx11)+hxx(i)*hh2x(xx11))
h(i+2,6) = -(h(i,1)*eemx(xx22)+h(i+3,1)*eepx(xx22)+hxx(i)*hh2x(xx22))
end do
else if (imod.eq.6) then ! SEM2
xx11SE2 = 0.0
hlin = h(:,1)
do i = -ib, ie+ib-2, 2
hlin(i+1) = (hlin(i)*eemSE2(xx11SE2) + hlin(i+2)*eepSE2(xx11SE2))
end do
h(:,3) = 0.
h(:,6) = 0.
hxxSE2(i) = 0.
do i = -ib, ie+ib-2, 2
del1 = h(i+1,1) - hlin(i+1)
hxxSE2(i) = - 1.0 / 1.0 * (2*h(i,1) - h(i-1,1) - h(i+1,1))
end do
do i = -ib, ie+ib-2, 2
h(i,3) = -(3.0/2.0)*(h(i+1,1)-h(i-1,1))/2.0 + 0.5*(h(i+2,1)-h(i-2,1))/4.0
end do
do i = -ib, ie+ib-2, 2
h(i+1,3)  =  -(h(i,1)*eemSE2x(xx11SE2)  +  h(i+2,1)*eepSE2x(xx11SE2)  +
hxxSE2(i)*hh2SE2x(xx11SE2))
end do
else if (imod .eq. 7) then
! spectral differencing at main nodes, spectral time stepping
hlin = h(:,1)
! linear interpolated h
do i = -ib, ie+ib-3, 3
hlin(i+1) = (hlin(i)*2./3. + hlin(i+3)/3.)
hlin(i+2) = (hlin(i)*1./3. + hlin(i+3)*2./3.)
end do
h(:,3) = 0.
do i = -ib, ie+ib-3, 3
hxl = (h(i+3,1)-h(i,1))/3.
h(i,3) = h(i,3)-.5*(hxl-ha2(i,1)*1.5+ha3(i,1)*3./4.)
h(i+3,3) = h(i+3,3)-.5*(hxl+ha2(i,1)*1.5+ha3(i,1)*3./4.)
end do
do i = -ib+3, ie+ib-3,3
hta2(i) = -(-3.*((h(i+3,1)-h(i,1))/3. + (h(i+3,3)+h(i,3))/2.))*4./9.
ha2(i,3) = hta2(i)
end do
do i = -ib+3, ie+ib-3, 3
hta3(i) = (hta2(i+3)-hta2(i-3))/6.
ha3(i,3) = hta3(i)
end do
```

```
endif
! RK weights
if (irk.eq.1) ftim1 = dt * .5
if (irk.eq.2) ftim1 = dt * .5
if (irk.eq.3) ftim1 = dt * 1.
if (irk.eq.4) ftim1 = dt * .0
if (irk.eq.1) ftim2 = dt / 6.
if (irk.eq.2) ftim2 = dt / 3.
if (irk.eq.3) ftim2 = dt / 3.
if (irk.eq.4) ftim2 = dt / 6.
h(:,1) = h(:,3) * ftim1 + h(:,2)
h(:,4) = h(:,3) * ftim2 + h(:,4)
h(- ib:0,1) = h(ie- ib:ie,1);
h(ie:ie+ ib,1) = h(0: ib,1) ! per bc
if (imod .ne. 5) then
ha2(:,1) = ha2(:,3) * ftim1 + h(:,2)
ha2(:,4) = ha2(:,3) * ftim2 + h(:,4)
ha2(- ib:0,1) = ha2(ie- ib:ie,1);
ha2(ie:ie+ ib,1) = ha2(0: ib,1) ! per bc !ha2 for imod=2
ha3(:,1) = ha3(:,3) * ftim1 + h(:,2)
ha3(:,4) = ha3(:,3) * ftim2 + h(:,4)
ha3(- ib:0,1) = ha3(ie- ib:ie,1);
ha3(ie:ie+ ib,1) = ha3(0: ib,1) ! per bc !ha3 for imod=2
do i=0, ie, 2
h(i,6) = h(i,1);
if (imod == 6) then
h(i+1,6) = h(i,1)*0.5+h(i+3,1)*0.5;
else
h(i+1,6) = h(i,1)*2./3.+h(i+3,1)/3.;
h(i+2,6) = h(i,1)/3.+h(i+3,1)*2./3.
end if
h(i,7) = h(i,3);
h(i+1,7) = (h(i,3)+h(i+3,3))*.5
h(i+1,7) = h(i+1,7)+hta2(i)*a21
end do
end if
end do ! end do irk=1,4
CALL CURVE( xp(0:ie) , h(0:ie,1), ie+1 )
h(:,1) = h(:,2) + h(:,4)
h(:,7) = h(:,7) * 5.
ipl = 300. / dt
iplot = iplot + 1
if (plot_together == 0) then
if (iplot .eq. nnend/line_number) then
CALL SET( 0. , 1. , 0. , 1. , -5.9 , 5.9, -5. , 5. , 1 )
```

```
call GSLN(1)
if (ntime.ne.1) CALL CURVE(xp(0:ie), h(0:ie,1)-iplotrep*.9-0.9, ie+1)
iplotrep=iplotrep+1
iplot=0
endif
else if (plot_together == 1) then
if (iplot.eq.nnend/line_number) then
CALL SET( 0., 1., 0., 1., -5.9, 5.9, -5., 5., 1 )
call GSLN(3) ! line style
iplot=0
endif
endif
! Plotting using NCAR_GRAFICS is used, which is a freely avalable package.
CALL SET( 0., 1., 0., 1., -5.9, 5.9, -5., 5., 1 )
if (ntime.eq.1) call GSLN(1)
if (ntime.eq.1) CALL CURVE( xp(0:ie), h(0:ie,1), ie+1 )
if (ntime.eq.300) call GSLN(2)
if (ntime.eq.300) CALL CURVE( xp(0:ie), h(0:ie,1), ie+1 )
if (ntime.eq.300) print*,'plot'
if (ntime.eq.29999) call GSLN(3)
if (ntime.eq.29999) print*,'plot'
if (matlab == 1) then ! output for matlab plotting
if ((ntime.eq.nnend/line_number)) then
write(filename_new_banyin2,*) "middle101.txt"
open(unit2Matlab,file=filename_new_banyin2)
do i = -ib, ie+ib
write(unit2Matlab,*) i, h(i,1), h(i,2), h(i,3), h(i,4), h(i,5), h(i,6), h(i,7), max-
val(abs(h(:,1) - h(:,6)))
end do
close(unit2Matlab)
else if ((ntime .eq. 2*nnend/line_number)) then
write(filename_new_banyin2,*) "middle202.txt"
open(unit2Matlab,file=filename_new_banyin2)
do i = -ib, ie+ib
write(unit2Matlab,*) i, h(i,1), h(i,2), h(i,3), h(i,4), h(i,5), h(i,6), h(i,7), max-
val(abs(h(:,1) - h(:,6)))
end do
close(unit2Matlab)
else if ((ntime.eq.3*nnend/line_number)) then
write(filename_new_banyin2,*) "middle303.txt"
open(unit2Matlab,file=filename_new_banyin2)
do i = -ib, ie+ib
write(unit2Matlab,*) i, h(i,1), h(i,2), h(i,3), h(i,4), h(i,5), h(i,6), h(i,7), max-
val(abs(h(:,1) - h(:,6)))
end do
```

```
close(unit2Matlab)
else if ((ntime.eq.4*nnend/line_number)) then
write(filename_new_banyin2,*) "middle404.txt"
open(unit2Matlab,file=filename_new_banyin2)
do i = -ib, ie+ib
write(unit2Matlab,*) i, h(i,1), h(i,2), h(i,3), h(i,4), h(i,5), h(i,6), h(i,7), max-
val(abs(h(:,1) - h(:,6)))
end do
close(unit2Matlab)
else if ((ntime.eq.300)) then
write(filename_new_banyin2,*) "middle606.txt"
open(unit2Matlab,file=filename_new_banyin2)
do i = -ib, ie+ib
write(unit2Matlab,*) i, h(i,1), h(i,2), h(i,3), h(i,4), h(i,5), h(i,6), h(i,7), max-
val(abs(h(:,1) - h(:,6)))
end do
close(unit2Matlab)
else if ((ntime.eq.5*nnend/line_number)) then
write(unit2Matlab,*) i, h(i,1), h(i,2), h(i,3), h(i,4), h(i,5), h(i,6), h(i,7), max-
val(abs(h(:,1) - h(:,6)))
endif
else if (matlab == 0) then
end if
enddo
CALL CURVE( xp(0:ie) , h(0:ie,1), ie+1 )
if (matlab == 1) then ! output for plotting by matlab
select case (ie+ib)
case (0:9)
write(form_banyin3,'(i1)') ie+ib
case (10:99)
write(form_banyin3,'(i2)') ie+ib
case (100:999)
write(form_banyin3,'(i3)') ie+ib
case (1000:9999)
write(form_banyin3,'(i4)') ie+ib
case (10000:99999)
write(form_banyin3,'(i5)') ie+ib
end select
write(filename_new_banyin3,*) "result",trim(form_banyin3),".txt"
open(unit3Matlab, file=filename_new_banyin3)
do i = -ib, ie+ib
write(unit3Matlab, *) i, h(i,1), h(i,2), h(i,3), h(i,4), h(i,5), h(i,6), h(i,7), max-
val(abs(h(:,1) - h(:,6)))
end do
close(unit3Matlab)
```

```
else if (matlab == 0) then
end if
CALL SET( 0. , 1. , 0. , 1. , -5.9 , 5.9, -5. , 5. , 1 )
call GSLN(2)
CALL CURVE( xp(-ib:ie+ib) , h(-ib:ie+ib,6), ie+1+2*ib )
CALL CLSGKS( )
END PROGRAM o3o3_2_vers_book
```

Appendix A

Neighborhood Relations for the Full Triangular Grid and a Compact Storage System

The grid and cells are shown in Fig. 1a, b, c, and f in Chapter "2D Basis Functions for Triangular and Rectangular Meshes". A quadrilateral cell is indexed by $\Omega_{i,j}$. Each cell contains two triangles, a lower $\Omega_{i,j}^{\Delta_1}$ and an upper triangle $\Omega_{i,j}^{\Delta_2}$. These are indexed by $i' \in 1, 2$, with 1 being the lower triangle and 2 the upper. The division of a quadrilateral cell into triangles is shown in Fig. 1b in Chapter "2D Basis Functions for Triangular and Rectangular Meshes". Each triangle has 4 kinds of amplitudes: 3 corner amplitudes, 3 edge amplitudes, second and third directional derivatives along the edges, and one center point amplitude as indicated in Fig. 2c in Chapter "2D Basis Functions for Triangular and Rectangular Meshes". As some of these amplitudes are shared between triangles, we need to define a compact storage where each storage is addressed only once. The locations of the amplitudes are indicated in Fig. 2c in Chapter "2D Basis Functions for Triangular and Rectangular Meshes".

Let the index of the lower left corner i, j be the cell index. We store amplitudes of the quadrilateral cell or the two triangles associated with it under the index i, j, m_p where p is a pointer index. Amplitudes at locations shown in Fig. 2c in Chapter "2D Basis Functions for Triangular and Rectangular Meshes" are stored with indices i, j, k. Let us as an example consider the collocation point representation. In Fig. 2c in Chapter "2D Basis Functions for Triangular and Rectangular Meshes", collocation points are indicated by small black or white points. There are 16 collocation points belonging to the cell i, j. This means that $m_p \in \{1, 2, \ldots, 16\}$. With the range of m_p, we do not have a compact storage, as some of the points belong also to other cells.

© Springer Nature Switzerland AG 2022   299
J. Steppeler, J. Li, *Mathematics of the Weather*, Springer Atmospheric Sciences,
https://doi.org/10.1007/978-3-031-07238-3

A compact storage is obtained by storing less than 16 points in the square $\Omega_{i,j}$. To achieve a compact storage, we must delete some of the 16 points in such a way that the points not represented with i, j are represented with some other i', j'. For the case of the collocation grid, a compact storage system is achieved by storing with the 4 inner points of cell $\Omega_{i,j}$ and the points $\mathbf{r}_{i,j}$, $\mathbf{r}_{i+\frac{1}{3},j}$, $\mathbf{r}_{i+\frac{2}{3},j}$, $\mathbf{r}_{i,j+\frac{1}{3}}$, $\mathbf{r}_{i,j+\frac{2}{3}}$. These are 9 points per quadrilateral cell $\Omega_{i,j}$ or on average $4\frac{1}{2}$ per triangular cell. The dimension of the discretization in phase space is 9 times the number of quadrilateral cells. The sparse grid is obtained by not using the 4 inner points of a cell. This reduces the dimension of the representation in phase space to 5 per cell. The unused points are obtained by interpolation, which for the schemes envisaged is done for plotting purposes only.

From the second and third derivatives, we store all edge locations of the lower triangle and the inner points of both triangles. These are 10 amplitudes and the dimension of the triangular spectral representation is 9 per $\Omega_{i,j}$. This is the triangular redundant representation. To obtain the compact representation, it is best to use the quadrilateral grid, which still allows to use triangle based numerical methods.

To obtain all amplitudes for computing the polynomial representation within a triangle, data associated with different $\Omega_{i,j}$ must be assembled. For example, corner point with the index $\mathbf{r}_{i,j+1}$ of the lower triangle $\Omega_{i,j}^{\Delta 1}$ is the corner amplitude at i, $j + 1$. The indices in spectral space of amplitudes are:

1. (i, j), $(i+1, j)$, $(i, j+1)$ for corner points. 3 points for the redundant representation and 4 for the quadrilateral cell.
2. $(i + \frac{1}{3}, j)$, $(i + \frac{2}{3}, j)$, $(i, j + \frac{1}{3})$, $(i, j + \frac{2}{3})$, $(i + \frac{1}{3}, j + \frac{2}{3})$, $(i + \frac{2}{3}, j + \frac{1}{3})$ for edge points. Each edge has two amplitudes: 6 points.
3. $(i + \frac{1}{3}, j + \frac{1}{3})$ for the inner point. These are 10 points for the redundant lower triangle representation and another 10 points for the upper triangle. As the 4 points on the diagonal are identical, we have 16 points for two triangles meaning 16 points for a quadrilateral. This is the same as we had obtained for the redundant quadrilateral cell representation. Note that it is possible to do a quadrilateral cell administration and combine it with triangular dynamics. The 16 redundant points of the two triangles can be transformed into the 9 points per double triangle cell of the compact grid system.

The 16 points of the redundant quadrilateral cell are not all stored with the cell $\Omega_{i,j}$. For the compact representation, 1 corner point, 4 edge points, and 4 inner points are stored with the cell $\Omega_{i,j}$. This makes 9 points per cell for the compact representation of the full grid. The sparse grid does not use the 4 inner point of a quadrilateral cell. Thus, there are only 16 points for the redundant representation and 9 points for the compact representation. In this way, there are 9 compact amplitudes associated with cell $\Omega_{i,j}$. To obtain a compact storage, one more dimension of index for the field h is required: h_{i,j,m_p}, where $m_p \in (1, \ldots, 5)$. For a given value of i, j, the set of 9 points is called the compact representation in cell $\Omega_{i,j}$. $m_p = 1$ contains the compact corner point, and $m_p = 2, 3, 4, 5$ store the amplitudes

$h_{xx,i+\frac{1}{2},j}, h_{xxx,i+\frac{1}{2},j}, h_{yy,i,j+\frac{1}{2}}, h_{yyy,i,j+\frac{1}{2}}$. According to Chapter "Local-Galerkin Schemes in 1D", there is an equivalent representation using one-sided derivatives. This means that for $m_p = 2, 3, 4, 5$, we can equivalently store the amplitudes $h^+_{x,i,j}, h^-_{x,i+1,j}, h^+_{y,i,j}, h^-_{y,i,j+1}$, which mean the one-sided derivative of h. There is another possibility, where all amplitudes stored are at position i, j: the stored amplitudes are $h^+_{x,i,j}, h^-_{x,i,j}, h^+_{y,i,j}, h^-_{y,i,j}$.

The Serendipity Interpolation on the Sphere

Consider a rhomboidal patch on the sphere, such as one of the rhomboids of the $R12$ or $R18$ polygonal partition on the sphere. We consider a grid numbered $I = 0, 1, \ldots, I_e$ and $J = 0, 1, \ldots, J_e$, and let $h_{I,J}$ be the grid points for the field $h(\mathbf{r})$: $h_{I,J} = h(\mathbf{r}_{I,J})$. As defined in Chapter "Platonic and Semi-Platonic Solids", the grid $\mathbf{r}_{I,J}$ can be a set of spherical grid points, $|\mathbf{r}^s| = 1$, or the corresponding point $\mathbf{r}^p_{I,J} = 1$ on the bilinear surface that is generated by linear interpolation from definition at the boundary. The projections of a λ or μ coordinate line to the sphere will result. The coordinate lines in λ or μ space are projected to great circles on the sphere. As in particular the boundaries of the rhomboid are coordinate lines, they are m great circles. Therefore on the boundaries, we assume the functional for m of h to be a piecewise polynomial of the arc, as measured from the initial point. The purpose of this appendix is to compute second derivatives along an edge of the computational cell. The computational cell is defined as the arc spanned by 3×3 points. The corner points are $\mathbf{r}_{I,J}, \mathbf{r}_{I+1,J}, \mathbf{r}_{I+1,J+1}, \mathbf{r}_{I,J+1}$. I and J are assumed to be divisible by 3: $I, J = 0, 1, 2, \ldots$. The corner point amplitudes $h_{I,J}, h_{I+1,J}, h_{I+1,J+1}, h_{I,J+1}$ are also amplitudes in spectral space. We consider the four edges, and for each of these, we must compute the second and third derivatives of the arches. These eight amplitudes in addition to the four corner amplitudes form the spectral space. It must be noted that the corner amplitudes are shared by four cells and the edge amplitudes by two. So we have five independent spectral amplitudes.

The aim is to compute the amplitude of the field $h(\mathbf{r})$ at any given point $\mathbf{r}$ in a given cell using the field representation defined by cell amplitudes on the cell boundaries. Any point $\mathbf{r}^r$ on the bilinear surface and the corresponding point $\mathbf{r}^s$ on the sphere are connected by the transformation:

$$\mathbf{r}^r = a^{red}\mathbf{r}^s. \tag{A.1}$$

The basis functions on the sphere and the bilinear basis functions are connected by this transformation, as defined in Section "Geometric Properties of Spherical Grids". Note that each point on the sphere and each spherical area have a one-to-one correspondence to a bilinear area. High-order and grid point amplitudes are related to each other. In particular, a high-order interpolation scheme is defined on the bilinear surface. The interpolation schemes on the sphere and the bilinear surface are equivalent by definition. So let a point $\mathbf{r}^s$ on the sphere be given. $\mathbf{r}^s$ may be one of the points in Fig. 11 in Chapter "Platonic and Semi-Platonic Solids" where field values need to be interpolated using the value $\mathbf{r}^r$. When now α_k are the relevant amplitudes for interpolation, then call $\mathbf{r}^{r,int}(\alpha_k)$ the interpolated value form α_k. The interpolation formula is the same as used with plane grids as described in Chapter "2D Basis Functions for Triangular and Rectangular Meshes". There is no difference of the interpolation formula between plane and bilinear grids. Then we define the interpolated value of h at $\mathbf{r}^s$ as

$$r^{s,int} = r^{r,int}(\alpha_k). \tag{A.2}$$

The Quasi-arithmetic Rendition QUASAR to Obtain a Sparse Field Representation

The principle of sparseness is that properties of a field are used to compute amplitudes diagnostically that do not have to be computed as dynamic points. In the case of Section "Rhomboidal Basis Functions and Sparse Grids for the Regular Grid Case" in Chapter "2D Basis Functions for Triangular and Rectangular Meshes", the arithmetic principle leading to a rendition of a field $h(\mathbf{r})$ was the principle of order of approximation. In Chapter "2D Basis Functions for Triangular and Rectangular Meshes", sparseness was obtained by observing that inner grid points are obtained diagnostically for analytic basis functions when they would define a higher order than four in a fourth-order scheme. Here we require the fields and basis functions to be differentiable. When we assume that the approximation methods applied have only the accuracy for differentiable fields, this neglection can heuristically be assumed not to decrease the accuracy of the solution. Therefore, we may assume that the reduction of dynamic amplitudes achieved in this way will not decrease the accuracy of representation. We limit ourselves to the approximation of analytic fields, meaning those for which a power series exists. Then it was shown in Section "Rhomboidal Basis Functions and Sparse Grids for the Regular Grid Case" in Chapter "2D Basis Functions for Triangular and Rectangular Meshes" that the sparse grid points are sufficient to compute the grid points of the whole grid up to a requested order of approximation. All loss of information encountered with this field reconstruction is of an order higher than the constructed order of approximation of the scheme.

In the case described in Section "Rhomboidal Basis Functions and Sparse Grids for the Regular Grid Case" in Chapter "2D Basis Functions for Triangular and Rectangular Meshes", the full system consists of the grid point values of the full grid, and the reduced system consists of the grid point values in the sparse grid. The arithmetic principle of reduction is the request that analytic functions are approximated by the reduced system to the same order of approximation as with the un-reduced system. Note that in this book, we do not go beyond order three.

An arithmetic rendition in this example is a high-order interpolation of the sparse system to the full system. The sparseness achieved with methods of Chapter "2D Basis Functions for Triangular and Rectangular Meshes" leads to sparseness in 2 dimensions and not in 1 dimension. The methods considered here, however, have the potential to create sparseness in one dimension as well.

The L-Galerkin scheme investigated in Steppeler (1976) shows a number of small-scale modes of the third-order piecewise interpolation. Any analytic function and any *sin* or *cos* function are approximated rather inaccurately by some of these modes. It may appear as a tempting idea to change the function system in such a way that analytic functions are better approximated by the smallest scale basis function representations. The QUASAR system uses the arithmetic principle of requesting the fields to be differentiable. This does not change the order of approximation and does not change the quality of approximation for smooth fields, even when fewer degrees of freedom need to be applied due to the arithmetic request. A QUASAR system generally leads to a sparse grid, compared to the original system.

A natural arithmetic principle to achieve sparseness is regularity at corner points. So far, we have considered continuous basis functions. Discontinuous basis functions and numerical L-Galerkin methods based on this have been the subject of research (see Steppeler (1987)). L-Galerkin method functions using basis functions with higher regularity, such as differentiable or two times differentiable, have been rarely used. Steppeler (1988) is an example.

Let us first consider one dimension. Let us define the C^n-space to consist of function of the form discussed in Section "Functional Representations, Amplitudes, and Basis Functions" in Chapter "Local-Galerkin Schemes in 1D" or analog in higher order, which are n times differentiable. In this book, we do not go beyond C^1-spaces. To form functions of higher regularity than one, we must use the discretization function spaces defined in Section "Functional Representations, Amplitudes, and Basis Functions" in Chapter "Local-Galerkin Schemes in 1D" to include polynomials of higher order than one. The functions defined in Section "Functional Representations, Amplitudes, and Basis Functions" in Chapter "Local-Galerkin Schemes in 1D" then form a C^0-space. Starting from the C^0-space, the requirement of admitting differentiable basis functions will lead to the use of fewer basis functions for the description of the C^1-space as compared to the C^0-space. So the C^1 requirement leads to sparseness in 1D. We call a corner point a QUASAR point, if all amplitudes are defined at the corner points. As compared to the grid and function system with QUASAR points, the amplitudes are no longer given for corner and edge points, but rather only at corner points, where at the corner points besides the grid point value also high-order amplitudes are stored.

In Chapter "Local-Galerkin Schemes in 1D", we have introduced a number of equivalent field representations. All representations use the corner point amplitudes h_I for the representation of the field $h(x)$ in one dimension. The corner point amplitudes determine the linear part $h^{lin}(x)$ of h, and we have: $h(x) = h^{lin}(x) + h^{ho}(x)$. The determination of $h^{lin}(x)$ and $h^{ho}(x)$ by amplitudes is described in

Chapter "Local-Galerkin Schemes in 1D". For the amplitudes to determine $h^{ho}(x)$, we have the following options:

1. High-order derivative representation: $h_{xx,I+\frac{1}{2}}$ and $h_{xxx,I+\frac{1}{2}}$
2. Collocation point representation: $h_{I+\frac{1}{3}}, h_{I+\frac{2}{3}} = h\left(x_{I+\frac{1}{3}}\right), \left(x_{I+\frac{1}{3}}\right)$
3. One-sided derivative representation: $h_{x,I}^{+}, h_{x,I+1}^{-}$

The representation of h is according to Chapter "Local-Galerkin Schemes in 1D":

$$h(x) = \sum_I h_I e_I(x) + h_{x,I}^{+} b_I^{+}(x) + h_{x,I+1}^{-} b_I^{-}(x), \qquad (A.3)$$

where the basis functions are defined in Chapter "Local-Galerkin Schemes in 1D". This means that all amplitudes needed to describe the field h are defined at corner points x_I.

For the QUASAR method, we use mainly representations defined in Chapters "Local-Galerkin Schemes in 1D" and "2D Basis Functions for Triangular and Rectangular Meshes". The three representations are equivalent and can be transformed into each other using methods explained in Chapter "Local-Galerkin Schemes in 1D". With QUASAR, we want to use one times differentiable fields h. As the representation Eq. (A.3) can potentially be non-differentiable at corner points, we obtain a QUASAR representation by requesting $h_{x,I}^{+} = h_{x,I}^{-}$. This means that the QUASAR representation is very similar to Eq. (A.3):

$$h(x) = \sum_I h_I e_I(x) + h_{x,I} b_I^{+}(x) + h_{x,I+1} b_I^{-}(x), \qquad (A.4)$$

which means that QUASAR amplitudes are field derivatives at corner points. Note that in two or three dimensions, the reduction of the dimension of the approximation space can be large. For example, in two dimensions, triangular grids often have 6 edges meeting at a point (see Chapter "2D Basis Functions for Triangular and Rectangular Meshes"), and a continuous derivative will require 2 rather than 6 amplitudes for the high-order part of the field.

For the test problem of homogeneous advection, this means that the flux is differentiable at corner points, and the time derivative $u_0 h_{x,I}$ is the corner point amplitude. No computations are necessary to compute the time derivative of h at corner points with the test problem of homogeneous advection. The differentiation of the cubic spline gives a continuous quadratic spline in the representation Eq. (A.3). This continuous representation is determined by two directional derivatives $h_{t,I}^{+}$ and $h_{t,I}^{-}$. This representation according to Eq. (A.3) must be approximated by the QUASAR representation Eq. (A.4). This means that $h_{t,x,I}$ and $h_{t,I}'$ must be computed. We use the notation $h_{t,I}'$ as the time derivative approximation must not necessarily be the $h_{t,I} = h_{x,I}$ computed from the representation Eq. (A.4). We employ the regularization operator defined in Section "An Example of a Regularization Operator" in Chapter "2D Basis Functions for Triangular and Rectangular Meshes". Note that for a differentiable representation Eq. (A.4) the mass of h is exclusively in the linear part of h:

$$\int h^{ho}(x)dx = 0, \tag{A.5}$$

for differentiable representations.

For continuous representations Eq. (A.3), $h^{ho}(x)$ has mass, and we obtain for the mass M_I of h_t associated with point I:

$$M_I = M_I^{lin} + M_I^+ + M_I^-. \tag{A.6}$$

To prepare the time step, we also compute the time derivative of h at mid interval $x_{I+\frac{1}{2}}$:

$$h_{t,I+\frac{1}{2}} = -\frac{u_0}{dx}(h_{I+1} - h_I) - u_0 h_{xxx} b_x^3 \left(x_{I+\frac{1}{2}} \right). \tag{A.7}$$

We can define the amplitudes for h_t. The formula is given for the regular grid case:

$$h_{t,I} = h_{x,I}, \tag{A.8}$$

$$h_{t,x,I} = w_{m2}h_{t,I-1}w_{m1}h_{t,I+\frac{1}{2}} + w_{p1}h_{t,I+\frac{1}{2}} + w_{p2}h_{t,I+1}, \tag{A.9}$$

$$h'_{t,I} = \frac{1}{\int e_I(x)dx}M_I. \tag{A.10}$$

In Eq. (A.9), $h'_{t,I}$ and $h_{t,x,I}$ are the amplitudes to be used for a time step. The time step using RK4 can be done using these amplitudes. An alternative that was used in all examples used so far in this book is to transform to grid point space, do the time step there, and transform back to the representation Eq. (A.4).

Before we discuss the 2D case, a few remarks for the use of non-constant velocity fields are given. When we transform to grid point space, we can form the fluxes for the non-constant velocity case in grid point space, which requires one multiplication per point just as in the case of spectral elements. There is the other option of never leaving the spectral space and performing the time step in spectral space. This method would save the cost of transformations but the fluxes, such as the flux in $x-$direction, uh must be done in spectral space. As the principle of the methods proposed in this book is to use small spectral spaces associated with each cell, the multiplication in spectral space is not such an extremely expensive operation as it is with the classical Galerkin method. Still, 2 rather than 1 operation per amplitude is necessary to do the product uh in spectral space. This increased cost of multiplication must be considered against the saving of the spectral to grid point transformations. Unfortunately, this chapter remains largely theoretical as few practical tests of modeling with QUASAR have been done. The application to two dimensions will not be described in detail, but rather the principle of this approximations is given.

In two dimensions, differentiability at corner point is requested for the field representations. At edges, just continuity is requested. In two dimensions, any directional derivative is defined if the derivatives in two directions are given. We may use the derivatives in two orthogonal directions as dynamic amplitudes. Using these, the derivatives in all other directions will become diagnostic. For example, the regular triangular mesh shown in Fig. 1c in Chapter "2D Basis Functions for Triangular and Rectangular Meshes" has 6 edges meeting at each corner, meaning 6 independent high-order amplitudes being the one-sided derivatives. Requiring differentiability at these corner points will reduce these to 2 per corner point. The triangular mesh will in addition require the center point amplitude as dynamic variable. By constructing auxiliary lines, it is possible to make this point diagnostic. The cube as discretization cell needs for the o3o3 discretization 7 amplitude points. With QUASAR, they are reduced to 4. For the hexagon in 3 dimensions, each hexagon needs 2 corner amplitudes and 2×3 derivatives. This means 8 independent amplitudes where the full grid has 81 amplitudes. Without QUASAR, we have 12 dynamic amplitudes per hexagon. This means that QUASAR brings the sparseness factor from 7.5 to 10.

For a simple example, see Section "1D Example for the QUASAR System" in chapter "Numerical Tests".

Glossary

Ansatz   Appeared in section "The Lorenz Paradigmatic Model" in chapter "Simple Finite Difference Procedures". A function system depending on parameters used to obtain the solution of equations.

Boundary condition   Appeared in section "Validations of Numerical Methods Using NWP Models" in chapter "Introduction". Near a boundary amplitudes being determined by other principles than the dynamic equations. An example is the requirement that the velocities are in the direction of the boundary.

Classical Galerkin method   Appeared in section "Numerical Methods" in chapter "Introduction". A polygonal spline method where the remapping procedure is based on a least square method.

Collocation points   Appeared in section "Numerical Methods" in chapter "Introduction". Grid points within cells that carry field amplitudes.

Compact grid (also called compact storage)   Appeared in chapter "2D Basis Functions for Triangular and Rectangular Meshes". A structured grid representation where all i, j correspond to a different point. Compact storage is important with distributed memory computers, as it allows efficient message passing.

Consistency   Appeared in section "Some Further Properties of Finite Difference Schemes" in chapter "Simple Finite Difference Procedures". A scheme is consistent, if for $dx - > 0$ the approximation yields the analytic solution.

Corner   Appeared in section "Numerical Methods" in chapter "Introduction". A point of a cell boundary, where several edges meet.

Corner point   Appeared in section "Numerical Methods" in chapter "Introduction". A collocation point being on the corner of a cell.

Diffusion   Appeared in section "Diffusion" in chapter "Simple Finite Difference Procedures". An operation to remove small scales from a field.

Edge   Appeared in section "Numerical Methods" in chapter "Introduction". In a plane, a line being part of a cell boundary. In this book, only straight line edges are considered.

J. Steppeler, J. Li, *Mathematics of the Weather*, Springer Atmospheric Sciences,
https://doi.org/10.1007/978-3-031-07238-3

Essential arithmetic feature   Appeared in chapter "Introduction". One scheme has this in comparison with another scheme, when it produces different results larger than roundoff error. The opposite is a superficial arithmetic feature. Superficial arithmetic features are obtained by arithmetic transformation from a given scheme. For example, when changing the collocation grid with SE schemes, the difference is an essential arithmetic feature. For the scheme o3o3, the change to another collocation grid results into a superficial arithmetic feature.

Full grid   Appeared in section "The Interface to Physics in High-Order L-Galerkin Schemes" in chapter "Local-Galerkin Schemes in 1D". A regular structured grid, where for each i, j, there exists a grid point.

Galerkin method   Appeared in section "Discretization on Spherical Grids" in chapter "Introduction". A method to approximate the time derivative of a field based on the least square method or weak formulation.

Grid imprinting   Appeared in section "Verifications of Numerical Methods for Climate Modeling" in chapter "Introduction". Patches of the grid are visible in the solution. This can be caused by a non-uniform approximation order.

Grid isotropy factor   Appeared in section "Isotropy of the Hexagonal Grid in Comparison to Rhomboidal Grid" in chapter "Full and Sparse Hexagonal Grids in the Plane". Relation of effective grid lengths for two directions. Grid isotropy factors different from 1 occur typically even on regular grids on the plane.

Homogeneous grid   Appeared in section Efficiency of the Computational Grid in chapter "Introduction". A grid on the sphere with nearly equal grid size everywhere.

Inhomogeneous finite differences   Appeared in section "Numerical Methods" in chapter "Introduction". Finite differences where within a group of points each point has a different finite difference equation. Such schemes typically occur with L-Galerkin schemes.

Internal boundary   Appeared in chapter "Simple Finite Difference Procedures". A line of grid irregularity, sometimes caused by a change in resolution. At such lines, care must be taken to keep the uniform approximation order.

Isotropic grid   Appeared in section "Isotropy of the Hexagonal Grid in Comparison to Rhomboidal Grid" in chapter "Full and Sparse Hexagonal Grids in the Plane". A 2D regular grid where waves are approximately resolved equal when they propagate in different directions.

L-Galerkin method   Appeared in section "Discretization on Spherical Grids" in chapter "Introduction". The classic Galerkin method or one of its variants. While the classic Galerkin method uses global communication of data, the L-Galerkin method can be designed to exchange data with neighbors only.

Multiple amplitude representation for discontinuous fields   Appeared in section "Triangular Basis Functions and Full Grids" in chapter "2D Basis Functions for Triangular and Rectangular Meshes". Some discontinuous functions achieve converged values if the point is approached from different directions. Such limit values are called multiple values, and the function is called multivalued. For example, for a cell structure, a non-differentiable field often has multiple valued derivatives. For the example of hexagonal cells treated in this book, it has triple valued derivatives with the standard C^0 representation.

Neighbor difference scheme Appeared in chapter "Introduction". A finite difference scheme involving only immediate neighbors, meaning $i - 1$ and $i + 1$ in one dimension. For regular grids, this scheme is called centered difference scheme or standard finite volume method.

Nonlinear Galerkin method Appeared in section "The Classic Galerkin Procedure" in chapter "Local-Galerkin Schemes in 1D". A generalization of the classic Galerkin method to the case where the test function may depend on the dynamic parameters in a nonlinear way. This does not mean that we have a nonlinear Galerkin method. The alternative is just called Galerkin method.

Order n of approximation Appeared in chapter Simple Finite Difference Procedures. An integer number n measuring how fast a method approaches the limiting value for decreasing dx.

Polygonal method Appeared in section "Discretization on Spherical Grids" in chapter "Introduction". Discretization based on a partition of the model area into cells of polygonal form.

Polynomial method Appeared in section "Geometric Properties of Spherical Grids". Discretization based on basis functions that are piecewise polynomial.

Polygonal spline Appeared in section "Numerical Methods" in chapter "Introduction". A field representation by a piecewise polynomial function.

Remapping Appeared in section "A Conserving Version of the Cut-Cell Scheme". Approximation of a spline by another with increased regularity. For example, with the CG finite element method, a discontinuous function is approximated by a continuous function.

Scalability Appeared in sections "Verifications of Numerical Methods for Climate Modeling" in chapter "Introduction" and "Computer Aspects of Parallel Processing". A computer program is scalable if the real time used for a prediction is inversely proportional to the number of processors used. Scalability may depend on the computer used, the programming, and the mathematical approximations applied. Some procedures, such as Gaussian elimination, have a reputation of preventing scalability.

Sparseness factor Appeared in section "Rhomboidal Basis Functions and Sparse Grids for the Regular Grid Case" in chapter "2D Basis Functions for Triangular and Rectangular Meshes". For regular grids, the relation of the number of points in the sparse grid to that of the full grid.

Sparse grid Appeared in section "Numerical Methods" in chapter "Introduction". A grid generated from the full grid by not using some of the points as dynamic points. A 2-d or 3-d regular grid that has approximately the same cell size for all cells. On the sphere, only platonic solids are exactly homogeneous.

Stability Appeared in section "The Runge–Kutta and Other Time Discretization Schemes" in chapter "Simple Finite Difference Procedures". A scheme is stable, if the absolute value of fields is limited (does not go to infinity).

Staggered grid Appeared in section "The Von Neumann Method of Stability Analysis" in chapter "Simple Finite Difference Procedures". A grid where velocities and densities are not defined at the same point of a cell. With polynomial methods, staggered methods are caused by approximations of different polynomial orders for densities and velocities.

Structured grid   Appeared in chapter "Simple Finite Difference Procedures". In 2D grid points having indices i, j, where neighboring indices indicate neighboring points.

Super-convergence   Appeared in chapter "Simple Finite Difference Procedures". A convergence in higher order than implied by the polynomial representation. For example, centered differences converge in second order, while being derived from first-order field representation.

Toy model   Appeared in section "Verifications of Numerical Methods for Climate Modeling" in chapter "Introduction". Test models for atmospheric models that are simpler than models of the real atmosphere.

Uniform approximation order   Appeared in chapter "Platonic and Semi-Platonic Solids". Finite differences where a minimum approximation order n^{un} is maintained at every grid point. A non-uniform order can lead to grid imprinting.

Uniform order n of approximation   Appeared in section "Baumgardner's Cloud Derivative Method". A number n where for each grid point of a 2D or 3D scheme the convergence is at least n.

Useful range of resolutions (a useful range of wavenumbers, also called effective resolution)   Appeared in Fig. 7 in chapter "Local-Galerkin Schemes in 1D". The range of wavenumbers, where the group velocity is positive.

References

AMCG (2014). Fluidity manual. In: Applied modelling and computation group. Imperial College London, London. Available via DIALOG. http://fluidityproject.github.io/support.html

Adams SV, Ford RW, Hambley M, Hobson JM, Kavcic I, Maynard CM, Melvin T, Muller EH, Mullerworth S, Porter AR, Rezny M, Shipway BJ, Wonga R (2019) LFRic: meeting the challenges of scalability and performance portability in weather and climate models. J Parallel Distrib Comput 132:383–396

Adcroft A, Hill C, Marshall J (1997) Representation of topography by shaved cells in a height coordinate ocean model. Mon Wea Rev 125(9):2293–2315

Ahlberg JH, Nilson EN, Walsh JL (1967) The theory of splines and their application, 1st edn. Academic, New York, pp. 45–56

Arakawa A, Jung JH (2011) Multiscale modeling of the moist-convective atmosphere: a review. Atmos Res 102(3):263–285 (2011)

Arakawa A, Lamb VR (1977) Computational design of the basic dynamical processes of the UCLA general circulation model, 1st edn. University of California, Los Angeles, pp. 173–265

Baumgardner JR, Frederickson PO (1985) Icosahedral discretization of the two-sphere. SIAM J Numer Anal 22:107–115

Benoit R, Desgagne M, Pellerin P, Pellerin S, Chartier Y, Desjardins S (1997) The Canadian MC2: a semi-implicit wideband atmospheric model suited for finescale process studies and simulation. Mon Wea Rev 125(10):2382–2415

Braun VG (1958) Schlichting Grenzschicht theorie, 1st edn. Braun, Karlsruhe

Cannon F, Carvalho LMV, Jones C, Norris J, Bookhagen B, Kiladis GN (2017) Effects of topographic smoothing on the simulation of winter precipitation in High Mountain Asia. JGR Atmos 122(3):1456–1474

Carson DJ, Cullen MJP (1977) Intercomparison of short-range numerical forecasts using finite difference and finite element models from UK Meteorological Office. Beitr Phys Atmos 50:1–15

Casanova EG (1964) Eduard und Elisabeth oder die Reise in das Innere unseres Erdballs, 1st edn. Verlag Ullstein, Frankfurt

Charney JG, Fjortoft R, Von Neumann J (1950) Numerical integration of the barotropic vorticity equation. Tellus 2:237–250

Clancy C, Pudykiewicz J (2013) On the use of exponential time integration methods in atmospheric models. Tellus A 65:1–16

Cotter J, Shipton J (2012) Mixed finite elements for numerical weather prediction. J Comp Phys 231(21):7076–7091

Daley R (1991) Atmospheric data analysis, 1st edn. Cambridge University Press, Cambridge

© Springer Nature Switzerland AG 2022
J. Steppeler, J. Li, *Mathematics of the Weather*, Springer Atmospheric Sciences,
https://doi.org/10.1007/978-3-031-07238-3

Davies DC (1976) A lateral boundary formulation for multi-level prediction models. Q J R Meteorol Soc 102:405–418

Davies T (2014) Lateral boundary conditions for limited area model. Quart J Roy Meteor Soc 140(678):185–196

Doms G, Schättler U (2002) A description of the nonhydrostatic regional model LM. Part I: dynamics and numerics. Consortium for small-scale modeling (COSMO) LM F90 2.18, Tech. Rep., DWD, Germany

Dudhia J (1993) A nonhydrostatic version of the Penn State-NCAR mesoscale model: validation tests. Mon Wea Rev 121:1439–1513

Durran DR (2010) Numerical methods for fluid dynamics: with applications to geophysics, 2nd edn. Springer, New York, pp. 35–146

Einstein A (1905) Zur Elektrodynamik bewegter Körper. Ann Phys Chem 17:891–921

ECMWF (2016). https://www.ecmwf.int

Gallus WA, Klemp JB (2000) Behavior of flow over step orography. Mon Wea Rev 128:1153–1164

Gao Y, Leung LR, Zhao C, Hagos S (2017) Sensitivity of U.S. summer precipitation to model resolution and convective parameterizations across gray zone resolutions. J Geophys Res Atmos 122:2714–2733

Gardner DJ, Guerra JE, Hamon FP, Reynolds DR, Ullrich PA, Woodward CS (2018) Implicit-explicit (IMEX) Runge-Kutta methods for non-hydrostatic atmospheric models. Geosci Model Dev 11(4):1497–1515

Giraldo FX (2001) A spectral element shallow water model on spherical geodesic grids. Num Meth Fluids 8:869–901

Giraldo FX, Perotb JB, Fisher PF (2003) A spectral element semi-Lagrangian (SESL) method for the spherical shallow water equations. J Comp Phys 190(2):623–650

Giraldo FX, Kelly JF, Constantinescu EM (2013) Implicit-explicit formulations of a three-dimensional nonhydrostatic unified model of the atmosphere (NUMA). SIAM 35:1162–1194

Gresho PM, Lee RL, Sani RI (1977) Advection-dominated flows with emphasis on the consequence of mass lumping. Finite elements in fluids, vol 3, 1st edn. Wiley, New York, pp. 335–350

Griffiths DF (1986) Finite element methods for time dependent problems. Astrophysical radiation hydrodynamics, 1st edn. Springer, Dordrecht, pp. 327–357

Gutowski WJ, Ullrich PA, Hall A, Leung LR, O'Brien TA, Patricola, CM, Arritt RW, Bukovsky MS, Calvin KV, Feng Z, Jones AD, Kooperman GJ, Monier E, Pritchard MS, Pryor SC, Qian Y, Rhoades AM, Roberts AF (2019) The ongoing need for high-resolution regional climate models: process understanding and stakeholder information. B Am Meteor Soc 12:875–882

Haltiner GJ, Williams RT (1983) Numerical prediction and dynamic meteorology, 2nd edn. Wiley, New York, pp. 477

Herrington AR, Lauritzen PH, Taylor MA, Goldhaber S, Eaton BE, Reed KA, Ullrich PA (2019) Physics-dynamics coupling with element-based high-order Galerkin methods: quasi equal-area physics grid. Mon Wea Rev 147:69–84

Honnert R, Efstathiou GA, Beare RJ, Ito J, Lock A, Neggers R, Plant RS, Shin HH, Tomassini L, Zhou B (2020) The atmospheric boundary layer and the gray zone of turbulence: a critical review. J Geophys Res (2020). https://doi.org/10.1029/2019JD030317

Huckert R, Steppeler J (2021) Galerkin compiler MOW_GC. www.hucket.com/Meteo/y.html

Jablonowski C, Williamson DL (2006) A baroclinic instability test case for atmospheric model dynamical cores. Quart J Roy Meteor Soc 132:2943–2975

Kageyama A (2005) Dissection of a sphere and Yin-Yang grids. J Earth Simulator 3:20–28

Kageyama A, Sato T (2004) "Yin-Yang grid": an overset grid in spherical geometry. Geochem Geophys Geosyst 5:Q09005

Kalnay E, Bayliss A, Storch J (1977) The fourth order GISS model of the global atmosphere. Beitrage Phys Atmos 50:299–311

Kopera M, Giraldo FX (2011) Analysis of adaptive mesh refinement for IMEX discontinuous Galerkin solutions of the compressible Euler equations with application to atmospheric simulations. J Comput Phys 275:92–117

Kreiss HO, Oliger J (1994) Comparison of accurate methods for the integration of hyperbolic equations. Tellus 24:199–215

Lander J, Hoskins J (1997) Believable scales and parameterizations in a spectral transform model. Mon Wea Rev 125:292–303

Li J, Steppeler J, Fang F, Pain CC, Zhu J, Peng X, Dong L, Li Y, Tao L, Leng W, Wang Y, Zheng J (2019) Potential numerical techniques and challenges for atmospheric modelling. B Am Meteor Soc 9:239–242

Li J, Fang F, Steppeler J, Zhu J, Cheng Y, Wu X (2021) Demonstration of a three-dimensional dynamically adaptive atmospheric dynamic framework for the simulation of mountain waves. Meteorol Atmos Phys 133:1627–1645

Lock SJ, Bitzse HW, Coals A, Gadian A, Mobbs S (2012) Demonstration of a cut-cell representation of 3D orography for studies of atmospheric flows over very steep hills. Mon Wea Rev 140:411–424

Lorenz EN (1963) Deterministic nonperiodic flow. J Atmos Sci 20:130–141

Marras S, Kelly JF, Moragues M, Muller A, Kopera MA, Vazquez M, Giraldo FX, Houzeaux G, Jorba O (2016) A review of element-based Galerkin methods for numerical weather prediction: finite element, spectral elements, and discontinuous Galerkin. Arch Computat Methods Eng 23:673–722

Melvin T, Benacchio T, Milano P, Shipway BJ, Wood N (2019) A mixed finite-element, finite-volume, semi-implicit discretisation for atmospheric dynamics: Cartesian geometry. Quart J Roy Meteor Soc 145(724):2835–2853

Mesinger F (1981) Horizontal advection schemes of a staggered gridââť an enstrophy and energy-conserving model. Mon Wea Rev 109(3):467–478

Mesinger F, Janjic ZI, Nickovic S, Gavrilov D, Deaven DG (1988) The step-mountain coordinate: model description and performance for cases of Alpine lee cyclogenesis and for a case of an Appalachian redevelopment. Mon Wea Rev 116:1493–1518

Mishra SK, Taylor MA, Nair RD, Tufo HM, Tribbia JJ (2010) Performance of the HOMME dynamical core in the aqua-planet configuration of NCAR CAM4: equatorial waves. Ann Geophys 29:221–227

Moncrieff MW, Liu C, Bogenschutz P (2017) Simulation, modeling, and dynamically based parameterization of organized tropical convection for global climate models. J Atmos Sci 74(5):1363–1380

Müller A, Kopera MA, Marras S, Wilcox LC, Isaac T, Giraldo FX (2018) Strong scaling for numerical weather prediction at petascale with the atmospheric model NUMA. Int J High Perform Comput Appl 33:411–426

Nakamura G, Potthast R (2015) Inverse modeling: an introduction to the theory and methods of inverse problems and data assimilation, 1st edn. IOP Publishing, Bristol, UK

Navon IM (1987) FEUDX, a two-stage high accuracy finite element Fortran program for solving the shallow water equations. Comput Geosci 13:255–285

Ogura Y, Yagihashi A (1970) A numerical study of finite-amplitude time-dependent convection induced by time-independent internal heating: truncated systems. J Meteorol Soc Jpn 48:1–17

Orlanski I (1975) A rational subdivision of scales for atmospheric processes. B Am Meteor Soc 56(5):527–534

Pain CC, Umpleby A, Oliveria CRE, Goddard AJH (2001) Tetrahedral mesh optimisation and adaptivity for steady-state and transient finite element calculations. Comput Method Appl M 190:3771–3796

Pain CC, Piggott MD, Goddard AJH, Fang F, Gorman GJ, Marshall DP, Eaton MD, Power PW, Oliveria CRE (2005) Three dimensional unstructured mesh ocean modelling. Ocean Model 10:5–33

Park SH, Klemp JB, Skamarock WC (2014) A comparison of mesh refinement in the global MPAS-A and WRF models using an idealized normal-mode baroclinic wave simulation. Mon Wea Rev 142:3614–3634

Pedlosky J (1987) Geophysical fluid dynamics, 1st edn. Springer, New York, pp. 745–792

Peng X, Xiao F, Takahashi K (2006) Conservative constraint for a quasi-uniform overset grid on the sphere. Quart J Roy Meteor Soc 132:979–996

Phillips NA (1957) A coordinate system having some special advantages for numerical forecasting. J Meteorol 14:184–185

Rancic M, Purser RJ, Mesinger F (1996) A global shallow water model using an expanded spherical cube: gnomonic versus conformal coordinate. Quart J Roy Meteor Soc 122:959–982

Robert A (1982) A semi-Lagrangian and semi-implicit numerical integration scheme for the primitive meteorological equations. J Meteorol Soc Jpn 60:319–325

Rocket B, Will A, Hense A (2008) The regional climate model COSMO-LM (CCLM). Meteor Zeitsch 17:347–348

Sadourny R (1972) Conservative finite-difference approximations of the primitive equations on quasi-uniform spherical grids. Mon Wea Rev 100:136–144

Sadourny R, Morel P (1969) A finite-difference approximation of the primitive equations for a hexagonal grid on a plane. Mon Wea Rev 97:439–445

Saito K, Doms G (1998) 3-D mountain waves by the Lokal-Modell of DWD and the MRI mesoscale nonhydrostatic model. Meteorol Geophys 49(1):7–19

Sani RL, Gresho PM, Lee RL, Griffiths DF (1981) The cause and cure of the spurious pressure generated by certain FEM solutions of the incompressible Navier-Stokes equations, Part I. Int J Numer Methods Fluids 1:17–43

Satoh M, Tomita H, Yashiro H, Miura H, Kodama C, Seiki T, Noda A, Yamada Y, Goto D, Sawada M, Miyoshi T, Niwa Y, Hara M, Ohno T, Iga S, Arakawa T, Inoue T, Kubokawa H (2014) The non-hydrostatic icosahedral atmospheric model: description and development. Progr Earth Planet Sci 18 1. Article number: 18

Savre J, Percival J, Herzog M, Pain CC (2016)Two-dimensional evaluation of ATHAM-fluidity, a nonhydrostatic atmospheric model using mixed continuous/discontinuous finite elements and anisotropic grid optimization. Mon Wea Rev 144:4349–4372

Schoenstadt AL (1979) On one-dimensional geotropic adjustment with finite differencing. Mon Wea Rev 107(2):211–214

Schoenstadt A (1980) A transfer function analysis of numerical schemes used to simulate geostrophic adjustment. Mon Wea Rev 108:1248–1259

Schwarz L (1950) Theorie des distributions, Hermann, 2 volumes, 1st edn. Neuauflage

Shapiro R (1970) Smoothing, filtering, and boundary effects. Rev Geophys 8(2):359–387

Shaw J, Weller H (2016) Comparison of terrain-following and cut-cell grids using a nonhydrostatic model. Mon Wea Rev 144:2085–2099

Skamarock WC, Klemp JB (2008) A time-split nonhydrostatic atmospheric model for weather research and forecasting applications. J Comput Phys 227:3465–3485

Skamaroc WCk, Klemp JB, Dudhia J, GIll D, Baker D, Wang W, Powers J (2007) A description of the advanced research WRF version 2. Consortium for small-scale modeling (COSMO) LM F90 2.18. Technical Report, NCAR TN STR (2007), P. 468

Skamarock WC, Klemp JB, Duda MG, Fowler LD, Park SH, Ringler T (2012) A multiscale nonhydrostatic atmospheric model using centroidal Voronoi Tesselations and C-grid staggering. Mon Wea Rev 140:3090–3105

Slifka MK, Whitton JL (2000) Clinical implications of dysregulated cytokine production. J Mol Med. https://doi.org/10.1007/s001090000086

Staniforth AN (1984) The application of the finite-element method to meteorological simulations: a review. Int J Numer Methods Fluids 4:1–22

Staniforth A, Thuburn J (2012) Horizontal grids for global weather and climate prediction models: a review. Quart J Roy Met Soc 138:1–16

Steppeler J (1976) A baroclinic model using a high accuracy horizontal discretization. Beitr Phys Atm 49:278–289

Steppeler J (1976) The application of the second and third degree methods. J Comput Phys 22:295–318

Steppeler J (1978) Fluid computations using infinitesimal functionals. Comput Fluid 6:241–258

Steppeler J (1987) Energy conserving Galerkin finite element schemes for the primitive equations of numerical weather prediction. J Comp Phys 69:158–264

Steppeler J (1987) Quadratic Galerkin finite element schemes for the vertical discretization of numerical forecast models. Mon Wea Rev 115:1575–1588

Steppeler J (1987) Galerkin and finite element methods in numerical weather prediction, 1st edn. Duemmler, Bonn, pp. 45–56

Steppeler J (1988) A Galerkin finite element-spectral weather forecast model in hybrid coordinates. Comput Math App 16:23–30

Steppeler J (1988) A cubic spline Galerkin method for vertical discretization. Quart J Roy Met Soc 114:1549–1549

Steppeler J (1989) Analysis of group velocities of various finite element schemes. Beitr Phys Atmos 62:151–161

Steppeler J (1989) Report on the seventh Int. Conference on finite element methods in flow problem (FEMIF7). B Am Meteor Soc 70:1428

Steppeler J (1990) Simple test calculations concerning finite element applications to numerical weather prediction. Numer Method Fluids 11:209–226

Steppeler J (1993) The southern oscillation as an example of a simple ordered subsystem of a complex chaotic system. J Climate 10:473–480

Steppeler J (1996) Meeting summary: first international short range numerical weather prediction 11–13 March 1996 Offenbach Germany. B Am Meteor Soc 78:470

Steppeler J, Klemp JB (2017) Advection on cut-cell grids for an idealized mountain of constant slope. Mon Wea Rev 145:1765–1777

Steppeler J, Prohl P (1996) Application of finite volume methods to atmospheric models. Beitr Phys Atmos 69:297–306

Steppeler J, Navon IM, Lu H (1990) Finite element schemes for extended integrations of atmospheric models. J Comput Phys 130:213–235

Steppeler J, Saito K, Kato T, Eito H, Seino N, Murata A (2001) Meeting summary: report on the third international SRNWP workshop on nonhydroâ‍ŋstatic modeling. Bull Am. Meteor. Soc. 82:2245–2250

Steppeler J, Bitzer HW, Minotte M, Bonaventura L (2002) Nonhydrostatic atmospheric modeling using a z-coordinate representation. Mon Wea Rev 130:2143–2149

Steppeler J, Doms G, Schättler U, Bitzer HW, Gassmann A, Damrath U, Gregoric G (2003) Meso-Gamma scale forecasts using the nonhydrostatic model LM. Meteor Atmos Phys 82:75–96

Steppeler J, Bitzer HW, Janjic Z, Schättler U, Prohl P, Gjertsen U, Torrisi L, Parfinievicz J, Avgoustoglou E, Damrath U (2006) Prediction of clouds and rain using a z-coordinate nonhydrostatic model. Mon Wea Rev 134:3625–3643

Steppeler J, Ripodas P, Thomas S (2008) Third order finite difference schemes on icosahedral-type grids on the sphere. Mon Wea Rev 136:2683–2698

Steppeler J, Park SH, Dobler A (2013) Forecasts covering one month using a cut cell model. Geosci Model Dev 6:875–882

Steppeler J, Li J, Fang F, Zhu J (2019) Test of a cubic spline interface for physical processes with a 1-D third-order spectral element model. Tellus A: Dynam Meteorol Oceanogr 71:1–6

Steppeler J, Li J, Fang F, Zhu J, Ullrich PA (2019) o3o3: a variant of spectral elements with a regular collocation grid. Mon Wea Rev 147:2067–2082

Steppeler J, Li J, Fang F, Navon IM (2019) Third-order sparse grid generalized spectral elements on hexagonal cells for homogeneous advection in a plane. Meteorol Atmos Phys 132(5):703–719

Steppeler J, Li J, Navon IM, Fang F, Xiao Z (2019) Medium range forecasts using cut-cells: a sensitivity study. Meteorol Atmos Phys. https://doi.org/10.1007/s00703-019-00681-w

Steppeler J, Li J, Fang F, Zhu J (2021) The o2o3 local Galerkin method using a differential flux representation. J Meteo Soc Jpn 99:49–65

Stocker TF, Qin D, Plattner G-K, Tingnor MMB, Allen SK, Boschung J, Nauels A, Xia Y, Bex V, Midgley PM (2013) Climate change 2013: the physical science basis (Working Group I Contribution to the Fifth Assessment Report of the Intergovernmental Panel on Climate Change), 1st edn. The Cambridge University Press, Cambridge, pp. 741–866

Stöcker H (2008) Taschenbuch mathematischer Formeln und moderner Verfahren, 1st edn. Verlag Harri, Deutsch

Stull RB (2018) Practical meteorology: an algebra-based survey of atmospheric science, 1st edn. The University of British Columbia, Vancouver, pp. 745–792

Szabo Z, Babuska I (1991) Finite element analysis, 1st edn. Wiley, New York

Taylor M, Tribbia J, Iskandarani M (1997) The spectral element method for the shallow water equations on the sphere. J Comput Phys 130:92–108

Tomita H, Tsugawa M, Satoh M, Goto K (2001) Shallow water model on a modified icosahedral geodesic grid by using spring dynamics. J Comput Phys 174:579–613

TOP500 (2019). www.top500.org/statistic/perfdevel/

Ullrich PA (2014) Understanding the treatment of waves in atmospheric models. Part I: The shortest resolved waves of the 1D linearized shallow-water equations. Quart J Roy Met Soc 140:1426–1440

Ullrich PA, Jablonowski C, Kent J, Lauritzen PH, Nair RD, Taylor MA (2012) Dynamical Core Model Intercomparison Project (DCMIP) test case document. Technical Report. http://earthsystemcog.org/site_media/docs/DCMIP-TestCaseDocument_v1.7.pdf

Ullrich PA, Reynolds DR, Guerra JE, Taylor MA (2018) Impacts and importance of diffusion on the spectral element method: a linear analysis. J Comput Phys 375:427–446

Williamson DL (1968) Integrations of the barotropic vorticity equation on a spherical geodesic grid. Tellus 20:643–653

Williams RT (1981) Zienkiewicz: improved finite element forms for the shallow water wave equations. Int J Numer Methods Fluids. 1:81–97

Williamson DL, Drake JB, Hack JJ, Jakob R, Swarztrauber PN (1992) A standard test set for numerical approximations to the shallow water equations in spherical geometry. J Comput Phys 102:211–224

Wilson DR, Ballard SP (1999) A microphysically based precipitation scheme for the UK meteorological office unified model. Quart J Roy Meteor Soc 125:1607–1636

Yamazaki H, Satomura T (2010) Nonhydrostatic atmospheric modeling using a combined Cartesian grid. Mon Wea Rev 132:3932–3945

Yang X, Hu J, Chen D, Zhang H, Shen X, Chen J, Ji L (2008) Verification of GRAPES unified global and reginal numerical weather prediction model dynamic core. Chin. Sci. Bull. 53:3458–3464

Zängl G, Daniel Reinert D, Ripodas P, Baldauf M (2015) The ICON (ICOsahedral Non-hydrostatic) modeling framework of DWD and MPI-M: description of the non-hydrostatic dynamical core. Quart J Roy Meteor Soc 141:563–579

Zheng J, Zhu J, Wang Z, Fang F, Pain CC, Xiang J (2015) Towards a new multiscale air quality transport model using the fully unstructured anisotropic adaptive mesh technology of Fluidity (version 4.1.9). Geosi Model Dev 8:3421–3440

Zheng J, Wu X, Fang F, Li J, Wang Z, Xiao H, Zhu J, Pain CC, Linden PF, Xiang B (2021) Numerical study of COVID-19 spatial-temporal spreading in London. Phys Fluids 33:046605

Index

© Springer Nature Switzerland AG 2022
J. Steppeler, J. Li, *Mathematics of the Weather*, Springer Atmospheric Sciences,
https://doi.org/10.1007/978-3-031-07238-3

Printed in the United States
by Baker & Taylor Publisher Services